Lecture Notes in Computer Science 8304

Commenced Publication in 1973
Founding and Former Series Editors:
Gerhard Goos, Juris Hartmanis, and Jan van Leeuwen

Roberto Baldoni   Nicolas Nisse
Maarten van Steen (Eds.)

Principles of Distributed Systems

17th International Conference, OPODIS 2013
Nice, France, December 16-18, 2013
Proceedings

Volume Editors

Roberto Baldoni
Sapienza Research Center of Cyber Intelligence and Information Security
and
Università degli Studi di Roma "La Sapienza"
Dipartimento di Ingegneria Informatica, Automatica e Gestionale "Antonio Ruberti"
Via Ariosto 25, 00185 Rome, Italy
E-mail: baldoni@dis.uniroma1.it

Nicolas Nisse
Inria, France
and
Université Nice Sophia Antipolis CNRS, 13S, UMR 7271
06900 Sophia Antipolis, France
E-mail: nicolas.nisse@inria.fr

Maarten van Steen
Vrije Universiteit Amsterdam
Department of Computer Science
De Boelelaan 1081a, 1081 HV Amsterdam, The Netherlands
E-mail: steen@cs.vu.nl

ISSN 0302-9743   e-ISSN 1611-3349
ISBN 978-3-319-03849-0   e-ISBN 978-3-319-03850-6
DOI 10.1007/978-3-319-03850-6
Springer Cham Heidelberg New York Dordrecht London

Library of Congress Control Number: 2013954562

CR Subject Classification (1998): C.2.4, C.2, F.2, D.2, I.2.11, G.2.2

LNCS Sublibrary: SL 1 – Theoretical Computer Science and General Issues

Typesetting: Camera-ready by author, data conversion by Scientific Publishing Services, Chennai, India

Printed on acid-free paper

Springer is part of Springer Science+Business Media (www.springer.com)

Preface

This volume contains the papers presented at OPODIS 2013. OPODIS, the International Conference on Principles of Distributed Systems, is an international forum for the exchange of state-of-the-art knowledge on distributed computing and systems among researchers from around the world. The 17th edition of OPODIS was held during December 16–18, 2013 in Nice, France.

Papers were sought soliciting original research contributions to the theory, specification, design and implementation of distributed systems. In response to the call for papers, 41 submissions were received, out of which 18 papers were accepted, after a rigorous reviewing process that involved 33 Program Committee members and at least three reviews per paper.

We would like to thank the Program Committee members, as well as the external reviewers, for their fundamental contribution in selecting the best papers.

In addition to the technical papers, the program included five invited presentations by: Marcos k. Aguilera (Microsoft Research, USA), Eitan Altman (Inria, France), Hein Meling (University of Stravanger, Norway), Nuno Preguica (Universidade Nova de Lisboa, Portugal) and Marc Shapiro (Inria, France).

September 2013

Roberto Baldoni
Nicolas Nisse
Maarten van Steen

Organization

Program Committee

Marco Aiello	University of Groningen, The Netherlands
Roberto Baldoni	Università di Roma "La Sapienza", Italy
Christian Cachin	IBM Research, Zurich, Switzerland
Antonio Carzaniga	University of Lugano, Switzerland
Gregory Chockler	IBM Haifa Research Laboratory, Israel
Allen Clement	Max Planck Institute for Software Systems, Germany
Paolo Costa	Microsoft Research Cambridge, UK
Dick Epema	Delft University of Technology, The Netherlands
Patrick Eugster	Purdue University, USA
Pascal Felber	Université de Neuchâtel, Switzerland
Antonio Fernandez Anta	Institute IMDEA Networks, Spain
Paola Flocchini	University of Ottawa, Canada
Ali Ghodsi	University of California at Berkeley, USA
Rachid Guerraoui	EPFL, Switzerland
Aaron Harwood	University of Merbourne, Australia
Konrad Iwanicki	University of Warsaw, Poland
Mark Jelasity	University of Szeged, Hungary
Ricardo Jimenez Peris	Universidad Politécnica de Madrid, Spain
Anne-Marie Kermarrec	Inria, France
Hein Meling	University of Stavanger, Norway
Alessia Milani	Bordeaux Institute of Technology, France
Alberto Montresor	University of Trento, Italy
Nicolas Nisse	Inria, France
Peter Pietzuch	Imperial College London, UK
Maria Potop-Butucaru	Université Pierre et Marie Curie-LIP6, France
Luis Rodrigues	Universidade de Lisboa, Portugal
Cristina Seceleanu	Malardalen University, Sweden
Marc Shapiro	Inria, Univ. Pierre et Marie Curie-LIP6, France
Alex Shraer	Google, USA
Peter Triantafillou	University of Glasgow, Scotland
Frits Vaandrager	Radboud University Nijmegen, The Netherlands
Maarten van Steen	VU University Amsterdam, The Netherlands
Paulo Verissimo	Universidade de Lisboa, Portugal

Roman Vitenberg   University of Oslo, Norway
Spyros Voulgaris   Vrije Universiteit Amsterdam, The Netherlands
Masafumi Yamashita   Kyushu University, Fukuoka, Japan

Additional Reviewers

Anagnostopoulos, Christos   Perelman, Dmitri
Ananthanarayanan, Ganesh   Schmid, Ulrich
Arad, Cosmin   Sens, Pierre
Culhane, William   Shafaat, Tallat M.
Dobre, Dan   Shavit, Nir
Georgievski, Ilche   Sutra, Pierre
Godard, Emmanuel   Taherkordi, Amir
Jehl, Leander   Travers, Corentin
Kogan, Kirill   Trehan, Amitabh
Lamani, Anissa   Tretmans, Jan
Lehmann, Anja   Tso, Posco
Marinescu, Raluca   Urdaneta, Guido
Mostefaoui, Achour   Verbeek, Freek
Pagani, Giuliano Andrea   Vilaca, Xavier

Invited Talks

Geo-Distributed Storage in Data Centers

Marcos K. Aguilera

Microsoft Research
Mountain View, CA, USA

Abstract. Data centers increasingly have a storage system that is *geo-distributed*, that is, distributed across several geographic locations. We explain the general characteristics of this setting and the challenges that it brings, chief among them the need to operate with low latency despite significant network delays. These challenges lead to many interesting problems: migrating data online, dealing with congestion, providing efficient transactions, and more. We discuss these problems and some recent solutions, which bring together techniques from distributed computing, distributed systems, and database systems. Despite much progress, however, several algorithmic and fundamental questions remain open and serve as inspiration for further investigation.

Dynamic Game Models in Complex Systems

Eitan Altman *

Inria, France
eitan.altman@inria.fr

Abstract. We begin the tutorial with a theoretic part that covers two areas: non-cooperative game theory, and population propagation models. In the game theory part, a particular attention will be given to potential games. We shall focus in particular on congestion games and on the game version of the generalized Kelly mechanism problem, both of which are known to be potential games. In our presentation of models for population propagation models, we shall present several models which we shall classify according to the size of population of potential interested destination nodes (which can be finite and constant, finite but non-constant or infinite), and the virality of the content. This will include branching and epidemic models. We shall then use these tools to study various applications to large networks. This will include (1) security issues related to e-virus attacks, (2) the question of what type of content should service providers specialize in, which will be solved by transforming it into an equivalent congestion game, (3) issues related to viral marketing and competition issues in social networks. In these problems the generalized Kelly mechanism will be frequently used. The game theoretic analysis will allow us to get insight on how much to spend on advertising products and on what product should we advertise. Both journal and conference papers as well as video presentations covering this tutorial are available at http://www-sop.inria.fr/members/Eitan.Altman/dodescaden.html

---

* This work was supported by CONGAS Project (FP7- ICT-2011-8-317672), see www.congas-project.eu

Table of Contents

Tutorial Summary: Paxos Explained from Scratch

Hein Meling and Leander Jehl

University of Stavanger, Norway

Abstract. Paxos is a flexible and fault tolerant protocol for solving the consensus problem, where participants in a distributed system need to agree on a common value. However, Paxos is reputed for being difficult to understand. This tutorial aims to address this difficulty by visualizing Paxos in a completely new way. Starting from a naive solution and strong assumptions, Paxos is derived in a step-wise fashion. In each step, minimal changes are made to the solution and assumptions, aimed at understanding why the solution fails. In this manner, a correct solution that corresponds to Paxos is eventually reached.

1 Introduction

Paxos is a flexible and fault tolerant consensus protocol that can be used in applications that need to agree on a common value among distributed participants. Paxos was proposed by Lamport in his seminal paper [1] and later gave a simplified description in [2]. Paxos can be used to solve the atomic commit problem in distributed transactions, or to order client requests sent to a replicated state machine (RSM). An RSM provides fault tolerance and high availability, by implementing a service as a deterministic state machine and replicating it on different machines. Paxos is relevant because it is often used in production systems such as Chubby and ZooKeeper [3, 4] among many others. Understanding Paxos is important because it reveals the distinction between a strongly consistent RSM and a primary-backup system.

Both before and after its publication in [1], Paxos attracted much attention for its unorthodox exposition in the form of a fictional parliamentary system, supposedly used by legislators at the Greek island of Paxos. But the scientific contribution was also significant; it provided a new way to implement RSMs, and proved that the protocol guarantees that participants make consistent decisions, irrespective of the number of failures. Clearly Paxos cannot always make progress, e.g. during network partitions, as was shown in [5]. But perhaps most important, Paxos was described in a flexible and general way, ignoring many technical details. This made it an excellent foundation for further research into RSM-based protocols [6–9], aimed at supporting different failure models, wide-area networking, to improve latency, and so on. The fact that these protocols build on the Paxos foundation, which has been formally proven, makes it much easier to reason about their correctness through step-wise modifications of Paxos.

With this powerful foundation that Paxos offered, came also a challenge: the flexible description made it harder to understand. This remains true to this day, even as numerous papers have been written aimed at explaining Paxos for system builders [10, 11] and more generally [12, 13]. These and other papers are still challenging for students and others to understand without significant efforts.

R. Baldoni, N. Nisse, and M. van Steen (Eds.): OPODIS 2013, LNCS 8304, pp. 1–10, 2013.

The aim of this tutorial is to explain Paxos from the bare fundamentals by deriving a Paxos-based RSM in a step-wise and pictorial manner. We start with a non-replicated service that we want to harden with fault tolerance and high availability. That is, the server must be replicated. Initially, we make unrealistic assumptions about the environment and propose the simplest protocol that we can imagine to coordinate the server replicas to ensure that they remain mutually consistent, and explain why the protocol is insufficient. Then in each step, minimal changes are introduced to the coordination protocol aimed at understanding why each protocol fails. Continuing, we finally reach a correct protocol that corresponds to a Paxos-based RSM.

Our objective is that you understand that many seemingly intuitive approaches do not work and why. Having read this tutorial, we hope that you will gain appreciation for Paxos' contribution, and perhaps put you in a better position to read the Paxos literature.

2   A Stateful Service: Assumptions and Notation

We will explain Paxos starting from a simple stateful service that should be made fault-tolerant and highly available. Initially the service is implemented by a single server that receives requests from a set of clients, processes the requests, updates its state, and replies back to the clients. This pattern is visualized as a message sequence diagram in Fig. 1, where server S_1 processes requests from clients, C_1 and C_2. Further notation is explained below.

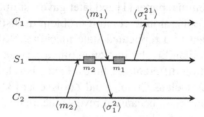

Fig. 1. Solution *SingleServer*: A single server can order and process requests from several clients

Notation. A request message received by the server causes a state transition affecting the current state of the server. The outcome of processing requests sent to the server depends on its current state. A box on the timeline is meant to illustrate that a state change has taken place, caused by the processing of some message m_i.

A common assumption also adopted here, is that requests from different clients are unrelated, and the order in which they are executed is irrelevant. Clearly, requests from the same client should be executed in sending order. We use σ_i^{kl} to denote the local state of server S_i after having processed messages $m_k m_l$, in that order. We ignore the server index and write σ^{kl}, when the origin is irrelevant. In our examples, the reply sent to clients is determined by the server's state, denoted $\langle \sigma_i^{kl} \rangle$. In practice, the reply is usually not the server's state, but rather some value computed from the server's state.

Single Server. In the single server case shown in Fig. 1, it is easy to see that the two clients observe a consistent reflection of the server's execution of the two requests. It is easy for the server to determine an ordering for the client requests that it receives. However, implementing the service with a single server is not fault-tolerant.

3   Fault Tolerance with Two Servers

As a first attempt at improving the fault tolerance and availability of our service, we can add one server to the system, under the assumption that if one of the servers fail, the other can take over and service client requests. This architecture is frequently used, and is called primary-backup. In our first naive solution we use two servers without coordination between them; i.e. the clients simply send their requests directly to the two servers, as shown in Fig. 2. However, as is apparent from this diagram, the two requests can be processed in different orders at the two servers, e.g. because of message delays: $m_1 m_2$ at S_1 and $m_2 m_1$ at S_2, resulting in deviating server states. We say that the servers become inconsistent. This inconsistency is also exposed to the clients: C_1 observes possibly inconsistent replies σ^1 and σ^{21}, while C_2 observes replies: $\sigma^2 \sigma^{12}$.

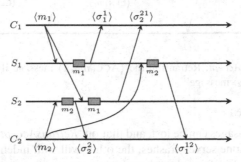

Fig. 2. Problem: Two servers cannot order messages from several clients without coordination

Our first solution to coordinate among the servers is to let one server be leader, also called the primary. The leader simply sends an *accept* message to the other server and executes the request locally. The accept message $\langle \text{ACC}, m_i, j \rangle$ is used to tell the other server that m_i should be executed as the jth request, where j is a *sequence number*. This approach is illustrated in Fig. 3. It is easy to see that both servers remain consistent and that replies to clients are also consistent, since σ^2 is a prefix of σ^{21}. Since the service is implemented as deterministic state machine, processing a request results in a unique state transition. Therefore $\sigma_1^2 = \sigma_2^2$ and $\sigma_1^{21} = \sigma_2^{21}$.

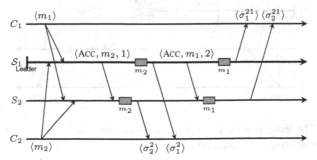

Fig. 3. Solution *SendAccept*: Leader (S_1) sends an accept message to the other server telling it the order in which the messages should be processed

This approach works fine as long as messages are not lost. However, if $\langle \text{ACC}, m_2, 1 \rangle$ in Fig. 3 is lost, then S_2 gets stuck and cannot process the next message, m_1.

The solution to this problem is simply to add a lost message detection mechanism. That is, let the leader retransmit its accept message until it is acknowledged by S_2. This solution is shown in Fig. 4, where a *learn* message $\langle \text{LRN}, m_i \rangle$ corresponds to an acknowledgement. This approach allows for the servers to eventually make progress as long as messages are not lost infinitely often.

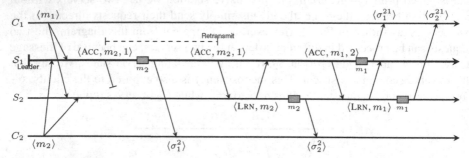

Fig. 4. Solution *RetransAccept*: Retransmit the $\langle \text{ACC}, m_2, 1 \rangle$ message if it does not receive a corresponding $\langle \text{LRN}, m_2 \rangle$ message

4   Server Crashes

We have seen that messages can be lost, and that our *RetransAccept* protocol can fix the problem. However, if one server crashes, the other will wait indefinitely for an accept or learn message. We therefore adopt the rule that once a server crashes, the remaining server continues to serve clients following the *SingleServer* protocol (Fig. 1). With this rule we can see from Fig. 5, that our *RetransAccept* protocol is insufficient. This is because the initial leader (S_1) replies to request m_1 before learning that S_2 has seen its $\langle \text{ACC}, m_1, 1 \rangle$ message, and because S_1 crashes before it can retransmit the accept. Instead S_2 takes over and decides to execute request m_2 before m_1, and thus the two clients observe inconsistent replies; σ^2 is seen by C_2 while σ^{21} is expected.

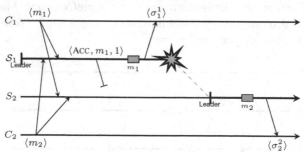

Fig. 5. Problem: The leader crashes after sending reply to client C_1, without ensuring that S_2 has learned about the ordering message, $\langle \text{ACC}, m_1, 1 \rangle$

To solve this problem, we require that the leader wait for the $\langle \text{LRN}, m_1 \rangle$ message before executing the request as shown in Fig. 6, and we also require a retransmission in case of message loss. If S_1 receives the learn message and replies to C_1 before crashing, the ordering information has already been propagated to S_2. If S_1 crashes before sending

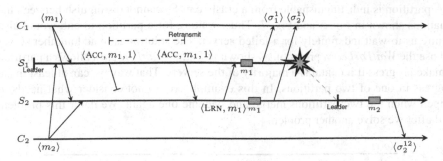

Fig. 6. Solution *WaitForLearn*: The leader waits for $\langle \text{LRN}, m_1 \rangle$ before executing the request m_1

its reply to the client, S_2 may or may not have seen the accept message. If S_2 has seen the accept, S_2 will obey it. If the accept didn't reach S_2, it can decide its own ordering.

5   Network Partitions

So far we have not specified how failures are actually detected. In practice a server is assumed to have failed if it is unresponsive for a given period of time. This is typically done using a timer mechanism, which upon a timeout triggers a *failure detection*. However, identifying a suitable timeout period is difficult in practice, and there is always a chance of false detections due to the stochastic nature of networked systems.

Recall that we adopted the rule to fall back to the *SingleServer* protocol when failure is detected. This rule was intended to allow the service to *make progress* after a server had failed. However, if we cannot reliably detect that the other server *really failed*, then we have a problem, as is illustrated in Fig. 7. Here we see that both servers remain operational, but are unable to communicate due to a network partition. After the failure detection time, both servers fall back to the *SingleServer* protocol and continue to process client requests, exposing the clients to different server states, σ^1 and σ^2. This state divergence violates our desire to remain consistent, especially towards clients. This is since reconciling the state divergence when communication is reestablished would involve rollback on multiple clients, and would quickly become unmanageable.

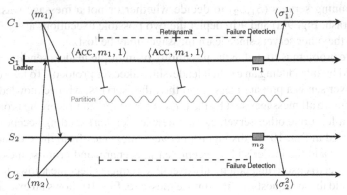

Fig. 7. Problem: Our *SingleServer* protocol can make progress in separate partitions, but it will lead to inconsistencies

A partition is indistinguishable from a crash, e.g. S_2 cannot distinguish between the situations shown in Fig. 5 and Fig. 7. Thus, waiting for a partition to end would also require us to wait indefinitely for a failed server. The solution is to add another server and use the *WaitForLearn* protocol, as shown in Fig. 8. *WaitForLearn* allows a partition to make progress if it contains a majority of the servers. That way we can at least make progress in one of two partitions. In this example we do not consider what needs to happen when the two partitions merge and become one again. We defer this problem until after we solve another problem.

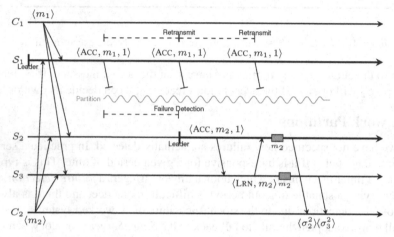

Fig. 8. Solution: Add another server and use the *WaitForLearn* protocol

6   Leader Change

Our *WaitForLearn* protocol tolerates either a crash or a partition. However, a concurrent partition and crash is not handled by our protocol. In cases of false detection, several servers may send out accepts concurrently. In Fig. 9 both S_1 and S_2 send accepts for different messages. If S_3 crashes shortly after receiving these accepts it might have executed one of the requests and sent a reply to the client. In this case it is impossible for the remaining servers (S_1,S_2) to decide whether or not a message was executed before the crash. Fig. 9(a) and 9(b) depict the two possible executions at S_3. These are unknown to the other servers since learn messages may be lost.

The above problem is rooted in the possibility of multiple leaders sending accepts. It can be solved by introducing an explicit leadership takeover protocol. To take over leadership, a server sends a prepare message to the other servers, who acknowledge with a promise to ignore all messages sent by the old leader. Only after receiving promise messages from at least one other server, can the new leader start sending accept messages. This is depicted in Fig. 10. To distinguish between messages from the old and the new leader, we now add the leader's id to accept, learn, prepare, and promise messages.

Furthermore, to ensure that potentially executed requests become known to the other servers, we add those requests to the promise message. Fig. 10 shows an example where no requests have been executed, indicated as $(-)$ in the promise. If the promise contains requests, the new leader resends the accept for these requests, as depicted in Fig. 11.

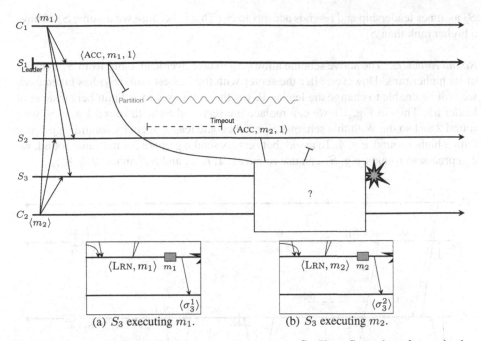

Fig. 9. Problem: Both S_1 and S_2 sent an accept message to S_3. Since S_3 crashes afterwards, the remaining servers cannot determine whether S_3 executed m_1 or m_2.

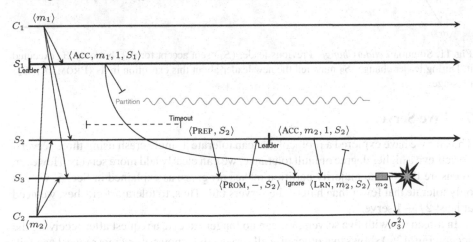

Fig. 10. Solution *LeaderChange*: S_2 announces its wish to become leader by sending a $\langle \mathrm{PREP}, S_2 \rangle$. S_3 replaces S_1 by S_2 as leader and confirms this with a $\langle \mathrm{PROM}, -, S_2 \rangle$ message. S_2 acts as leader after receiving this promise.

Merging Partitions. When two partitions merge, the leader resends accept messages to servers that missed them. However, the merged partition may now have several leaders. For example, when the partition in Fig. 11 ends, both S_1 and S_2 consider themselves leaders. To establish a single leader, we assume a predefined ranking. In Fig. 11,

S_2 assumes leadership and resends accepts to S_1. That is because we assume S_2 to have a higher rank than S_1.

Round Numbers. The above scheme allows S_2 to take over leadership from S_1 because of its higher rank. However, after the server with the highest rank (S_3) has taken over, we will be unable to change the leader. Paxos therefore uses round numbers instead of leader ids. Thus in Fig. 10, we can replace the server id S_1 with round 1 and S_2 with round 2 and so on. With this scheme, S_1 can become leader again by sending a prepare with a higher round, e.g. 4. To avoid that servers send a prepare for the same round, we can preassign rounds, e.g. S_1 can use rounds $1, 4, 7, \ldots$ and S_2 can use $2, 5, 8, \ldots$

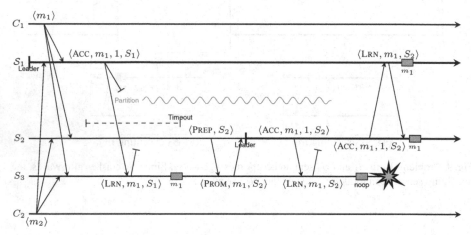

Fig. 11. Solution *LeaderChange*: Previous leader (S_1) sent accept for m_1, but only S_3 executed it. During leader change, S_3 must tell the new leader about this execution in its $\langle \text{PROM}, m_1, S_2 \rangle$ message.

7   Five Servers

Thus far we have explored a protocol that can tolerate a single crash using three servers. To achieve a higher degree of fault tolerance, we can clearly add more servers. However, to ensure that only a majority partition makes progress, as explained in Sec. 5, we can only tolerate that fewer than half of the servers fail. Thus, to tolerate f crashes, we need at least $2f + 1$ servers.

In a scenario with five servers, we can no longer execute a request after receiving the accept. Fig. 12(a) shows that otherwise all servers that knows about this request can fail. We therefore adjust our protocol to send learns to all servers, as depicted in Fig. 12(b), and only execute after receiving three learns for one message. Note that the accept is an implicit learn from the leader, and every server can also send a learn to itself. Therefore a follower can, in practice, execute after receiving one accept and one learn, while the leader can execute after receiving two learns.

Similarly, the new leader needs to collect two promises from the other servers to begin its leader role. Also here the new leader makes an implicit promise to itself as the third one. After multiple, successive leader changes it is possible to receive promises

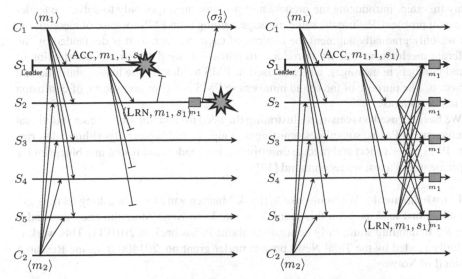

(a) With five servers we can no longer execute immediately after receiving an accept message.

(b) Execute after receiving one accept and one learn, or two learn messages.

Fig. 12. Paxos with five servers requires additional messages

including different values, sent by different leaders. E.g. a leader receiving promises $\langle\text{PROM}, m_1, S_3\rangle$ and $\langle\text{PROM}, m_2, S_3\rangle$ has to choose wether to send an accept for m_1 or m_2. We solve this by adding the identity used in the accept to the promise. Our promises now look like $\langle\text{PROM}, (S_1, m_1), S_3\rangle$ and $\langle\text{PROM}, (S_2, m_2), S_3\rangle$. The new leader S_3 sends $\langle\text{ACC}, m_2, S_3\rangle$, since S_2 has a higher rank than S_1. As in Sec. 6, we can also here replace the server identity with round numbers.

8 Summary

We have presented Paxos, aiming to understand the fundamental mechanisms. Our presentation differs significantly from previous attempts to explain Paxos, and in this section we explain how it relates to the presentation in *Paxos made Simple* (PMS) [2].

The first distinction is that PMS introduces separate agent roles: proposers, acceptors, and learners. These roles are at the heart of Paxos' flexibility, and allows one to structure a Paxos system in different ways. While this is very useful for formal reasoning over a wide variety of structures, it can be challenging to comprehend at first. Our servers each combine these three roles. Another difference is that PMS presents the protocol for agreeing on a single client request, among several requests seen by the servers. Thus, one instance of Paxos is used to agree on the next request to be executed. PMS then explains how multiple Paxos instances can be combined to build a Paxos-based RSM. These instances are numbered sequentially, and corresponds to our sequence numbers. In PMS, Lamport also explains that Paxos instances can be optimized to run with only the accept and learn messages, when the leader is stable. Instead we delay this step, introducing the prepare and promise messages only to solve the leader take over problem. We use the same message naming as in PMS for ease of recognition, but we only gradually augment the content of each message as it is demanded by the different mechanisms that we introduce. In particular, we deferred the introduction of round numbers in messages, which is used in PMS to identify the leader, until the end of Sec. 6. The purpose of the round numbers in PMS is a common source of confusion for many students.

We have focused on scenarios illustrating the need for and function of each individual mechanism in Paxos, sometimes omitting a complete and precise algorithmic description. PMS gives a short and precise description. For readers interested in a blueprint for implementing Paxos, we recommend [13].

Acknowledgements. We would like to thank Maarten van Steen for asking all the right questions that lead us down this path. Also thanks to Keith Marzullo and Alessandro Mei for untangling some early confusions about Paxos back in 2010/11. This work is partially funded by the Tidal News project under grant no. 201406 from the Research Council of Norway.

References

1. Lamport, L.: The part-time parliment. ACM Trans. on Comp. Syst. 16(2), 133–169 (1998)
2. Lamport, L.: Paxos made simple. ACM SIGACT News 32(4), 18–25 (2001)
3. Burrows, M.: The chubby lock service for loosely-coupled distributed systems. In: Proc. OSDI, pp. 335–350 (2006)
4. Hunt, P., Konar, M., Junqueira, F.P., Reed, B.: Zookeeper: wait-free coordination for internet-scale systems. In: Proc. USENIX ATC (2010)
5. Fischer, M.J., Lynch, N.A., Paterson, M.S.: Impossibility of distributed consensus with one faulty process. J. ACM 32(2), 374–382 (1985)
6. Martin, J.P., Alvisi, L.: Fast byzantine consensus. IEEE Trans. Dependable Secur. Comput. 3(3), 202–215 (2006)
7. Mao, Y., Junqueira, F.P., Marzullo, K.: Mencius: building efficient replicated state machines for wans. In: Proc. OSDI, pp. 369–384 (2008)
8. Meling, H., Marzullo, K., Mei, A.: When you don't trust clients: Byzantine proposer fast paxos. In: Proc. ICDCS, pp. 193–202 (2012)
9. Lamport, L.: Fast paxos. Distributed Computing 19(2), 79–103 (2006)
10. Ongaro, D., Ousterhout, J.: In search of an understandable consensus algorithm. Technical report, Stanford University (2013)
11. Chandra, T.D., Griesemer, R., Redstone, J.: Paxos made live: an engineering perspective. In: Proc. PODC, pp. 398–407 (2007)
12. De Prisco, R., Lampson, B., Lynch, N.: Revisiting the paxos algorithm. Theor. Comput. Sci. 243(1-2), 35–91 (2000)
13. van Renesse, R.: Paxos made moderately complex. Technical report, Cornell University (2011)

On Two-Party Communication through Dynamic Networks*

Sebastian Abshoff, Markus Benter, Manuel Malatyali, and
Friedhelm Meyer auf der Heide

Heinz Nixdorf Institute & Computer Science Department,
University of Paderborn, Fürstenallee 11, 33102 Paderborn, Germany
{abshoff,benter,malatya,fmadh}@hni.upb.de

Abstract. We study two-party communication in the context of directed dynamic networks that are controlled by an adaptive adversary. This adversary is able to change all edges as long as the networks stay strongly-connected in each round. In this work, we establish a relation between counting the total number of nodes in the network and the problem of exchanging tokens between two communication partners which communicate through a dynamic network. We show that the communication problem for a constant fraction of n tokens in a dynamic network with n nodes is at most as hard as counting the number of nodes in a dynamic network with at most $4n + 3$ nodes. For the proof, we construct a family of directed dynamic networks and apply a lower bound from two-party communication complexity.

Keywords: Directed Dynamic Networks, Two-Party Communication, Counting, Token Dissemination, Communication Complexity.

1 Introduction

Many networks, such as wireless sensor and mobile ad-hoc networks, tend to be highly dynamic in the sense that their topologies could change very fast and in an unpredictable way. All these kinds of dynamics in networks pose a major challenge for the design of distributed algorithms and they may even render the computation of non-trivial functions impossible, especially for arbitrarily adversarial dynamics.

Kuhn et al. [7,9,13] study dynamic networks from a worst-case perspective. These networks may change almost completely from round to round and may even be controlled by an adversary that tries to interfere with the algorithms executed on the nodes by changing edges in the network. Under the restriction that messages must be as small as $\Theta(\log(n))$ bits and that the adversary must

---

* This work was partially supported by the German Research Foundation (DFG) within the Collaborative Research Centre "On-The-Fly Computing" (SFB 901), by the EU within FET project MULTIPLEX under contract no. 317532, and the International Graduate School "Dynamic Intelligent Systems".

R. Baldoni, N. Nisse, and M. van Steen (Eds.): OPODIS 2013, LNCS 8304, pp. 11–22, 2013.

give a strongly-connected network in each round, they investigate fundamental problems such as counting the total number of nodes and k-token dissemination.

In the k-token dissemination problem, k pieces of information (or tokens), that are initially stored somewhere in the network, have to be disseminated to all nodes by exchanging small messages. Kuhn et al. [7] present an algorithm for k-token dissemination that requires $\mathcal{O}(n(n + k))$ rounds even if the number of nodes is not known beforehand. They and other authors give matching lower bounds for k-token dissemination for restricted classes of algorithms [3].

The all-to-all token dissemination problem is a special instance of the k-token dissemination problem with $k = n$, where each node initially holds exactly one token. It is known that the counting problem can be solved if the nodes solve the all-to-all token dissemination problem where they disseminate their unique IDs. Using the algorithm by Kuhn et al., this gives an upper bound of $\mathcal{O}(n^2)$ rounds for counting which is currently the best known upper bound for deterministic counting algorithms. It is an open question whether this bound can be improved.

1.1 Our Contribution

We are interested in a two-party version of the k-token dissemination problem, where k tokens, that are initially stored in one node, have to be sent to a given other node in the network. In this work, we show an interesting relation between the counting problem and this two-party version for dynamic networks. It turns out that if there is a fast counting algorithm, then this two-party version can also be solved fast. Or stated differently, any lower bound for the two-party version is also a lower bound for counting algorithms. We would like to point out that our construction holds for any deterministic algorithm that may send arbitrary messages of size $\Theta(\log(n))$ and that the algorithms are not restricted to token-forwarding. To the best of our knowledge, this is the first approach of applying classical results from two-party communication complexity to the domain of dynamic networks.

1.2 Organization of the Paper

Our paper is organized as follows. In Section 2, we introduce the directed dynamic network model and the problems we are interested in. In Section 3, we discuss related work on dynamic networks and on applying lower bounds from communication complexity on static networks. After that, we state our main result in Section 4. Our construction starts in Section 4.1 with a directed dynamic graph G for which we know the round complexity of the two-party version of k-token dissemination. Then, copies of this dynamic network G are used to create special dynamic networks, where the problem of deciding whether the predecessors of some nodes are the same reduces to counting. Next, in Section 4.2, we show that any algorithm for this decision problem can be simulated in a two-party communication network where the channels are replaced by copies of G to decide the set equality problem. By applying a result from communication complexity in Section 4.3, we prove that many messages need to be exchanged in order to solve the set equality problem. From this, we reason that counting cannot be solved faster than the two-party version of k-token dissemination in G. Finally, Section 5 concludes the paper.

2   Models and Problems

A *directed dynamic network* is a directed dynamic graph $G = (V, E)$ where V is a static set of n nodes and $E : \mathbb{N}_{\geq 0} \to V \times V$ is a round-dependent set of directed edges. We define the set of successors of node v in round r as $N^+(v, r) := \{(v, u) \subseteq E(r)\}$, and accordingly, the set of predecessors as $N^-(v, r) := \{(u, v) \subseteq E(r)\}$ in the graph of each round r. We may omit the round parameter r if it is clear from context. Furthermore, we assume that the graph $G(r)$ is strongly-connected in each round r, i.e., there must be a path from each node to every other node in the graph. Apart from that, an adaptive adversary may choose an arbitrary set of edges.

Throughout this paper, we assume that the nodes of the network have unique IDs from a fixed range $\{1, \ldots, n^d\}$ for a fixed constant d. Further, in each round r, each node v can broadcast an arbitrary message of (bit-)size at most $c \log(n)$ to its successors $N^+(v, r+1)$ in the following round $r+1$. c is also a fixed constant and we assume $c > d$. Thus, a node can broadcast, e.g., a constant number of IDs in one round, but it may also broadcast any other type of information encoded in at most $c \log(n)$ bits. As we cannot simulate the transmission of a long message by transmission of several short messages in our dynamic setting, we have to assume fixed values for the lengths of messages, IDs, and tokens. Note that the nodes never know who their successors are due to the dynamic of the network. We say an algorithm has terminated when all nodes have terminated and output the results of their computation.

In particular, we are interested in the following fundamental problems that have also been studied by Kuhn et al. [7,9,13].

Definition 1 (Counting Problem). *All nodes should terminate and output the number n of nodes in the network.*

Definition 2 (k-Token Dissemination Problem). *Each node u in the network receives as input $I(u)$ a possibly empty subset from some token universe $\mathcal{T}$ such that $\left|\bigcup_{v \in V} I(v)\right| = k$. The nodes should disseminate the tokens such that each node knows all k tokens when it terminates. Parameter k is not known by the nodes beforehand.*

Definition 3 (All-to-all Token Dissemination Problem). *This problem is an instance of the k-token dissemination problem with $k = n$, where each node holds exactly one token in the beginning.*

A typical application of the all-to-all token dissemination problem is to disseminate IDs of nodes. Accordingly, we define a two-party version of the k-token dissemination problem which we want to relate to the counting problem.

Definition 4 (Two-Party k-Token Dissemination Problem). *One node v_s in the network receives as input $I(v_s)$ a subset of cardinality k from some token universe $\mathcal{T}$. All other nodes do not receive any input. The nodes should send the tokens through the network such that a given v_t knows all k tokens when it terminates. The nodes may know parameter k beforehand.*

Throughout this paper, we assume that $\mathcal{T} = \{1, \ldots, n^{d'}\}$, for some fixed constant d'.

3   Related Work

Dynamic networks, where the set of bidirectional edges in the network may change arbitrarily and in an adversarial way from round to round as long as the graph is strongly-connected in each round, were introduced by Kuhn et al. [7,9,13]. If there is a strongly-connected subgraph in every sequence of T graphs, they call the dynamic network T-interval connected. On the one hand, for the k-token dissemination problem, Kuhn et al. present a deterministic $\mathcal{O}(n(n + k)/T)$ token-forwarding algorithm which is a randomized algorithm that is only allowed to store and forward one token without modification. This algorithm can be used to obtain an $\mathcal{O}(n^2/T)$ algorithm for the counting problem. On the other hand, they give an $\Omega(nk/T)$ lower bound for k-token dissemination for a restricted class of knowledge-based token-forwarding algorithms[1] and they provide an $\Omega(n \log(k))$ lower bound for deterministic (even centralized) token-forwarding algorithms.

Dutta et al. [3] improve the latter lower bound by Kuhn et al. to $\Omega(nk/\log(n)+n)$ that holds for any randomized, centralized token-forwarding algorithm. Moreover, they show that k-token dissemination can be done in $\mathcal{O}((n+k)\log(n)\log(k))$ w.h.p. if the adversary is only weakly-adaptive. Furthermore, they provide two polynomial-time, randomized and centralized offline algorithms, where one of them returns an $\mathcal{O}(n, \min\{k, \sqrt{k\log(n)}\})$ schedule w.h.p. and the other one an $\mathcal{O}((n+k)\log^2 n)$ schedule w.h.p. if nodes can send a different token along every edge.

Haeupler et al. [5] go beyond the class of token forwarding algorithms and send linear combinations of tokens. They solve the k-token dissemination problem in $\mathcal{O}(nk/\log(n))$ rounds w.h.p. and their technique can be derandomized such that a deterministic centralized algorithm is able to disseminate k tokens in $\mathcal{O}(n \cdot \min\{k, \frac{n}{T}\})$ rounds in a T-interval connected dynamic network.

Brandes et al. [2] develop algorithms for counting if additionally to the worst-case dynamic every edge in the network fail with some probability.

Michail et al. [10] study counting and naming in anonymous unknown dynamic networks. Here, the nodes do not have IDs and naming refers to the

---

[1] A token-forwarding algorithm is called *knowledge-based* if the probability distribution that determines which token is sent by some node in some round depends only on its unique ID, the set of token it is aware of in each round and the sequence of its coin tosses up to this round.

problem of finding unique IDs for the nodes. Interestingly, they also introduce a different communication model where the nodes in the network can send different, individual messages to their neighbors but without any information about their state. In a different paper [11], they study worst-case dynamic networks under a temporal connectivity assumption, where the network is not necessarily strongly-connected in each round.

Kuhn et al. [8] study consensus problems in dynamic networks where all nodes have to agree on the input of one node, and they also study clock synchronization problems in a different time model [6]. Consensus problems were also considered by Biely et al. [1] in directed dynamic networks.

Das Sarma et al. [15] describe randomized algorithms based on random walks on dynamic networks. Wattenhofer et al. [12] analyzed information dissemination in a different but worst-case adversarial model. The field of gossiping algorithms as well as some previous work on evolving graphs is related but different since the communication graph is usually chosen randomly.

In static networks, Das Sarma et al. [14] apply techniques similar to the one we use in this paper. They utilize lower bounds from communication complexity to obtain lower bounds for distributed algorithms for many fundamental graph problems such as the verification of a minimum spanning tree. Subsequently, Frischknecht et al. [4] show that (static) networks require $\Omega(n/\log(n))$ time to compute their diameter.

4 Relating Counting to Communication through a Dynamic Network

In this section, we state and prove our main result.

Theorem 1. *Assume there is a distributed counting algorithm that needs at most $T_{Counting}(n)$ rounds for counting in directed dynamic networks of size at most n. Then, there is an algorithm that can solve the two-party αn-token dissemination problem in an arbitrary directed dynamic network of size n using at most $T_{Counting}(4n + 3)$ rounds for some fixed constant $\alpha > 0$.*

Remark 1. To be precise, we have to assume that the message size for the two-party αn-token dissemination problem is somewhat longer than $c \log(n)$, the one used for counting, but it is still logarithmic. Recall that the (bit-)sizes of messages, IDs, and tokens are logarithmically bounded.

To prove this theorem, we define a family of directed dynamic networks of size at most $4n + 3$ where we embed two directed dynamic networks of size n. For these, we show that many bits must have passed through them to solve a problem defined on the family of directed dynamic networks.

4.1   Special Dynamic Networks and the Same Predecessor Problem

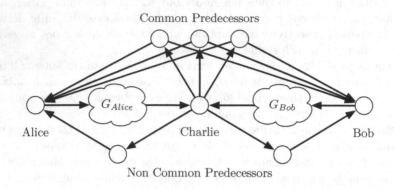

Fig. 1. Construction of the Special Dynamic Network G'

We start with any two directed dynamic networks G_{Alice} and G_{Bob} with n nodes and embed them into another dynamic network G' with n' nodes where $3n + 3 \leq n' \leq 4n + 3$. Beside G_{Alice} and G_{Bob}, this dynamic network consists of three special nodes Alice, Bob, and Charlie, and n to $2n$ other nodes. Both Alice and Bob have one outgoing edge to one fixed node $v_{s,Alice}$ in G_{Alice} or $v_{s,Bob}$ in G_{Bob} respectively. Both G_{Alice} and G_{Bob} have one fixed outgoing edge from one node $v_{t,Alice}$ in G_{Alice} or $v_{t,Bob}$ in G_{Bob} to Charlie. All other n to $2n$ nodes become predecessors of Alice or Bob such that the in-degrees of Alice and Bob are equal to n. Charlie is connected to all these predecessor nodes. G_{Alice} and G_{Bob} are dynamic, all other edges are static. The construction is depicted in Figure 1.

The set of all these directed dynamic graphs G' that can be constructed based on any two directed dynamic graphs G_{Alice} and G_{Bob} with n nodes will be referred to as *special dynamic network* $\mathcal{SDN}_n$. We define $\mathcal{SDN} = \bigcup_{n \in \mathbb{N}^+} \mathcal{SDN}_n$. We allow that all nodes know that the instance is in $\mathcal{SDN}_n$. Also, the nodes Alice and Bob know that they are Alice and Bob respectively but all other nodes do not know whether they are Charlie, a common or a non common predecessor or a node in G_{Alice}, G_{Bob}. Furthermore, all nodes do not know the total number of nodes n' in $\mathcal{SDN}_n$.

Next, we define a problem defined on graphs from $\mathcal{SDN}_n$ which we will reduce to counting and prove afterwards that it is hard to solve.

Definition 5 (Same Predecessor Problem SP_n). *Let U_n be the set of all subsets of $\{1, \ldots, n'^d\}$ with cardinality n. Furthermore, let A and B be the set of predecessors $N^-(Alice)$ and $N^-(Bob)$ of Alice and Bob respectively. The same predecessor function $SP_n : U_n \times U_n \to \{0, 1\}$ is defined as*

$$SP_n(A, B) = \begin{cases} 1 & \text{if } A = B \\ 0 & \text{otherwise} \end{cases}.$$

Although both Alice and Bob can easily obtain their own fixed set of predecessors if the predecessors send their unique IDs, Alice and Bob have to communicate to determine if they have the same set of predecessors.

Lemma 1. *Let $T_{SP_n}(n')$ be the number of rounds a deterministic distributed algorithm needs to decide SP_n in a special dynamic network SDN_n of size at most n', where $3n + 3 \leq n' \leq 4n + 3$. Then,*

$$T_{SP_n}(n') \leq T_{Counting}(n') + 1.$$

Proof. A distributed algorithm for SP_n can determine the in-degree d of Alice and Bob if each node sends its own ID in the first round. Then, given an algorithm for counting, Alice and Bob can count the total number of nodes n' in the network G'. In any graph G', there are Alice, Bob, Charlie, G_{Alice} with n nodes, G_{Bob} with n nodes, and between n and $2n$ predecessors of Alice and Bob. All predecessors are the same iff the number of predecessors is equal to n. Therefore, iff $n' = 3 + 3d = 3 + 3n$, then the predecessors of Alice and Bob are equal. □

4.2 Dynamic Channel Networks and the Set Equality Problem

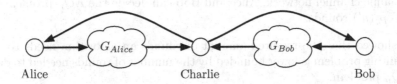

Alice   Charlie   Bob

Fig. 2. Construction of the Dynamic Channel Network G''

To bound $T_{SP_n}(n')$, we define another model which we will refer to as the *dynamic channel network* DCN_n and define $DCN := \bigcup_{n \in \mathbb{N}^+} DCN_n$. Similar to the construction of the graphs in SDN_n, $G'' \in DCN_n$ with $n'' = 2n + 3$ nodes is defined for any two networks G_{Alice} and G_{Bob} with n nodes and three special nodes Alice, Bob, and Charlie. Again, both Alice and Bob have one outgoing edge to one node $v_{s,Alice}$ in G_{Alice} or $v_{s,Bob}$ in G_{Bob} respectively, and both G_{Alice} and G_{Bob} have one fixed outgoing edge from one node $v_{t,Alice}$ in G_{Alice} or $v_{t,Bob}$ in G_{Bob} to Charlie. Charlie is connected to both Alice and Bob via a directed edge. The construction of G'' is depicted in Figure 2.

We can define a function where Alice and Bob are given two sets A and B respectively from some universe $\mathcal{T}$ as input.

Definition 6 (Set Equality Problem EQ_n). *Let T_n be the set of all subsets of $\mathcal{T}$ with cardinality n. For $A, B \in T_n$, the set equality function $EQ_n : T_n \times T_n \to \{0, 1\}$ is defined as*

$$EQ_n(A, B) = \begin{cases} 1 & \text{if } A = B \\ 0 & \text{otherwise} \end{cases}.$$

Note the similarity between SP_n and EQ_n: While SP_n is defined on subsets of n IDs of the predecessors of Alice and Bob respectively, EQ_n is defined on subsets of cardinality n of a token universe $\mathcal{T}$ given to Alice and Bob as input.

Lemma 2. *Let $T_{EQ_n}(n'')$ be the number of rounds a deterministic distributed algorithm needs to decide EQ_n in a dynamic channel network $\mathcal{DCN}_n$ of size n'', where $n'' = 2n + 3$ and $3n + 3 \leq n' \leq 4n + 3$. Then,*

$$T_{EQ_n}(n'') \leq T_{SP_n}(n').$$

Proof. Assume we are given an algorithm that solves SP_n in $T_{SP_n}(n')$ communication rounds. Since every node has unbounded computational power, in one round, Alice in the dynamic channel network can simulate the algorithm executed on Alice and on a predecessor for each token in the special dynamic network. Alice in the dynamic channel network gets all the messages Charlie is broadcasting. So, she is able to simulate each of her predecessors in the special dynamic network and sends, based on this simulation, the message Alice in the special dynamic network would send to G_{Alice}. Bob can do the same with a predecessor for each of his tokens. Since Alice and Bob have the same predecessors in the special dynamic network iff Alice and Bob have the same sets in the dynamic channel network, Alice and Bob can decide the EQ_n problem in at most $T_{SP_n}(n')$ rounds.   □

We showed that the required amount of rounds an algorithm needs to solve the counting problem is lower bounded by the number of rounds needed to decide the EQ_n problem.

4.3   Two-Party Communication and the Set Equality Problem

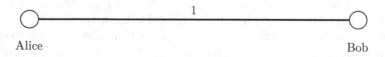

Fig. 3. Two-Party Communication

In this section, we show that Alice and Bob have to exchange at least $\Omega(n \log(n))$ bits through the dynamic channels in the dynamic channel network to bound $T_{EQ_n}(n'')$. We refer to the two-party communication model as depicted in Figure 3.

In the two-party communication model, two nodes Alice and Bob are given inputs A, B. Alice and Bob are connected by an undirected edge, which they can use to alternately exchange messages to compute a function $f(A, B)$. In this model, two rounds correspond to at least one round if Alice and Bob are allowed to send messages at the same time. Yao [16] introduced a lower bounding technique which can be used to get a lower bound on the number of bits that need to be exchanged in our model.

Lemma 3. *In the two-party communication model by Yao, any deterministic algorithm for EQ_n requires at least*

$$C'(EQ_n) := (d'-1) \cdot n \cdot \log_2(n) - 4 = \Omega(n \log(n))$$

rounds for at least $\frac{1}{2} \cdot \binom{n^{d'}}{n}$ inputs.

Proof. EQ_n is a Boolean function on $T_n \times T_n$. Let $\tilde{A}, \tilde{B}$ be subsets of T_n containing sets with cardinality n from universe $\mathcal{T} = \{1, \ldots, n^{d'}\}$. To compute the minimum depth of a decision tree for EQ_n, and thus the number of communication rounds required for deciding EQ_n, we need the minimum number of leaves or equivalently the minimum number of EQ_n-monochromatic rectangles[2] that decomposes EQ_n. Since each monochromatic rectangle has to be constant over $\tilde{A} \times \tilde{B}$, we state that $|\tilde{A} \times \tilde{B}| = 1$ for arguments evaluating to 1, otherwise two different arguments $(X, X), (Y, Y) \in \tilde{A} \times \tilde{B}$, with $X \neq Y$ would lead to $EQ_n(X, Y) = 1$. Hence, the number of EQ_n monochromatic rectangles with result 1 is exactly the number of inputs (X, X) with $X \in T_n$. We conclude,

$$\#EQ_n\text{-monochromatic rectangles}$$
$$\geq \#EQ_n\text{-monochromatic rectangles with result 1}$$
$$= |T_n| = \binom{n^{d'}}{n} \geq (n^{d'-1} - 1)^n.$$

Applying Yao's lower bound,

$$C(EQ_n) \geq \log_2 \left((n^{d'-1} - 1)^n \right) - 2 \geq (d'-1) \cdot n \cdot \log_2(n) - 3,$$

yields that $\Omega(n \log(n))$ rounds are necessary for Alice and Bob to decide EQ_n.

Following a similar argument, it can be shown that for at least $\frac{1}{2} \cdot \binom{n^{d'}}{n}$ inputs at least $(d'-1) \cdot n \cdot \log_2(n) - 4$ rounds are required: If we cut the decision tree at level $(d'-1) \cdot n \cdot \log_2(n) - 4$, then it contains less than $\frac{1}{2} \cdot \binom{n^{d'}}{n}$ leaves. Hence, all other leaves must have height greater than $C'(EQ_n)$. $\square$

Next, we would like to transfer the round complexity from Yao's model to message complexity in terms of transferred bits between Alice and Bob in the dynamic network model. While in Yao's model, Alice and Bob have to send one bit alternately, nodes are able to exchange messages logarithmic size simultaneously in the dynamic network model.

---

[2] For a function $f : M \times N \to \{0, 1\}$, the Cartesian product $S \times T$ (where $S \subseteq M$ and $T \subseteq N$) is called an *f-monochromatic rectangle* if f is constant over $S \times T$. We refer to [16] for more details.

Lemma 4. *In order to decide EQ_n, either Alice has to send at least*

$$\frac{1}{4} \cdot C'(EQ_n) = \frac{1}{4} \cdot (d' - 1) \cdot n \cdot \log_2(n) - 1 = \Omega(n \log(n))$$

bits to Bob or vice versa for at least $\frac{1}{2} \cdot \binom{n^{d'}}{n}$ inputs.

Proof. Assume that both Alice and Bob have to send less than $\frac{1}{4} \cdot C'(EQ_n)$ bits to Bob or Alice respectively in order to solve EQ_n. So, in total, strictly less than $\frac{1}{2} \cdot C'(EQ_n)$ bits are needed. Consequently, strictly less than $C'(EQ_n)$ rounds are required if bits have to be sent alternately as in Yao's model.

Since this model defines that the next bit some communication partner sends is based on its given input and the complete history of received bits, there is a strong interaction between Alice and Bob. Assume Alice and Bob are not allowed to use their complete history to decide what reply to send. Clearly, this could only increase the number of messages necessary to evaluate any function. Thus, the communication complexity for a lower level of interaction is also lower bounded by $C'(EQ_n)$. It follows, that Alice and Bob could have solved the problem in less than $C'(EQ_n)$ rounds which contradicts Lemma 3.   □

Lemma 5. *Let $\alpha = \frac{1}{5}$. Let $T_{Dissemination}(n)$ be the number of rounds a deterministic distributed algorithm needs to solve the two-party αn-token dissemination problem in any directed dynamic network of size n, where $n'' = 2n + 3$. Then,*

$$T_{Dissemination}(n) \leq T_{EQ_n}(n'').$$

Proof. If $T_{Dissemination}(n)$ rounds are required to solve the two-party αn-token dissemination problem in G_{Alice} and G_{Bob}, then it is in particular not possible to send any

$$\left\lceil \log_2 \binom{n^{d'}}{\alpha n} \right\rceil < \alpha \cdot n \cdot d' \log_2(n)$$

bit string from $v_{s,Alice}$ to $v_{t,Alice}$ (or from $v_{s,Bob}$ to $v_{t,Bob}$) in strictly less than $T_{Dissemination}(n)$ rounds since this string could be used to encode the αn tokens given to $v_{s,Alice}$ (or $v_{s,Bob}$). However, by Lemma 4, at least $\frac{1}{4} \cdot C'(EQ_n)$ bits have to be sent from Alice to Bob or vice versa, in order to solve EQ_n. Hence, for $\alpha = \frac{1}{5}$ and $d' > 5$, more bits have to be sent through the dynamic network for solving the equality problem EQ_n than for solving the two-party αn-token dissemination problem. Thus, we conclude $T_{Dissemination}(n) \leq T_{EQ_n}(n'')$.   □

Now, we want to sum up the steps to our main result stated in Theorem 1. In a directed dynamic network of size n, a counting algorithm needs $T_{Counting}(n)$ rounds. Thus, in our special dynamic network $\mathcal{SDN}_n$ of size $n' \leq 4n + 3$, the counting algorithm needs at most $T_{Counting}(n')$ rounds. By Lemma 1, we know that if there is an algorithm which solves the counting problem in the special dynamic network, then there is an algorithm which solves SP_n in at most $T_{Counting}(n') + 1$ rounds. By Lemma 2, we know the EQ_n problem in the dynamic channel network can be decided by a counting algorithm in at most $T_{Counting}(n') + 1$ rounds. Finally by Lemma 5, the two-party αn-token dissemination problem can be solved in time $T_{Counting}(n') + 1$ which proves the claim.

5   Conclusion and Future Prospects

We showed that the number of rounds necessary to count the nodes in a directed dynamic network can be lower bounded by the number of rounds necessary to solve a two-party token dissemination problem. While our construction heavily depends on directed edges, it would be interesting to see if the construction can be extended for undirected dynamic networks. Also, since EQ_n can be solved faster if randomization is used, we would like to know whether another function such as the set disjointness function can be used to derive a stronger lower bound for randomized algorithms.

We think that the two-party k-token dissemination problem is a challenging problem. On the one hand, finding a non-trivial upper bound for this problem seems to be difficult since it is not clear how a dissemination algorithm could exploit the knowledge that v_t is the only node that has to receive all tokens. On the other hand, any non-trivial lower bound for this problem using our construction yields a non-trivial lower bound for the counting problem in directed dynamic networks which at this time does not exist. This poses an interesting open question.

References

1. Biely, M., Robinson, P., Schmid, U.: Agreement in directed dynamic networks. In: SIROCCO, pp. 73–84 (2012)
2. Brandes, P., Meyer auf der Heide, F.: Distributed computing in fault-prone dynamic networks. In: TADDS, pp. 9–14 (2012)
3. Dutta, C., Pandurangan, G., Rajaraman, R., Sun, Z., Viola, E.: On the complexity of information spreading in dynamic networks. In: SODA, pp. 717–736 (2013)
4. Frischknecht, S., Holzer, S., Wattenhofer, R.: Networks cannot compute their diameter in sublinear time. In: SODA, pp. 1150–1162 (2012)
5. Haeupler, B., Karger, D.R.: Faster information dissemination in dynamic networks via network coding. In: PODC, pp. 381–390 (2011)
6. Kuhn, F., Lenzen, C., Locher, T., Oshman, R.: Optimal gradient clock synchronization in dynamic networks. In: PODC, pp. 430–439 (2010)
7. Kuhn, F., Lynch, N.A., Oshman, R.: Distributed computation in dynamic networks. In: STOC, pp. 513–522 (2010)
8. Kuhn, F., Moses, Y., Oshman, R.: Coordinated consensus in dynamic networks. In: PODC, pp. 1–10 (2011)
9. Kuhn, F., Oshman, R.: Dynamic networks: models and algorithms. SIGACT News 42(1), 82–96 (2011)
10. Michail, O., Chatzigiannakis, I., Spirakis, P.G.: Brief announcement: Naming and counting in anonymous unknown dynamic networks. In: Aguilera, M.K. (ed.) DISC 2012. LNCS, vol. 7611, pp. 437–438. Springer, Heidelberg (2012)
11. Michail, O., Chatzigiannakis, I., Spirakis, P.G.: Causality, influence, and computation in possibly disconnected synchronous dynamic networks. In: OPODIS, pp. 269–283 (2012)
12. O'Dell, R., Wattenhofer, R.: Information dissemination in highly dynamic graphs. In: DIALM-POMC, pp. 104–110 (2005)

13. Oshman, R.: Distributed Computation in Wireless and Dynamic Networks. PhD thesis, Department of Electrical Engineering and Computer Science, Massachusetts Institute of Technology, Cambridge, MA 02139 (September 2012)
14. Sarma, A.D., Holzer, S., Kor, L., Korman, A., Nanongkai, D., Pandurangan, G., Peleg, D., Wattenhofer, R.: Distributed verification and hardness of distributed approximation. SIAM J. Comput. 41(5), 1235–1265 (2012)
15. Das Sarma, A., Molla, A.R., Pandurangan, G.: Fast distributed computation in dynamic networks via random walks. In: Aguilera, M.K. (ed.) DISC 2012. LNCS, vol. 7611, pp. 136–150. Springer, Heidelberg (2012)
16. Yao, A.C.-C.: Some complexity questions related to distributive computing (preliminary report). In: STOC, pp. 209–213 (1979)

Eventual Leader Election
in Evolving Mobile Networks

Luciana Arantes[1], Fabíola Greve[2], Pierre Sens[1], and Véronique Simon[1]

[1] LIP6, Université Pierre et Marie Curie, Inria, CNRS, France
firstname.lastname@lip6.fr
[2] DCC - Computer Science Department / Federal University of Bahia, Brazil
fabiola@dcc.ufba.br

Abstract. Many reliable distributed services rely on an eventual leader election to coordinate actions. The eventual leader detector has been proposed as a way to implement such an abstraction. It ensures that, eventually, each process in the system will be provided by an unique leader, elected among the set of correct processes in spite of crashes and uncertainties. A number of eventual leader election protocols were suggested. Nonetheless, as far as we are aware of, no one of these protocols tolerates a free pattern of node mobility. This paper proposes a new protocol for this scenario of dynamic and mobile unknown networks.

Keywords: Fault-tolerant leader election, dynamic networks, process mobility, asynchronous systems.

1 Introduction

Dynamic distributed systems based on ad-hoc collections of distributed computing devices, wireless and mobile networks, unstructured peer to peer networks, opportunistic grids, or clouds are supposed to allow participants to access services and information regardless of their location, topology or mobility pattern. Nonetheless, the issue of designing reliable services which can cope with the high dynamics of these systems is a challenge.

Many reliable distributed services rely on an eventual leader election to coordinate actions. The Ω *leader detector* has been proposed as a way to implement such an abstraction [1]. It ensures that, eventually, each process in the system will be provided by an unique leader, elected among the set of correct processes, in spite of crashes, uncertainties and dynamics. However, the Ω detector cannot be implemented in a purely asynchronous system [1]. Thus, some additional assumptions on the underlying system should be made in order to implement it. With this aim, two orthogonal approaches can be distinguished: *timer-based* and *message-pattern*. The timer-based is the traditional approach and supposes that channels are eventually timely; the system may be described as being partially synchronous. An alternative approach assumes that the system satisfies a message exchange pattern on the execution of a communication mechanism. While the timer-based approach imposes a constraint on the physical time (to satisfy message transfer delays), the message-pattern approach imposes a constraint on the logical time (to satisfy a message delivery order) [2].

R. Baldoni, N. Nisse, and M. van Steen (Eds.): OPODIS 2013, LNCS 8304, pp. 23–37, 2013.
© Springer International Publishing Switzerland 2013

A number of leadership protocols were proposed to implement Ω. The first timer-based solutions adopted strong assumptions concerning time and channel reliability [1, 3]; afterwards, they seek to find weaker and weaker conditions regarding synchrony and reliability [4–8]. Nonetheless, the totality of these protocols adopts a classical model of "known" networks in which the set of participants (Π), its cardinality (n), and the maximum number of faults (f) are known.

It happens that the inherent dynamics of the new environments prevent processes from gathering a global knowledge of the system properties. That is why recent solutions, aiming to implement Ω in a dynamic system of "unknown" networks have emerged [9, 10]. They seek for models and solutions with the possible weakest assumptions, regarding the knowledge of the topology, the communication graph, as well as the channel connectivity and reliability, trying to get as close as possible to reality. Although these proposals lead to a breakthrough in the implementation of the leader abstraction with dynamics requirements, none of them tolerate node mobility.

Very few papers deal with node mobility [11–14]. However, to the best of our knowledge, none of them consider a system with an arbitrary graph topology that changes over time. In this paper we provide a first Ω algorithm to tolerate a generic pattern of node mobility in an unknown network, subject to messages losses and a topology that changes over time. [14] is perhaps the work with most similarity with ours. However, differently from our solution which follows a message-pattern approach, it considers a timer-based one and the existence of stable periods that should last long enough to elect a leader.

The current paper brings thus two main contributions: *(i)* The proposition of a *model to solve the leader election problem in mobile dynamic systems*. This model, although simple, captures the requirements to solve the problem and represents the network by a communication graph with a dynamic topology. *(ii)* A *leadership algorithm that implements the Ω class under the proposed model*. It follows the message-pattern approach and does not assume timely links.

2   Related Work

Leadership Protocols for "known" Networks. A number of leadership protocols were proposed to implement Ω in an asynchronous system prone to crash failures and taking into account the classical model of "known" networks in which Π, n and f are known and moreover the communication graph is complete.

The first solutions [1, 3] adopted strong assumptions concerning reliability and time. They consider that *all* links were reliable and eventually timely. Further solutions seek to find weaker and weaker conditions regarding synchrony and channel reliability. Aguilera et al. relax the necessity regarding the time constraints of all links, firstly proposing an algorithm in which only one process should maintain an *eventual timely link* to all the other processes [4]. Afterwards, they weaken the condition to an outgoing link, in such a way that one node (namely, the $\Diamond$-*source* process) should have an eventually outgoing timely link to all the other processes, while the other links may still lose messages [5, 6]. These conditions ensure that after some time only the current leader sends message forever.

Another important work in this line is due to Malkhi et al. [7] that proposes a solution without having any eventual timely links, but which considers *eventual accessible links*. Their algorithm assumes that eventually one process (namely, the $\Diamond$-*accessible* process) can send messages such that every message obtains f timely responses. One very practical interest of this assumption is that the links are moving, that is, the f responders need not to be the same and may change from one message to another. Most recently, [8] presents a solution with a weaker model that unifies the assumptions made in [5, 6] and [7]. It shows that Ω can be implemented with at least one process with f outgoing moving eventual timely links, assuming either unicast or broadcast steps.

An orthogonal and totally different approach for implementing Ω is based on the satisfaction of a *message exchange pattern* in the system. It has been proposed by [15] to implement a $\Diamond S$ failure detector and exploited so far by [2, 12] to implement Ω. They show that Ω can be built as soon as the following process behavior property is satisfied: there is a correct process p and a set Q of $(f+1)$ processes, such that eventually the response of p from any query issued by one process $q \in Q$ is always received by q among the first $(n-f)$ responses.

Leadership Protocols for "unknown" Networks. Some recent works aiming to implement Ω in dynamic systems with the possible weakest assumptions, regarding the knowledge and communication graph have emerged. In common, they share a reachability communication assumption between every pair of correct processes. Jimenez et al. [16] show that it is possible to implement Ω with no knowledge about the membership of the system, even under the minimal conditions regarding link synchrony and reliability. They provide an algorithm for Ω considering an unknown network, a complete communication graph and links that are fair-lossy, but timely.

Fernandez et al. [9] propose two Ω algorithms with weakest assumptions. The first algorithm considers a partial unknown network, with a global knowledge about the lower bound on the number of correct processes (represented by $\alpha = n - f$) and fair-lossy timely links. The communication graph is not complete but there are direct links between a correct process p and a set of correct processes. The second algorithm considers unknown networks and a complete communication graph. Links are fair-lossy and timely composed of output direct links between a correct process p and every correct process in the system.

Tucci et al. [10] study the Ω abstraction in a system with bounded concurrency. It assumes an unknown network, but a fully connected dynamic graph. It provides the first proposal for Ω algorithms for the infinite arrival message-passing mode [17], in which an infinite number of processes may arrive and depart over time, but the number of processes which are simultaneously up is finite (including the correct processes).

Leadership Protocols with Node Mobility. Masum et al. [11] present an Ω algorithm which, contrarily to ours, assumes totally reliable and timely channels. Cao et al. [12] provide an implementation of Ω for a network composed of mobile hosts (MH) and mobile support stations (MSS). The eventual leader is an MH, but it is elected by the MSSs. Differently from our work, the set of MSS forms a static distributed system of reliable channels in a "known" network. Melit et al. [13] propose both a model and an Ω algorithm that tolerate node mobility and partitions. But, to converge, their approach requires that the topology eventually does not change. Unlike ours, this last requirement prevents arbitrary changes in the topology along the system existence.

Larrea et al. [18] propose an Ω specification and algorithm suited to dynamic systems, where processes join and leave, relying on eventually timely links. Gomez-Calzado et al. [14] extended the previous works so as to take into account graph joins/fragmentations and process mobility. They make a stability assumption in which there is no graph partitioning and the existence of bidirectional connectivity among processes. Differently from our solution, they adopt a timely based assumption during stable periods and some other conditions in the graph.

3 Model for Eventual Leader Election in Mobile Systems

The system is a collection of mobile nodes which communicate by sending and receiving messages via a network with broadcast facilities. The system is *asynchronous* and there are no assumptions on the relative speed of processes or on message transfer delays. We take the range $\mathcal{T}$ of the clock's tick to be the set of natural numbers. There is no global clock and processes do not have access to $\mathcal{T}$: it is introduced for the convenience of the presentation and proofs.

3.1 Communication Model

Time-Varying Communication Graph. Following [19], we consider that the dynamics of the network is represented by a *time-varying graph*, denoted TVG.

Definition 1. *[Time-varying graph]. A TVG is a tuple* $\mathcal{G} = (V, E, \mathcal{T}, \rho, \zeta, \psi)$, *where: (1)* $V = \Pi$ *represents the set of nodes, (2)* $E \subseteq V \times V$ *represents the set of communication links between nodes, (3)* $\mathcal{T} \subseteq \mathbb{N}$ *is a time span, (4)* $\rho :$ $E \times \mathcal{T} \to \{0, 1\}$ *is an edge presence function: an edge* $e \in E$ *is available at a given time* $t \in \mathcal{T}$, *such that* $\rho(e, t) = 1$ *iff* e *is present at* t, *otherwise* $\rho(e, t) = 0$, *(5)* $\zeta : E \times \mathcal{T} \to \mathbb{N}$ *is a latency function, indicating the time taken to cross a given edge* e *if starting at a given time* t [1]; *(6)* $\psi : V \times \mathcal{T} \to \{0, 1\}$ *is a node presence function, indicating whether a given process* $p_i \in V$ *is up at a given time* $t \in \mathcal{T}$, *such that* $\psi(p_i, t) = 1$ *iff node* p_i *is up at* t, *otherwise* $\psi(p, t) = 0$.

Links are directed and the edge from p_i to p_j is denoted $e_{i,j} = (p_i, p_j)$. N_i^t is the set of 1-hop *neighbors* of p_i and E_i^t is the set of edges that connect p_i to these neighbors at time t. The neighborhood relationship establishes the edge set, in such a way that $p_j \in N_i^t$ iff $e_{i,j} \in E_i^t$, such that $\rho(e_{i,j}, t) = 1$. The *degree* of p_i at time t is defined to be $Deg_i^t = |E_i^t|$. Given a TVG $\mathcal{G}$, the graph $G = (V, E)$ is called the *underlying graph* of $\mathcal{G}$. G should be considered as a sort of *footprint*

---

[1] Note that the effective delivery of a message sent at time t on an edge could be subjected to further constraints regarding the latency function, such as the condition that $\rho(e)$ returns 1 for the whole interval $[t; t + \zeta(e, t))]$.

of $\mathcal{G}$ which flattens the time dimension and indicates only the pair of nodes that have relations at some time in $\mathcal{T}$. Journeys can be thought of as paths over time from a source to a destination.

Definition 2. *[Journey] A sequence of couples* $\mathcal{J} = \{(e^1, t_1), (e^2, t_2), \ldots, (e^k, t_k)\}$, *such that* $\{e^1, e^2, \ldots, e^k\}$ *is a walk in* G, *is a journey in* $\mathcal{G}$ *if and only if* $\rho(e^i, t_i) = 1$ *and* $t_{i+1} \geq t_i + \zeta(e^i, t_i)$ *for all* $i < k$. *Let departure*$(\mathcal{J}) = t_1$ *be the starting date and arrival*$(\mathcal{J}) = t_k + \zeta(e^k, t_k)$ *be the last date of* $\mathcal{J}$. *Let* $\mathcal{J}_{(i,j)}$ *be a journey starting at node* p_i *and ending at node* p_j; *in this case, we say that* p_i *reaches* p_j *or more simply,* $p_i \rightsquigarrow p_j$. *Let us denote by* $\mathcal{J}_{\mathcal{G}}^*$ *the set of all possible journeys in* $\mathcal{G}$, *and by* $\mathcal{J}_{(i,j)}^* \subseteq \mathcal{J}_{\mathcal{G}}^*$ *those journeys starting at* p_i *and ending at* p_j.

Channels. Local broadcast between 1-hop neighbors is *fair-lossy*. This means that messages may be lost, but, if a correct p_i broadcasts m an infinite number of times, then every p_j permanently in its neighborhood receives m from p_i an infinite number of times, otherwise p_j is faulty or out of p_i's neighborhood. That is, if p_i starts to send m at time t an infinite number of times, then, if $\rho(e_{i,j}, t') = 1, \forall t' \in [t; +\infty)$, p_j receives m an infinite number of times if p_j is a correct neighbor of p_i. In the case of a wireless network, this condition is e.g. attained if the MAC layer reliably delivers broadcast data, even in presence of unpredictable behaviors, such as fading, collisions, and interference; solutions in this sense were proposed in [20, 21].

3.2 Process Model

We consider the *finite arrival model* [17]: The network is a dynamic system composed of infinitely many mobile processes, but each run consists of a finite set Π of n nodes, namely, $\Pi = \{p_1, \ldots, p_n\}$. *The membership is unknown.* Processes are not aware of Π or n, because, moreover, these values can vary from run to run [17]. There is one process per node; each process knows its own identity, but it does not necessarily know the identities of the others. A process may fail by *crashing*, i.e., by prematurely or by deliberately halting (switched off); a crashed process does not recover. Indeed, a process can re-connect to the system, but with a new identity, thus, it is considered as a new process. Processes may re-connect as they wish, but the number of re-entries is bounded, due to the finite arrival assumption. Until it possibly crashes, a process behaves according to its specification. A process that does follow its algorithm specification and never crashes is said to be *correct*.

Let us thus define the status that a process may exhibit along the system execution. Informally, a *stable* process is a correct process that never leaves the system; otherwise, it is *faulty*.

Definition 3. *[Process status]. Let* $t \in \mathcal{T}$. *A process* p_i *may assume the following status.*

$stable^t(p_i) \Leftrightarrow \forall t' \geq t, \ \psi(p_i, t') = 1$

$faulty^t(p_i) \Leftrightarrow (\exists s, s < t, \ \psi(p_i, s) = 1) \wedge (\forall t' \geq t, \psi(p_i, t') = 0)$

The *failure pattern* of the system, namely $F(t)$, is the set of processes that have failed in the system by time t. That is, $F(t) = \{p_i : faulty^t(p_i)\}$. Similarly, $S(t)$, is the set of processes that are stable in the system by time t. That is, $S(t) = \{p_i : stable^t(p_i)\}$.

Definition 4. *[Process sets]. The set of processes in the system may be divided into:* $\text{STABLE} \stackrel{def}{=} \bigcup_{t \in \mathcal{T}} S(t)$ *and* $\text{FAULTY} \stackrel{def}{=} \bigcup_{t \in \mathcal{T}} F(t)$

3.3   Network Connectivity

To solve the eventual leader abstraction, we are mostly interested in the *transmission* TVG induced by the stable nodes in the system.

Definition 5. *[Transmission TVG]. The transmission TVG is a tuple* $\mathcal{G}_S^{tr} = (V_S, E_S, \mathcal{T}, \rho_{tr}, \zeta, \psi)$, *in which* $V_S = \text{STABLE}$; $E_S \subseteq V_S \times V_S$ *and* ρ_{tr} *is a transmission edge presence function:* $\rho_{tr}(e_{i,j}, t) = 1$ *iff a message sent from* p_i *at time* t *is delivered to* p_j *at time* $t + \zeta(e_{i,j}, t)$.

We can identify classes of TVG based on the temporal properties established by the entities. The classes are important because they imply necessary conditions and impossibility results for distributed computations. Notably, Class 5 (Recurrent connectivity) [19] is important to our study.

Assumption 1. *[Network recurrent connectivity]. In the subsystem of stable nodes, represented by TVG* $\mathcal{G}_S^{tr}$, $\forall p_i, p_j \in V_S$, $\forall t \in \mathcal{T}$, $\exists \mathcal{J} \in \mathcal{J}_{(p_i,p_j)}^*$: *departure* $(\mathcal{J}) > t$.

The recurrent connectivity is a fundamental assumption, mandatory to ensure reliable dissemination of messages to all stable processes in a dynamic network [19] and thus to ensure the properties of the leader oracle [1, 16, 22]. It means that, at any point t in time, the TVG $\mathcal{G}_S^{tr}$ remains connected over time. Thus, for all stable nodes p_i, p_j, at any time, $p_i \rightsquigarrow p_j$.

3.4   The Ω Class

A *leader* oracle is a distributed entity that provides processes with a function *leader()* that when invoked by any process p outputs a single process q, denoted the leader. In the context of a dynamic system, a leader oracle of the Ω class satisfies the following Eventual leadership property: *There is a time after which every invocation of leader() by any stable process returns the same stable process.* Therefore, a unique leader is eventually elected but no processes knows when the leader is elected.

4   An Eventual Leader Oracle for Mobile Systems

4.1   Stable Query-Response Communication Mechanism

Our eventual leader oracle solution is based on the *message pattern approach* [15] and uses, to this end, a local QUERY-RESPONSE communication mechanism [22] adapted to a network with unknown membership. At each *query-response* round, a node systematically broadcasts a QUERY message to the nodes in its neighborhood until it possibly crashes or leaves the system. The interval between two consecutive queries is finite but arbitrary. Each couple of QUERY-RESPONSE messages is uniquely identified in the system. A process p_i launches the primitive by sending a QUERY(m) with a message m. When a process p_j delivers this query, it updates its local state and answers by sending back a RESPONSE(m') with a message m' to p_i. Then, when p_i has received at least α_i responses from different processes, the current QUERY-RESPONSE terminates. Without loss of generality, the response for p_i itself is among the α_i responses.

Formally, the QUERY–RESPONSE primitive has the following properties: (i) QR-Validity: If a QUERY(m) is delivered by process p_j, it has been sent by process p_i; (ii) QR-Uniformity: A QUERY(m) is delivered at most once by a process; (iii) QR-Termination: Let t be the time at which a process p_i terminates to send a query. If $faulty^t(p_i)$ does not hold, then that query generates at least α_i RESPONSE(m') messages from a subset of X_i processes, $|X_i| \geq \alpha_i$.

An implementation of a couple of QUERY-RESPONSE communication over fair-lossy local channels can be done by the repeated broadcast of the query by the sender p_i until it has received at least α_i responses from its neighbors. Since the communication pattern followed is local, α_i is defined locally as a function of the expected number of stable known neighbors with whom p_i may communicate at the time t in which the QUERY is issued. We consider that f_i is the maximum number of faulty processes in p_i's neighborhood. Thus, since the set of responses received by p_i includes its own response, $\alpha_i = |N_i^t| - f_i + 1$, which guarantees the liveness of QUERY-RESPONSE rounds. To ensure that at least one stable node p_j $(p_j \neq p_i)$ receives the QUERY and sends a response to p_i, $\alpha_i > f_i + 1$.

The local choice for α_i changes from existing solutions which consider a global value proportional to the number of processes [12, 15, 22]. Moreover, it follows recent works on fault tolerant communication in radio networks which propose a "local" fault model, instead of a "global" fault model, as an adequate strategy to deal with the dynamics and unreliability of wireless channels in spite of failures [21]. To reliably deliver data in spite of crashes, the maximum number of local failures should be $f_i < Deg_i^t/2$ [23].

To implement Ω, a *stable termination property* is necessary for the reliable dissemination of the information to the whole network and consequent election.

Property 1. Stable Termination Property ($\mathcal{S}at\mathcal{P}$). Let p_i be a node which issues a QUERY. Thus, $\exists p_j \in$ STABLE, $p_j \neq p_i$, which receives that QUERY.

$\mathcal{S}at\mathcal{P}$ is a guarantee that information from/to p_i is going to be sent/received to/from at least a stable p_j in its neighborhood.

4.2 Behavioral Property

Instead of synchrony assumptions, to ensure the accuracy of the election, we have adopted a *message pattern* model which establishes conditions on the logical time the messages are delivered by processes. These are unified in the *stabilized responsiveness property* or $\mathcal{SRP}$.

Property 2. Stabilized Responsiveness Property ($\mathcal{SRP}$). Let X_j^t be the set of processes whose respective response to the latest QUERY of p_j before t is among the first α_j responses received by p_j. Process p_i satisfies $\mathcal{SRP}$ at time t:

$$\mathcal{SRP}^t(p_i) \ \textit{iff} \ \textit{stable}^t(p_i) \wedge \forall p_j \in \Pi \ ((\exists e_{i,j}, \exists t' \geq t, \rho_{tr}(e_{i,j}, t') = 1)$$

$$\Rightarrow \forall t'' \geq t' + \zeta(e_{i,j}, t'), p_i \in X_j^{t''})$$

$\mathcal{SRP}^t(p_i)$ states that there exists a time t after which all nodes of p_i's neighborhood receive, to every of their queries, a response from p_i which is always among the first α_j responses to the query. It denotes the ability of a stable node p_i to eventually always reply, among the first α_j nodes, to a QUERY sent by p_j. Moreover, as nodes may move, the $\mathcal{SRP}^t(p_i)$ states as well that neighbors of p_i eventually stop moving outside p_i's neighborhood.

To solve Ω, the $\mathcal{SRP}(p_i)$ property should hold for one stable node p_i in the system, thus preventing a probable leader p_i to be permanently demoted. As a matter of comparison, in the timer-based model, this property would be: there is a time t after which the output channels from a stable node p_i to every other neighbor p_j that communicates with p_i are eventually timely.

4.3   An Eventual Leader Election Algorithm

Algorithm 1 describes a protocol for implementing Ω in a mobile system satisfying the model, properties, and assumptions stated in Sections 3 and 4.

Notations. The algorithm uses the following variables and functions:

- mid_i: a counter used to timestamp every couple of QUERY-RESPONSE messages. Before broadcasting a new QUERY, p_i increments mid_i. These two operations are atomically performed.
- $local\_known_i$: the current knowledge of p_i about its neighborhood, i.e., the set of nodes that communicated directly with p_i. It is composed of tuples of the form $\langle mid_j, p_j \rangle$: mid_j is associated with the greatest timestamp value of a QUERY or RESPONSE message received by p_i from p_j.
- $global\_known_i$: the current knowledge of p_i about the membership of the system. Similarly to $local\_known_i$, it is composed of tuples of the form $\langle mid_j, p_j \rangle$.
- $punish_i$: a set of tuples of the form $\langle ct, p \rangle$ where ct is a punish counter and p the identity of the punished node.
- $recvfrom_i$: the set of processes that replied to the latest QUERY of p_i.
- $MaxKnown()$: a boolean function that checks if p_i has the greatest timestamp associated to a message received from p_j. It is used to verify if a given neighbor process has moved or not.
- $UnionMax(set_1, set_2, ...)$: a function that performs the union of sets whose tuple elements have the form $\langle ct, p \rangle$. If $\langle -, p \rangle$ belongs to several sets, the function considers the one whose value ct is the greatest one.
- $Update\_State()$: a function used to update the state of p_i's sets with the most recent information. It keeps the tuples $\langle ct, p \rangle$ with the greatest counters in these sets. It is used to evaluate the contents of a receiving message (QUERY or RESPONSE).
- $leader()$: function that returns the current leader.

Algorithm 1. Eventual Leader Election for Mobile Networks

```
1    Init:
2
3        punish_i ← {⟨0, p_i⟩}
4        local_known_i ← {⟨mid_i, p_i⟩}
5        global_known_i ← {⟨mid_i, p_i⟩}
6        recvfrom_i ← ∅
7        mid_i ← 1
8        broadcast QUERY(mid_i, punish_i, global_known_i)
9
10   Task T1: [Punishment]
11   Repeat forever
12        Wait until |recvfrom_i| ≥ α_i
13        { Punishing known processes which did not responded }
14        If ∀p_j : ⟨−, p_j⟩ ∈ local_known_i ∧ p_j ∉ recvfrom_i ∧ MaxKnown(p_j) then
15            If ⟨0, p_j⟩ ∈ punish_i then
16                c_min ← min c : ⟨c, −⟩ ∈ punish_i
17                replace in punish_i ⟨0, p_j⟩ by ⟨c_min + 1, p_j⟩
18            Else
19                replace in punish_i ⟨v, p_j⟩ by ⟨v + 1, p_j⟩
20        recvfrom_i ← ∅
21        mid_i ← mid_i + 1
22        broadcast QUERY(mid_i, punish_i, global_known_i)
23   End repeat
24   Task T2: [Response]
25   upon reception of RESPONSE (mid_j, punish_j, global_known_j) from p_j
26
27        UpdateState(mid_j, punish_j, global_known_j, p_j)
28        recvfrom_i ← recvfrom_i ∪ {p_j}
29
30   Task T3 [Query]
31   upon reception of QUERY (mid_j, punish_j, global_known_j) from p_j
32
33        UpdateState(mid_j, punish_j, global_known_j, p_j)
34        send RESPONSE (mid_i, punish_i, global_known_i) to p_j
35
36   Task T4 [Leader Election]
37   upon the invocation of leader()
38
39        return l such that ⟨c, l⟩ = Min(punish_i)
40
41   MaxKnown (p) [Max counter associated with p]
42
43        If x : ⟨x, p⟩ ∈ local_known_i ≥ y : ⟨y, p⟩ ∈ global_known_i then
44            return true
45        Else
46            return false
47
48   UpdateState (mid_j, punish_j, global_known_j, p_j) [Union of states]
49
50        local_known_i ← UnionMax(local_known_i, {⟨mid_j, p_j⟩})
51        global_known_i ← UnionMax(global_known_i, global_known_j, {⟨mid_j, p_j⟩})
52        punish_i ← UnionMax(punish_i, punish_j)
53
```

Underlying Principle. The algorithm elects the leader on a basis of a punishment procedure and on the periodic exchange of QUERY-RESPONSE messages. Processes exchange these messages to know each other, to show that they are alive, as well as to exchange information to elect the leader. If a QUERY sent by process p_i is not responded by a process p_j that p_i locally knows, then p_j is punished by p_i. Each time p_i punishes p_j, it increments the counter ct_j associated to p_j in $punish_i$.

The rationale behind the punishment procedure is that a process that fail will be infinitely often punished. The algorithm thus will eventually elect a stable process that has the smallest punish counter. To ensure that all the nodes will elect the same leader, processes should exchange their information regarding locally known processes and their respective punishment counters. Thus, each QUERY or RESPONSE message sent by p_i, beyond the message id (mid_i), carries the sets $punish_i$ and $global\_known_i$. Since the network remains connected over time (Assumption 1), the information will spread over all stable processes.

To tolerate the mobility of nodes, the algorithm makes use of the message counters. The timestamp of the latest message received from processes is used to avoid false suspicions in case of mobility. If p_j is in $local\_known_i$ and if it moves from p_i's neighborhood, then it will be punished by p_i according to the latest message received. But, as soon as p_i gets the information (by the contents of a received message) that another node has received a message from p_j with a greater timestamp, p_i stops to punish p_j. In this case, p_i suspects p_j to have moved from its neighborhood and considers that it is still alive in the network.

Description. Initially, p_i sends a first QUERY to introduce itself to the network (line 8). Then, four tasks are launched: $T1$, $T2$, $T3$ and $T4$.

Task $T1$ [Punishment]: This task is made up of an infinite loop. At each round, process p_i waits for at least α_i responses, which includes p_i's own response. For each RESPONSE$(mid_j, punish_j, global\_known_j)$ not received from p_j such that p_j is considered as a current neighbor of p_i (i.e., it belongs to $local\_known_i$) and p_j is not suspected to have moved from p_i's neighborhood (i.e., p_i has a greater message timestamp received from p_j than the other processes of which p_i is aware), then p_j will be punished by p_i (lines 15 – 19). Notice that if it is the first time that p_j is punished by p_i, then, its punish counter will be equal to $\langle c_{min} + 1, p_j \rangle$ (line 17). Then, mid_i counter is incremented and a QUERY$(mid_i, punish_i, global\_known_i)$ message is sent to all nodes in p_i's neighborhood.

Task $T2$ [Response]: In this task, node p_i handles the reception of a RESPONSE message sent by p_j containing mid_j, as well as the sets $punish_j$ and $global\_known_j$. Process p_i calls upon $Update\_State()$ to update its state about punishment of processes, membership, and neighborhood with more recent information coming from p_j. It also includes p_j with the respective mid_j in the set of processes that answered to its latest query $(local\_know_i)$, as well as in the set that keeps its membership knowledge $(global\_known_i)$.

Task $T3$ [Query]: In this task, node p_i handles the reception of a QUERY message sent by p_j containing mid_j, as well as the sets $punish_j$ and $global\_known_j$. Similarly to $T2$, node p_i updates its state about punishment of processes, membership, and neighborhood with more recent information coming from p_j. It also answers p_j's QUERY with a RESPONSE message timestamped with its mid_i counter.

Task $T4$ [Leader]: This task handles the invocation of $leader()$. Whenever called, the $leader()$ function returns the process with the smallest counter in $punish_i$, thanks to the $Min(punish_i)$ function (line 39). In the case that more than a node satisfies such a condition, the identities of the nodes break the tie. Eventually, all nodes will elect the same leader, as proved in the next section.

5 Proof of Correctness

We present a sketch of proof that Algorithm 1 ensures the *eventual leadership* property. We consider a mobile network of unknown membership that satisfies the model and assumptions stated in Sections 3 and 4.

Notations

(i) The *state* of a process p_i in time t is represented by the contents of its variables at t. (ii) Let set_i be one of the sets *local_known$_i$*, *global_known$_i$* or *punish$_i$* of process p_i. We denote set_i^t this set at time t. Moreover, $set_i^t(p_j) = c$ if the value $\langle c, p_j \rangle \in set_i$ at time t; otherwise $set_i^t(p_j) = \bot$. We denote $set_i^*(p_j)$ as the set of all values of $set_i^t(p_j)$ such that $t \in \mathcal{T}$ and $set_*^*(p_j)$ as the set of all $set_i^*(p_j)$, $i \in \Pi$. (iii) Let m be a message sent by p_i. Then, m is either a QUERY or a RESPONSE message and it contains mid_i and the sets *punish$_i$* and *global_known$_i$*. (iv) We consider that process p_j *punishes* p_i if it executes lines 17 or 19 increasing the counter of p_i in its *punish$_j$* set. (v) Let us denote the set SBP as the subset of processes that have a bounded value on the punish set of all processes, $SBP = \{p_i \in \Pi \mid punish_*^*(pi)\ is\ bounded\}$.

Lemma 1. *Let $\mathcal{J}_{(i,j)}$ be a journey from p_i to p_j in the TVG $\mathcal{G}_S^{tr}$. Let t_0 be the departure and t_f be the arrival of $\mathcal{J}_{(i,j)}$. Let set be either punish or global_known. For any process p_k, if $set_i^{t_0}(p_k) \neq \bot$ then $set_j^{t_f}(p_k) \neq \bot \wedge set_j^{t_f}(p_k) \geq set_i^{t_0}(p_k)$.*

Proof. We first show that the lemma holds for the one-step journey $\mathcal{J}_{(i,j)} = \{(e_{i,j}, t_0)\}$, i.e., there is a message m sent by p_i at time t_0 which is delivered and handled by p_j at time $t_f = Arrival(\mathcal{J}_{(i,j)})$. We denote $punish_m$ and $global\_known_m$ the sets $punish_i^{t_0}$ and $global\_known_i^{t_0}$ carried by m respectively. Upon reception of m, p_j calls the function $UpdateState()$ and the result of $UnionMax(set_j, set_m, ...)$ is stored in set_j. Thus, after m is handled, if $set_i^{t_0}(p_k) \neq \bot$, then $set_j^{t_f}(p_k) \neq \bot \wedge set_j^{t_f}(p_k) \geq set_i^{t_0}(p_k)$. Moreover, $punish_i$ is modified either (i) when p_i punishes some process or (ii) upon reception of m. In (i), $punish_i$ is updated in line 17 and 19. In both cases, the associated counter values are increased by at least one. In (ii), the result of $UnionMax(punish_i, punish_k)$ is stored in $punish_i$. Therefore, values in $punish_i$ never decrease locally. On the other hand, $global\_known_i$ is only updated on reception of a message and, thus, similarly to $punish_i$, values in $global\_known_i$ never decrease as well. Since in a journey, the arrival of a message precedes the departure of the message that follows, by induction and transitivity of inequality, the lemma holds for a journey of any step size.

Observation 1. Let $mid_i = c$ at time t. If a process p_j does not receive any message sent by p_i after t then $local\_known_j^{t'}(p_i) \leq c$ or $local\_known_j^{t'}(p_i) = \perp$, $\forall t' \geq t$. This follows since $local\_known_j(p_i)$ is updated by p_j upon the reception of a message from p_i and, from assumption, the value mid_m carried by this message is such that $mid_m \leq c$.

Lemma 2. *Let p_i be a stable process and $t \in \mathcal{T}$. If $\mathcal{SRP}^t(p_i)$ then there is a time $u \geq t$ after which no process punishes p_i.*

Proof. Let p_j be a process. Three cases are possible.

Case 1: p_j is faulty. If $faulty^u(p_j)$, $u \geq t$, then p_j will not punish p_i after u.

Case 2: p_j is stable and it receives a message sent from p_i at time $t' > t$. Since $\mathcal{SRP}^t(p_i)$ holds and $t' > t$, $\forall u \geq t' + \zeta(e_{i,j}, t')$ $p_i \in X_j^u$. Thus, after u, because $p_i \in recvfrom_j$, the predicate of line 14 will always return false and p_j will never punish p_i after u.

Case 3: p_j is stable and it never receives a message from p_i, sent after t. In this case, (i) either p_j does not receive any message from p_i or (ii) p_j receives at least one message from p_i. In (i), if p_j never receives a message from p_i at any time, the latter will never be added to the set $local\_known_j$. Therefore, the predicate of line 14 always returns false and p_j never punishes p_i. In (ii), if p_j receives at least one message from p_i, then p_i sent this message at time t at the latest. Let $mid_i = c$ at time t. Due to Observation 1, $local\_known_j^t(p_i) \leq c$. As p_i is stable, there is a time $t' > t$ such that $mid_i = c + 1$ and p_i broadcasts a QUERY. Upon reception of its own RESPONSE at time $t'' > t'$, p_i updates its local state. In particular $global\_known_i^{t''}(p_i)$ is updated to $c+1$ (line 51). Furthermore, Assumption 1 (recurrent connectivity) ensures that there is a journey $\mathcal{J}_{(i,j)}$ from p_i to p_j, such that $departure(\mathcal{J}_{(i,j)}) > t''$ and $arrival(\mathcal{J}_{(i,j)}) = u$. According to Lemma 1, $global\_known_j^u(p_i) \geq global\_known_i^{t''}(p_i) = c + 1$. Thus, $\forall u' \in T$, $u' > u \Rightarrow global\_known_j^{u'}(p_i) > c \geq local\_known_j^t(p_i)$ and, thus, every call to $Maxknown()$ will always return false. It follows then that after u, p_j never punishes p_i.

We have shown that for any process p_j, there is a time u after which p_j never punishes p_i. As there is a finite number of processes, there is a finite time after which no process punishes p_i.

Lemma 3. *Let p_i be a process such that no process punishes p_i after a finite time t. Thus, $p_i \in SBP$.*

Proof. Since after t, no process punishes p_i, a process p_j punishes p_i at most the number of times p_j broadcasts a query till t. As there is a finite number of processes (from the finite arrival assumption), over all processes, the overall total number of times p_i is punished is finite. Let pun_i be this number and let max_pun_i be the maximum value by which the punish counter of p_i is incremented or updated $\forall p_j \in \Pi$ (note that at each punish step, the counter associated to p_i is either incremented by 1 at line 19 or set to $c_{min} + 1$ at line 17). Then, as the initial value of every punish counter is 0, we have $\forall s \in T, \forall p_j \in \Pi, punish_j^s(p_i) \leq pun_i * max\_pun_i \lor punish_j^s(p_i) = \bot$; and, by definition of SBP, $p_i \in SBP$.

Lemma 4. *Let $p_i \in SBP$. There is a time t after which p_i is not punished by any process.*

Proof. The proof is by contradiction. Let us assume that $\forall t \in T, \exists (t', p_j) \in T \times \Pi$, such that $t' > t$ and p_j punishes p_i at time t'. Hence, process p_i is punished infinitely often and, as the number of processes is finite, there is a process p_j that punishes p_i infinitely often. It follows, therefore, that $punish_j^*(p_i)$ is not bounded, which is a contradiction.

Theorem 1. *SBP is the set of processes that are eventually not punished.*

Proof. Theorem 1 follows directly from lemma 3 and lemma 4.

Lemma 5. *Let $p_j \in$ FAULTY. p_j will be punished an infinite number of times by at least one process $p_i \in$ STABLE. Thus, it follows that $SBP \subset$ STABLE.*

Proof. When p_j connects to the system, it broadcasts at least one QUERY, corresponding to the first message sent upon execution of line 8. Let $faulty^t(p_j)$ and $last\_mid_j$ be the last value of mid_j before t. Since the increment of variable mid_j and the QUERY (lines 7–8 or 21–22) are performed atomically (i.e., p_j does not crash between these two operations), p_j broadcasts a query with $mid_j = last\_mid_j$ before crashing. Furthermore, due to the stable termination Property 1 ($\mathcal{S}at\mathcal{P}$), there is at least one process $p_i \in$ STABLE that receives this query. Thus, there is a time t' such that $local\_known_i^{t'}(p_j)$ and $global\_known_i^{t'}(p_j)$ equal to $last\_mid_j$.

We remark (lines 50 and 51) that no process p_k inserts in its $global\_known_k$ set neither in its $local\_known_k$ set the tuple $\langle mid_j, p_j \rangle$, such that $mid_j > last\_mid_j$, since $last\_mid_j$ is the greatest value of mid_j of any message received from p_j. Thus, after t', each call by process p_i to the function $MaxKnown(p_j)$ returns true. Let be $t'' = max(t, t')$. Since $stable^{t''}(p_i)$, the number of queries sent by p_i after t'' is infinite. Moreover, since p_j crashed at time $t \leq t''$, p_j does not respond to any of those queries. Therefore, p_i will punish p_j infinitely often.

Lemma 6. *Let $p_j \notin SBP$ be a process such that $\exists p_i, p_i \in$ STABLE which punishes p_j infinitely often. Then, $\forall p_k \in$ STABLE, $punish_k^*(p_j)$ is unbounded.*

Proof. Since p_i punishes p_j infinitely often, $punish_i^*(p_j)$ is unbounded. Let $p_k \in$ STABLE, $p_k \neq p_i$. Let us show that $punish_k^*(p_j)$ is unbounded as well. Let $b \in \mathbb{N}$, since $punish_i^*(p_j)$ is unbounded, there is a time $t \in T$ such as $punish_i^t(p_j) \geq b$. From Assumption 1 (recurrent connectivity) there is a journey $\mathcal{J}_{(i,k)}$ from p_i to p_k, such that $t' = departure(\mathcal{J}_{(i,k)}) > t$ and $arrival(\mathcal{J}_{(i,k)}) = t''$. As punish values increase over time and according to Lemma 1, $punish_k^{t''}(p_j) \geq punish_i^t(p_j) > b$. We conclude then that $punish_k^*(p_j)$ is unbounded.

Lemma 7. *Let $p_i \in SBP$. There is a time t after which $\forall p_j, p_j \in$ STABLE will carry the same $punish_j^t(p_i)$ value for p_i and this value never changes after t.*

Proof. Since $p_i \in SBP$, $\exists b \in \mathbb{N}$, such that $\forall s \in T$, $\forall p_j \in \Pi$, $punish_j^s(p_i) < b \vee$ $punish_j^s(p_i) = \perp$. This remains true if $p_j \in$ STABLE. Furthermore, there is a time when p_i adds itself to $punish_i$ (line 3). Thus, $punish_*^*(p_i) \neq \emptyset$ and it is bounded. As $punish_*^*(p_i)$ is composed of integer values, there exists a maximum value; let $max\_punish(p_i)$ be such a maximum value. Let p_j be the stable process such that $punish_j^s(p_i) = max\_punish(p_i)$. Due to Assumption 1 (recurrent connectivity), there is a journey $\mathcal{J}_{(j,k)}$ from p_j to p_k, such that $departure(\mathcal{J}_{(j,k)}) > s$ and $arrival(\mathcal{J}_{(j,k)}) = s'$, $s' > s$. On the one hand, following Lemma 1, $punish_k^{s'}(p_i) \geq max\_punish(p_i)$. On the other hand, since $punish_k^{s'}(p_i) \leq max\_punish(p_i)$, we conclude that $punish_k^{s'}(p_i) = max\_punish(p_i)$. Moreover, the punish values increase over time. Thus, $\forall s''$, $s'' \geq s' \Rightarrow punish_k^{s''}(p_i) = max\_punish(p_i)$. Since there is a finite number of stable processes, $\forall p_k \in$ STABLE, there is a time s_k' where $punish_k^{s_k'}(p_i) = max\_punish(p_i)$. Let be $t = max(s_k'|p_k \in$ STABLE$)$ then $\forall p_k \in$ STABLE, $\forall t' \geq t$, $punish_k^{t'}(p_i) = max\_punish(p_i)$.

Theorem 2. *Algorithm 1 satisfies the eventual leadership property.*

Proof. From assumption, there is at least one process $p_i \in$ STABLE satisfying $\mathcal{SRP}^s(p_i)$ at time s. According to Lemma 2, $p_i \in SBP$; thus, $SBP \neq \emptyset$. According to Lemma 7 and the finite arrival assumption, $\exists t \in T, \forall t' > t, \forall p_i \in SBP, \forall p_j \in$ STABLE, $punish_j^{t'}(p_i) = max\_punish(p_i)$. Let $maxSBP = Max(max\_punish(p_k))$, $p_k \in SBP$. From Lemma 6, the finite arrival assumption and the fact that the punish values never decrease, $\exists t'' \in T$, $\forall p_j \in$ STABLE, $\forall p_k \notin SBP, \forall t' > t''$ $maxSBP < punish_j^{t'}(p_k)$. Thus, there exists a time $u = max(t, t'')$ after which $Min(punish_j)$ will return the same tuple $\langle c, p_i \rangle$, $\forall p_j$, such that $p_i \in SBP$. Hence, upon invoking the $leader()$ function after u, all stable processes will return the same process identity as the leader.

6   Conclusion

This paper has provided a model and an algorithm to solve the eventual leader election problem in mobile dynamic systems, in which both the network topology and relations between mobile nodes evolve over time. The algorithm implements the Ω class, following the message-pattern approach and exploiting the TVG framework to represent the dynamics of the network topology. As a future research, the timer-based approach will be considered.

References

1. Chandra, T., Toueg, S.: Unreliable failure detectors for reliable distributed systems. JACM 43(2), 225–267 (1996)
2. Mostefaoui, A., Raynal, M., Travers, C.: Time-free and timer-based assumptions can be combined to obtain eventual leadership. IEEE TPDS 17(7), 656–666 (2006)
3. Larrea, M., Fernandez, A., Arévalo, S.: Optimal implementation of the weakest failure detector for solving consensus. In: SRDS 2000, pp. 334–334 (2000)

4. Aguilera, M.K., Delporte-Gallet, C., Fauconnier, H., Toueg, S.: Stable leader election. In: Welch, J.L. (ed.) DISC 2001. LNCS, vol. 2180, pp. 108–122. Springer, Heidelberg (2001)
5. Aguilera, M.K., Delporte-Gallet, C., Fauconnier, H., Toueg, S.: On implementing omega with weak reliability and synchrony assumptions. In: PODC 2003, pp. 306–314. ACM Press (2003)
6. Aguilera, M.K., Delporte-Gallet, C., Fauconnier, H., Toueg, S.: Communication-efficient leader election and consensus with limited link synchrony. In: PODC 2004, pp. 328–337 (July 2004)
7. Malkhi, D., Oprea, F., Zhou, L.: O meets paxos: Leader election and stability without eventual timely links. In: Fraigniaud, P. (ed.) DISC 2005. LNCS, vol. 3724, pp. 199–213. Springer, Heidelberg (2005)
8. Hutle, M., Malkhi, D., Schmid, U., Zhou, L.: Chasing the weakest system model for implementing omega and consensus. IEEE Transactions on Dependable and Secure Computing 6, 269–281 (2009)
9. Fernandez, A., Jimenez, E., Raynal, M.: Eventual leader election with weak assumptions on initial knowledge, communication reliability, and synchrony. Journal of Computer Science and Technology 25(6), 1267–1281 (2010)
10. Tucci-Piergiovanni, S., Baldoni, R.: Eventual leader election in infinite arrival message-passing system model with bounded concurrency. In: EDCC 2010, pp. 127–134 (2010)
11. Masum, S.M., Ali, A.A., Touhid-youl Islam Bhuiyan, M.: Asynchronous leader election in mobile ad hoc networks. In: AINA Conference, pp. 827–831 (2006)
12. Cao, J., Raynal, M., Travers, C., Wu, W.: The eventual leadership in dynamic mobile networking environments. In: PRDC Conference, pp. 123–130 (2007)
13. Melit, L., Badache, N.: An Ω-based leader election algorithm for mobile ad hoc networks. In: 4th Networked Digital Technologies Conf., pp. 483–490 (2012)
14. Gomez-Calzado, C., Larrea, M., Raynal, M.: Fault-tolerant leader election in mobile dynamic distributed systems. Technical report, University of the Basque Country UPV/EHU (2013)
15. Mostefaoui, A., Mourgaya, E., Raynal, M.: Asynchronous implementation of failure detectors. In: DSN Conference, pp. 351–360 (2003)
16. Jiménez, E., Arévalo, S., Fernandez, A.: Implementing unreliable failure detectors with unknown membership. Inf. Process. Lett. 100(2), 60–63 (2006)
17. Aguilera, M.K.: A pleasant stroll through the land of infinitely many creatures. SIGACT News 35(2), 36–59 (2004)
18. Larrea, M., Raynal, M., Soraluze, I., Cortiñas, R.: Specifying and implementing an eventual leader service for dynamic systems. Int. J. Web Grid Serv. 8(3), 204–224 (2012)
19. Casteigts, A., Flocchini, P., Quattrociocchi, W., Santoro, N.: Time-varying graphs and dynamic networks. In: Adhoc-Now Conference, pp. 346–359 (2011)
20. Min-Te, S., Lifei, H., Arora, A., Ten-Hwang, A.L.: Reliable mac layer multicast in ieee 802.11 wireless networks. In: ICPP Conference, pp. 527–536 (2002)
21. Koo, C.Y.: Broadcast in radio networks tolerating byzantine adversarial behavior. In: PODC 2004, pp. 275–282 (2004)
22. Mostefaoui, A., Raynal, M., Travers, C., Patterson, S., Agrawal, D., Abbadi, A.: From static distributed systems to dynamic systems. In: SRDS 2005, pp. 109–118 (2005)
23. Bhandari, V., Vaidya, N.H.: Reliable broadcast in radio networks with locally bounded failures. IEEE TPDS 21, 801–811 (2010)

Self-stabilizing Leader Election in Population Protocols over Arbitrary Communication Graphs

Joffroy Beauquier[1], Peva Blanchard[2,*], and Janna Burman[1]

[1] LRI, Paris-South 11 University, Orsay, France
{jb,blanchard,burman}@lri.fr
[2] LRI, Bât. 650, Université Paris-Sud 11, 91405 Orsay Cedex France

Abstract. This paper considers the fundamental problem of *self-stabilizing leader election* ($\mathcal{SSLE}$) in the model of *population protocols*. In this model, an unknown number of asynchronous, anonymous and finite state mobile agents interact in pairs over a given communication graph. $\mathcal{SSLE}$ has been shown to be impossible in the original model. This impossibility can been circumvented by a modular technique augmenting the system with an *oracle* - an external module abstracting the added assumption about the system. Fischer and Jiang have proposed solutions to $\mathcal{SSLE}$, for complete communication graphs and rings, using an oracle Ω?, called the *eventual leader detector*. In this work, we present a solution for arbitrary graphs, using a *composition* of two copies of Ω?. We also prove that the difficulty comes from the requirement of self-stabilization, by giving a solution without oracle for arbitrary graphs, when an uniform initialization is allowed. Finally, we prove that there is no self-stabilizing *implementation* of Ω? using $\mathcal{SSLE}$, in a sense we define precisely.

Keywords: leader election, self-stabilization, population protocols, global fairness, oracles.

1 Introduction

Leader election and consensus are among the most fundamental problems in distributed computing. Both have been formally proven not to admit any solution under some assumptions and especially under the presence of faults. Consensus is impossible in asynchronous message passing or shared memory systems, even with a single crash fault [15]. Leader election is impossible each time the system is completely symmetrical, involving no identifiers, or is required to be *self-stabilizing* [13], i.e., withstand state-corrupting transient failures (see, e.g., [7,4]). To circumvent these impossibilities, a lot of studies have been performed for devising and defining the (minimum) supplementary information or assumptions needed to solve these problems. Such information generally should be available or possible to retrieve in real systems, allowing practical implementations.

Devising such necessary supplementary information in a modular way can be done using *oracles*. An oracle can be viewed as a black box, which, when asked

---

* Corresponding author.

R. Baldoni, N. Nisse, and M. van Steen (Eds.): OPODIS 2013, LNCS 8304, pp. 38–52, 2013.

by the system, provides some type of information, hopefully useful to solve a given problem. A great number of studies, following Chandra and Toueg [11], have been devoted to a specific type of oracles, named *failure detectors*, and allowing to solve consensus with crashes in asynchronous networks. Generally, failure detectors provide a quite precise type of information. It is a list of process identifiers (estimated to have crashed). Obviously, the oracle that gives as few information as possible, that is the *weakest oracle*, is both of theoretical and practical interest. For instance, in their framework, Chandra et al. [10] exhibit the weakest failure detector necessary to solve consensus. This oracle is called the *eventual leader elector* and is denoted by Ω.

Fischer and Jiang [14] introduced a different type of oracles, for solving the leader election problem in the model of tiny, asynchronously mobile and pairwise communicating agents called *population protocols* [3]. In particular, this model was introduced in order to characterize what can be computed with only minimal assumptions in a network of mobile agents. The agents are assumed to be undistinguishable (no identifiers and the same algorithm for all) and memory bounded (actually, constant memory). An agent cannot know with which agent it communicates, nor if the agent it communicates with presently is the same as the agent it communicated with just before. Moreover, no knowledge or an upper bound on the number of agents is available. Such characteristics, make the classical failure detectors, or any variant involving a list or the number of identifiers, not applicable to population protocols. This is one of the reasons why Fischer and Jiang introduced a totally different type of oracle. Their oracle is able to detect the presence or the absence of (at least) one leader. It is denoted by $\Omega?$, in reference to Ω, though it is quite different from a failure detector in the sense that it provides information taken from a global configuration of a system.

Fischer and Jiang studied the possibility to solve *self-stabilizing leader election* ($\mathcal{SSLE}$) over specific communication graphs. They prove that $\Omega?$ helps to solve $\mathcal{SSLE}$ in complete graphs and on rings, while the same problem in complete graphs is proven impossible without oracles [4,8]. After the introduction of $\Omega?$, other oracles for leader election in population protocols appeared in the literature, all based on some information related to global configurations. Michail et al. [17] introduced the *absence detector*, an oracle that indicates which agent states are not present in a configuration, as well as a *covering service* which informs an agent that it has met (communicated with) all the other agents. Intuitively, both are much stronger than $\Omega?$. In [6], we solve $\mathcal{SSLE}$ in arbitrary graphs with $\Omega\$$, an oracle which distinguishes between the presence of zero, one or more leaders in a configuration (in the way that $\Omega?$ does for zero or at least one leader). Additional oracle $W\Omega?$ is introduced in [6]. It is a weaker version of $\Omega?$ that can be used to solve $\mathcal{SSLE}$ over oriented or bounded degree trees.

Our Contribution

Comparing precisely and relating all these different oracles seemed necessary. That is why the first contribution of this paper is to provide a formal framework for dealing with oracles related to $\mathcal{SSLE}$ and encompassing all the particular oracles described above. Although it may seem complicated at a first glance, this framework is necessary for two reasons. First, it provides a unified formalism, taking into account both oracles that interact with a protocol (like Ω?), and problems, which are independent of any protocol. A second important feature of the framework is a formal definition of the *implementation* of an oracle by another oracle. This step goes through the definition of *compositions* (*sequential, parallel, self*), which, e.g., allows to express that two copies of Ω?, are *stronger* than a single one, or that an oracle that provides information on a three value variable is stronger than an oracle that provides only information on two. Then, based on the notion of implementation, this framework allows to classify some class of leader election oracles under the form of a double hierarchy, which leads to a lattice.

We then show that one of the elements in the lattice, Ω?$(2,1)$ (a notation which we define in the sequel and which represents two instances of Ω?, giving independently two different outputs), allows to solve $\mathcal{SSLE}$ over any connected communication graph (Sec. 6). The protocol is non trivial and, with its correctness proof, may be considered as the major contribution of this paper. On the contrary, we prove that if the property of self-stabilization is not mandatory, that is if some (uniform) initialization is allowed, leader election can be solved without oracle in any communication graph (Sec. 5). This result confirms the fact that the difficulties for solving $\mathcal{SSLE}$ do come from the tolerance to (transient) failures, modeled by the framework of self-stabilization. In addition, to the best of our knowledge, this is the first leader election population protocol over arbitrary graphs.

All the protocols proposed in the paper assume and require the original *global fairness* of population protocols. We show that, with only local fairness, leader election in arbitrary graphs is impossible even with (uniform) initialization (Sec. 4).

Finally we show that Ω? cannot be implemented using $\mathcal{SSLE}$ over the family of all graphs, even with multiple copies of $\mathcal{SSLE}$ (Sec. 7). This result is an illustration of what can be done in the proposed framework. It should be put in relation with a result in [6], stating that, over rings, Ω? and $\mathcal{SSLE}$ are *equivalent*. The paper ends with some open problems (Sec. 8).

Due to the lack of space, most of the proofs are sketched. All complete proofs appear in [5].

Related Work

Self-stabilization was introduced by Dijkstra [13]. A self-stabilizing protocol does not depend on initialization of process states and converges towards a correct behavior from arbitrary starting configurations. Self-stabilization is intended to deal with transient failures, that hit a system punctually, corrupting memory and channel contents. It also deals with dynamic networks, where the topology changes during an execution.

Being an important primitive in distributed computing, leader election has been extensively studied. Below, we mention only the most relevant literature.

Since the introduction of population protocols by Anglin et al. in [2], several studies have been devoted to self-stabilizing leader election in this model.

Angluin et al. [4] present a non-uniform $SSLE$ algorithm for rings in the population protocol model. They also show in the same paper that there does not exist a $SSLE$ protocol for general connected networks.

Fischer and Jiang [14] propose the eventual leader detector $\Omega?$ and, using it, present uniform $SSLE$ protocols for complete graphs and rings. The first protocol works under either a local or global fairness condition, whereas the second requires global fairness. It is also shown that with only local fairness, uniform self-stabilizing leader election in rings is impossible, even with the help of $\Omega?$. Canepa and Potop-Butucaru [9] propose deterministic and probabilistic protocols in arbitrary graphs, assuming $\Omega?$ and different types of local fairness conditions.

Cai et al. [8] show that, in complete communication graphs, n agent-states are necessary and sufficient to solve $SSLE$, where n is the population size. This result involves that an oracle is necessary for solving $SSLE$ in population protocols. For the enhanced model of *mediated population protocols* - MPP (allowing an extra memory on every agent pair) [16], the work of Mizoguchi et al. [18] shows that $(2/3)n$ agent states and a single bit memory on every agent pair are sufficient to solve $SSLE$. They also show that there is no MPP that solves $SSLE$ with any constant agent-states and any constant size memory on each agent-pair, for general n.

Michail et al. [17] introduce the *absence detector*, an oracle for population protocols that indicates which agent states are not present in a configuration, as well as a *covering service* which informs an agent that it has met (communicated with) all the other agents. Intuitively, both are much stronger than $\Omega?$.

Finally, in [6] we define $\Omega\$$ and $W\Omega?$, two oracles respectively stronger and weaker than $\Omega?$, and prove that $SSLE$ can be solved with $\Omega\$$ over weakly connected communication graphs, with $W\Omega?$ over oriented trees and with $\Omega?$ over weakly connected communication graphs of bounded degree.

2   Model and Definitions

2.1   Population Protocol

We use the same definitions as in [14] with some slight modifications. A network is represented by a directed graph $G = (V, \mathcal{E})$ with n vertices and no multi-edges nor self-loops. Each vertex represents a finite-state sensing device called an *agent*, and an edge $(u, v) \in \mathcal{E}$ indicates the possibility of a communication between two distinct nodes u and v in which u plays the role of the *initiator* and v of the *responder*. The orientation of an edge corresponds to this asymmetry in roles. In this paper, we consider weakly connected networks.

A *population protocol* $\mathcal{A}(D, \mathcal{Q}, Init, X, Y, O, \delta)$ consists of a family of graphs D (the *domain* of the protocol), a finite *state space* $\mathcal{Q}$, a function $Init$ that associates every graph $G(V, \mathcal{E})$ in D with a set $Init(G)$ of *initial configurations* (see below) on G , a finite *input alphabet* X, a finite *output alphabet* Y, an *output function* $O : \mathcal{Q} \to Y$ and a transition function $\delta : (\mathcal{Q} \times X)^2 \to \mathcal{P}(\mathcal{Q}^2)$ that maps any tuple (q_1, x_1, q_2, x_2) to a non-empty (finite) subset $\delta(q_1, x_1, q_2, x_2)$ in

Q^2. A *(transition) rule* of the protocol is a tuple $(q_1, x_1, q_2, x_2, q_1', q_2')$ such that $(q_1', q_2') \in \delta(q_1, x_1, q_2, x_2)$ and is denoted by $(q_1, x_1)(q_2, x_2) \to (q_1', q_2')$. The population protocol $\mathcal{A}$ is *deterministic* if the set $\delta(q_1, x_1, q_2, x_2)$ always has exactly one element.

Given a graph $G(V, \mathcal{E})$ in D and a set Z, an *assignment with values in Z* is a function from V to Z. A *configuration* C is an assignment with values in the state space $\mathcal{Q}$. An *input assignment* (resp. *output assignment*) is an assignment with values in the input alphabet X (resp. *output alphabet* Y). Each configuration C induces an output assignment $O \circ C$ where O is the output function of the protocol. A *trace T with values in Z* on the graph $G(V, \mathcal{E})$ is an infinite sequence of assignments with values in Z, i.e., $T = \alpha_0 \alpha_1 \ldots$ where $\alpha_i : V \to Z$. An *input trace* (resp. *output trace*) is a trace with values in the input alphabet X (resp. the output alphabet Y). The trace $\alpha_0 \alpha_1 \ldots$ is *constant* if $\alpha_0 = \alpha_1 = \ldots$, and it is *uniform constant* if it is constant and for every $u, v \in V$, $\alpha(u) = \alpha(v)$.

Given a graph $G(V, \mathcal{E})$ in D, an *action* is a pair $\sigma = (e, r)$ where r is a rule $(q_1, x_1)(q_2, x_2) \to (q_1', q_2')$ and $e = (u, v)$ an edge of G. Let C, C' be configurations and α be an input assignment. We say that σ *is enabled in* (C, α) if $C(u) = q_1, C(v) = q_2$ and $\alpha(u) = x_1, \alpha(v) = x_2$. We say that (C, α) *goes to C' via σ in one step*, denoted $(C, \alpha) \xrightarrow{\sigma} C'$, if σ is enabled in (C, α), $C'(u) = q_1', C'(v) = q_2'$ and $C'(w) = C(w)$ for all $w \in V - \{u, v\}$. In other words, C' is the configuration that results from C by applying the transition rule r to the node pair e. We also denote by $(C, \alpha) \to C'$ when $(C, \alpha) \xrightarrow{\sigma} C'$ for some action σ. Given an input trace $T_{in} = \alpha_0 \alpha_1 \ldots$, we write $C \xrightarrow{*} C'$ if there is a sequence of configurations $C_0 C_1 \ldots C_k$ such that $C = C_0$, $C' = C_k$ and $(C_i, \alpha_i) \to C_{i+1}$ for all $0 \leq i < k$, in which case we say that C' is *reachable* from C given the input trace T_{in}.

Given a graph G in D, a *virtual execution E* is an infinite sequence of configurations, input assignments and actions $E = (C_0, \alpha_0, \sigma_0)(C_1, \alpha_1, \sigma_1) \ldots$ such that $C_0 \in Init(G)$ and for each i, $(C_i, \alpha_i) \xrightarrow{\sigma_i} C_{i+1}$. Such a virtual execution induces an output trace denoted by $O(E)$ defined as $(O \circ C_0)(O \circ C_1) \ldots$ where O is the output function of the protocol. We denote by SE the (infinite) suffix of E such that each couple (C, α) (C being a configuration, and α an input assignment) in SE appears infinitely often in SE. This suffix is well-defined because the number of couples (C, α) that occurs finitely often in E is bounded.

We now define fair executions. We first recall two fairness conditions used with population protocols [14]:

(Local Fairness) a virtual execution $(C_0, \alpha_0, \sigma_0)(C_1, \alpha_1, \sigma_1) \ldots$ is *locally fair* when, for every action σ, if σ is enabled in (C_i, α_i) for infinitely many i, then $(C_j, \alpha_j) \xrightarrow{\sigma} C_{j+1}$ for infinitely many j.

(Global Fairness) a virtual execution $(C_0, \alpha_0, \sigma_0)(C_1, \alpha_1, \sigma_1) \ldots$ is *globally fair* when, for every C, C', α such that $(C, \alpha) \to C'$, if $(C, \alpha) = (C_i, \alpha_i)$ for infinitely many i, then $C' = C_j$ for infinitely many j.

In this paper, unless stated otherwise, an *execution* is a virtual execution that is globally fair. Finally we consider two types of population protocols. A population protocol is *uniformly initialized* if there exists a state q_0 such that every initial configuration is an assignment with values in $\{q_0\}$. In a *non-initialized population protocol*, the set of initial configurations is the set of all possible configurations.

2.2   Run, Behaviour, Oracle and Implementation

The definitions of runs, behaviours and oracles that we give below, are different from those in [4,14] and are required to obtain a proper framework for defining oracles and establishing relations between them. For instance, in this framework, the oracles are self-implementable, in contrast with the traditional failure detectors' frameworks [12].

A *schedule* on a network $G(V, \mathcal{E})$ is a sequence of edges $S = e_1 e_2 \ldots$, i.e., $e_i \in \mathcal{E}$ for all i. The *schedule S associated with an execution E* is the sequence S of edges that appear in the sequence of actions in E; we also say that E is an *execution with schedule S*.

The following notion of *compatibility of a trace with a schedule* involves that the changes in a trace are only caused by the interactions. A trace $T = \alpha_0 \alpha_1 \ldots$ on G is said to be *compatible* with the schedule $S = (u_0, v_0)(u_1, v_1) \ldots$ on G if, for every i, for every $w \in V - \{u_i, v_i\}$, $\alpha_i(w) = \alpha_{i+1}(w)$. That is, two consecutive assignments can differ only in the assignment values of the two agents in the corresponding edge in the schedule. Note that, by definition, the output trace induced by an execution with schedule S of a population protocol on G, is compatible with S.

(**Run**). A *run $R(X, Y)$ with an input alphabet X and output alphabet Y* on a network $G(V, \mathcal{E})$ is a triple (T_{in}, T_{out}, S), where T_{in} is a trace with alphabet X on G, T_{out} is a trace with alphabet Y on G and S is a schedule on G such that T_{in} and T_{out} are both compatible with S. The trace T_{in} (resp. T_{out}) is referred to as the *input trace* (resp. *output trace*) *of the run*.

(**Behaviour**). A *behaviour B* is given by a family D of graphs (the domain of B), an input alphabet X, an output alphabet Y and a function that maps any graph G in D to a set $B(G)$ of runs with input alphabet X and output alphabet Y. Given a population protocol $\mathcal{A}$ with domain D, input alphabet X and output alphabet Y, we define the *behaviour $Beh(\mathcal{A})$ associated with the protocol $\mathcal{A}$* as follows. The domain is D, the input alphabet is X, the output alphabet is Y, and, for any graph G in D, for any run (T_{in}, T_{out}, S) on G, $(T_{in}, T_{out}, S) \in Beh(\mathcal{A})(G)$ if and only if there exists an execution of $\mathcal{A}$ on G with the input trace T_{in}, the output trace T_{out} and the schedule S.

In the following paragraph, we define the notion of *composition* of behaviours. Informally, a *serial* composition uses the output of one behaviour as the input of another behaviour. A *parallel* composition consists in two behaviours being used independently. Finally, a *self* composition uses (a part of) the output of a behaviour as the input to the same behaviour, producing a sort of "feedback". In [14], the self composition is implicitly used, when the oracle $\Omega?$ produces a new input to a protocol based on the output of the same protocol.

Formally, consider two behaviours B_1, B_2 with (respectively) domains D_1, D_2 such that $D_1 \cap D_2 \neq \emptyset$, input alphabets X_1, X_2, and output alphabets Y_1, Y_2.

We denote by T_X a trace with values in X. The *parallel composition* $B = B_1 \otimes B_2$ is the behaviour with domain $D_1 \cap D_2$, alphabets $X_1 \times X_2$, $Y_1 \times Y_2$ such that, for every $G \in \mathcal{F}$, $B(G)$ is the set of runs $((T_{X_1}, T_{X_2}), (T_{Y_1}, T_{Y_2}), S)$ with $(T_{X_1}, T_{Y_1}, S) \in B_1(G)$ and $(T_{X_2}, T_{Y_2}, S) \in B_2(G)$. If $Y_1 = X_2 = U$, the *serial composition* $B = B_2 \circ B_1$ is the behaviour with domain $D_1 \cap D_2$ and alphabets X_1, Y_2 defined as follows. For every $G \in \mathcal{F}$, $B(G)$ is the set of runs (T_{X_1}, T_{Y_2}, S) such that there exists a trace T_U satisfying $(T_{X_1}, T_U, S) \in B_1$ and $(T_U, T_{Y_2}, S) \in B_2$. If $X_1 = U \times V$ and $Y_1 = U \times W$, the *self composition* $B = Self^U(B_1)$ *on* U is the behaviour with domain D_1, alphabets V, W, where, for every $G \in \mathcal{F}$, $B(G)$ is the set of runs $((T_U^{in}, T_V^{in}), (T_U^{out}, T_W^{out}), S) \in B$ such that $T_U^{in} = T_U^{out}$.

Given a family H of behaviours, a behaviour B is a *composition of behaviours from H* if it is a combination of serial, parallel and self composition of behaviours in H.

(Implementation, Comparison). A behaviour B_2 is an *implementation of a behaviour B_1 over a family $\mathcal{F}$ of graphs* when $\mathcal{F} \subset D_1 \cap D_2$, and for every graph $G \in \mathcal{F}$, $B_2(G) \subset B_1(G)$.

Consider a family $\mathcal{H}$ of behaviours and a family $\mathcal{F}$ of graphs. We say that a behaviour B_1 is *weaker* than a behaviour B_2 over $(\mathcal{F}, \mathcal{H})$, denoted by $B_1 \preccurlyeq B_2$ mod $(\mathcal{F}, \mathcal{H})$, when there exists a composition B involving the behaviour B_2 and behaviours from $\mathcal{H}$ that implements B_1 over $\mathcal{F}$. In other words, if we can compose behaviours from $\mathcal{H}$ with one copy of B_2 to implement B_1, then B_1 is said to be weaker than B_2. This is analogous to the definition in [11] of an oracle being weaker than another one.

The two behaviours are *equivalent* if $B_1 \preccurlyeq B_2$ mod $(\mathcal{F}, \mathcal{H})$ and $B_2 \preccurlyeq B_1$ mod $(\mathcal{F}, \mathcal{H})$. We denote this case by $B_1 \simeq B_2$ mod $(\mathcal{F}, \mathcal{H})$. When $\mathcal{F}$ and $\mathcal{H}$ are clear from the context, we write $B_1 \preccurlyeq B_2$ and $B_1 \simeq B_2$.

A *problem* and an *oracle* are defined as behaviours. A population protocol $\mathcal{A}$ is a *solution to a problem P* (resp. an *implementation of an oracle Θ*) *using a behaviour B over a family $\mathcal{F}$ of graphs* if there exists a composition involving the behaviours $Beh(\mathcal{A})$ and B that implements the behaviour P (resp. Θ) over $\mathcal{F}$. Note that with these definitions, if there exists a population protocol in some family $\mathcal{H}$ of protocols that solves the problem P_1 using the problem P_2 over a family $\mathcal{F}$, then P_1 is weaker than P_2 over $(\mathcal{F}, \mathcal{H}^*)$, where $\mathcal{H}^*$ is the family of the behaviours associated with the protocols in $\mathcal{H}$.

Given a behaviour B, we define *the stabilizing behaviour B_s associated with B* as follows. It has the same domain D, the same input and output alphabets as B, and for any graph G in D, the set of runs $B_s(G)$ comprises the runs having a suffix[1] belonging to $B(G)$. Given a problem P (resp. an oracle Θ), a population protocol $\mathcal{A}$ is a *self-stabilizing solution to P* (resp. *self-stabilizing implementation of Θ*) if it is non-initialized and it is a solution to the stabilizing problem P_s associated with P (resp. an implementation of the stabilizing oracle Θ_s associated with Θ).

---

[1] A run can be seen as a sequence of triples $(\alpha_s, \beta_s, e_s)_{s \in \mathbb{N}}$ where α_s (resp. β_s) is an input (resp. output) assignment and e_s is an edge.

Remark 1. The results in the paper concern the family $\mathcal{F}_{all}$ of all (weakly connected) graphs. Note however that in Sec. 5 and 6, we present protocols that solve the leader election problem in the family of all *strongly* connected graphs. The extension of these protocols to the family of all *weakly* connected graphs is detailed in [5]. Roughly speaking, given a weakly connected graph G, one can simulate an execution over the symmetric closure G' of G, which is strongly connected. This can be done by performing, at each interaction, a non-deterministic choice to select which agent plays the role of the initiator and which agent plays the role of the responder. Then, it can be shown that such a non-deterministic execution on G is an execution on G'. It is possible to get a deterministic version of this simulation using the transformer in [4].

3   Specific Behaviours and Oracles

3.1   Eventual Leader Election Behaviour $\mathcal{ELE}$

The domain of the behaviour $\mathcal{ELE}$ is the family $\mathcal{F}_{all}$ of all the graphs, the input alphabet is $\{\bot\}$ (no input), the output alphabet is $\{0, 1\}$ and, for any graph $G \in \mathcal{F}_{all}$, a run $(\bot, T, S)$ belongs to $\mathcal{ELE}(G)$ if and only if T has a constant suffix $T' = \alpha\alpha\alpha\ldots$ and there exists a node λ such that $\alpha(\lambda) = 1$ and $\alpha(u) = 0$ for every $u \neq \lambda$. In other words, λ is the unique leader. Note that for all our protocols, there is an implicit output function that maps a state to 1 if it is a leader state, and to 0 otherwise.

In our settings, the (informal) problem of Self-Stabilizing Leader Election ($\mathcal{SSLE}$) is reformulated as the problem of constructing a population protocol that is a self-stabilizing solution to the $\mathcal{ELE}$ problem (using some oracle, if necessary).

3.2   Oracles $\Omega?(k, d)$

We first define, for each $d \geq 1$, an oracle $\Omega?(1, d)$. Its input alphabet is $\{0, 1\}$, and its output alphabet is $\{0, \ldots, d\}$. The domain of $\Omega?(1, d)$ is all the graphs. Given an assignment α, we denote by $l(\alpha)$ the number of vertices that are assigned the value 1 by α. Informally, if $l(\alpha) = c$ or $l(\alpha) \geq c$ for all α in an (infinite) execution suffix, then the oracle will eventually permanently output values in $\{c\}$ in the former case, and in $\{c, \ldots, d\}$ in the latter. When $l(\alpha) = 0$ for all α in an (infinite) execution suffix, it is only required that the oracle permanently outputs 0 at one agent at least.

Given a graph G and a run (T_{in}, T_{out}, S) on G, $(T_{in}, T_{out}, S) \in \Omega?(1, d)(G)$ when the following conditions hold. If T_{in} has a suffix $\alpha_0 \alpha_1 \ldots$ such that $\forall s, l(\alpha_s) = 0$, then T_{out} has a suffix in which at least one agent is permanently assigned the value 0. For every $1 \leq r \leq d-1$, if T_{in} has a suffix $\alpha_0 \alpha_1 \ldots$ such that $\forall s, l(\alpha_s) = r$, then T_{out} has a suffix equal to the uniform constant trace r. For every $0 \leq r \leq d$, if T_{in} has a suffix $\alpha_0 \alpha_1 \ldots$ such that $\forall s, l(\alpha_s) \geq r$, then T_{out} has a suffix with values in $\{r, r+1, \ldots, d\}$. Otherwise, any T_{out} (compatible with S) is valid.

For any $k, d \geq 1$, we formally define $\Omega?(k, d) = \bigotimes^k \Omega?(1, d)$. In other words, $\Omega?(k, d)$ is the parallel composition of k copies of $\Omega?(1, d)$. Thus, the input alphabet of $\Omega?(k, d)$ is $\{0, 1\}^k$, and the output alphabet is $\{0, \ldots, d\}^k$.

Note that $\Omega?(1, 1)$ corresponds to the Fischer and Jiang's oracle $\Omega?$ in [14], while $\Omega?(1, 2)$ corresponds to the oracle $\Omega\$$ in [6], except that in the case of absence of a leader, it is only required that at least one agent reports the fact. It is easy to see that the oracles $\Omega?(k, d)$ form a lattice, i.e., if $k \leq k'$ and $d \leq d'$, then $\Omega?(k, d) \preccurlyeq \Omega?(k', d')$ over any graph and behaviour families.

4   Impossibility of Leader Election under Local Fairness with Uniform Initialization

In this section, we show that the eventual leader election problem cannot be solved by any uniformly initialized population protocol under the local fairness assumption.

We first recall the notion of *graph covering* [1,7]. A *fibration* (resp. *opfibration*) between graphs G and B is a graph morphism $\phi : G \to B$ such that for every node b in B, for every node y satisfying $\phi(y) = b$, ϕ induces a bijection between the set of incoming (resp. outgoing) edges at y and the set of incoming (resp. outgoing) edges at b. A *covering* from G to B is a graph morphism from G to B that is both a fibration and an opfibration. The graph G is called the *total* graph, and B is the *base* graph. The *fiber* over a node b in B is the set of nodes in G that are mapped to b via ϕ, which we denote by $\phi^{-1}(b)$. A fiber is *trivial* if it is a singleton. A covering is a k-*covering* if every fiber has k elements, i.e., $\forall b, |\phi^{-1}(b)| = k$. For instance, there is a covering from a ring of size $2 \cdot n$ to a ring of size n obtained by mapping two diametrically opposite nodes to the same node.

The following theorem is inspired by the impossibility result of leader election in the family of rings under local fairness [14] and the ideas developed in [1,7]. Note that the models considered in [1,7] are different from the population protocols. Hence, the results do not directly apply to our case.

Theorem 1. *Let $\mathcal{F}$ be a family of graphs that contains graphs G and B such that there exists a k-covering $\phi : G \to B$ with $k \geq 2$. There is no uniformly initialized population protocol that solves the $\mathcal{ELE}$ problem over the family $\mathcal{F}$ under the local fairness assumption.*

Proof (Sketch). The result is proved by contradiction. Assume that such a protocol exists, and consider a locally fair execution E_B on B with $\gamma_0 \gamma_1 \ldots$ being the corresponding sequence of configurations. Thanks to the property of covering, we can lift E_B to get a locally fair execution E_G on G containing configurations g_s such that $g_s = \gamma_s \circ \phi$ for every $s \in \mathbb{N}$. Hence, since ϕ is a k-covering, and since E_B has a suffix during which there is a unique leader, E_G contains infinitely many configurations with $k \geq 2$ leaders; whence a contradiction. $\square$

5   Leader Election under Global Fairness with Uniform Initialization

We establish that, under global fairness, solving the leader election problem on arbitrary communication graphs is possible without oracle, when an uniform initialization is possible (Alg. 1). In other words, there exists a uniformly initialized population protocol that solves the $\mathcal{ELE}$ problem over the family of all graphs under the global fairness assumption. This result highlights the difference between global and local fairness. It also shows that the necessity to use an oracle comes from the requirement of self-stabilization. As explained in Remark 1, our protocol considers *strongly* connected graphs. Each agent x can be leader or non-leader (implemented with a variable $leader_x$) and can hold a white or black token (implemented with a variable $token_x$). Initially, every agent is a leader and holds a black token (uniform initialization). The tokens move through the network by swapping between two agents during an interaction. When two black tokens meet, one of them turns white. When a white token interacts with a leader x, x becomes a non-leader and the token is destroyed.

Algorithm 1. Leader Election with Uniform Initialization

```
 1  variables for every agent x:
 2      leader_x : 0 (non-leader) or 1 (leader)
 3      token_x : ⊥ (no token), white or black
 4  initialization: ∀x, (leader_x, token_x) = (1, black)        /* uniform */
 5  protocol (initiator x, responder y):
 6      if token_x = token_y = black then
 7          token_y ← white
 8      if token_x = white ∧ leader_y = 1 then
 9          leader_y ← 0                      /* y becomes a non-leader */
10          token_x ← ⊥                       /* the token is destroyed */
11      token_x ↔ token_y                     /* swap the tokens */
```

We consider an execution E of Alg. 1 and prove that there is eventually a unique leader. Recall that SE denotes the infinite suffix of E such that each couple (C, α) in SE occurs infinitely often in SE (see Sec. 2.1). Given a configuration C, let $b(C)$ be the number of black tokens, $w(C)$ the number of white tokens and $l(C)$ the number of leaders in C. In addition, for every agent x, we denote by $C.leader_x$ (resp. $C.token_x$) the value of the variable $leader_x$ (resp. $token_x$) in the configuration C.

Lemma 1. *In each configuration C in every execution E of Alg. 1, $b(C) + w(C) = l(C)$ and $b(C) \geq 1$.*

Proof (Sketch). The initial configuration satisfies this relation. During an interaction, if no leader is turned into a non-leader, then the total number of tokens remains constant. When a leader is turned into a non-leader (by a white token), the corresponding token is also destroyed. □

Lemma 2. *For every configuration C in SE, $b(C) = 1$.*

Proof (Sketch). The global fairness and the fact that two colliding black tokens yield one black token and one white token involves that eventually in E, there is always a unique black token. □

Theorem 2. *In every execution E of Alg. 1, there exists exactly one agent λ such that for every configuration C in SE, $C.leader_\lambda = 1$ and for every agent $\mu \neq \lambda$, $C.leader_\mu = 0$.*

Proof (Sketch). By the previous lemmas, for every configuration C in SE, $l(C) = w(C) + 1$. If a configuration C in SE has $l \geq 2$ leaders, then C also has $l - 1$ white tokens. Thus there is a sequence of steps during which each white token is moved to turn one leader into a non-leader, then reaching a configuration C' with one leader. By global fairness, C' occurs in SE. The configuration C' has exactly one leader, one black token and no white token, thus every subsequent configuration has the same unique leader. □

6   Self-stabilizing Leader Election Using $\Omega?(2,1)$ under Global Fairness

In this section, we exhibit a self-stabilizing solution to $\mathcal{ELE}$ using $\Omega?(2,1)$, i.e., two copies of the Fischer and Jiang's oracle, over the family $\mathcal{F}_{all}$ of all graphs under the global fairness assumption. Alg. 2 below, referred to as the protocol $\mathcal{A}$, is a self-stabilizing solution[2] to $\mathcal{ELE}$ using $\Omega?(2,1)$ over $\mathcal{F}_{all}$. In this protocol, each agent can be a leader or not, and a leader can be either black or white. An agent can also hold a token, and a token can be either black or white. We denote by $\Omega?^l$, resp. $\Omega?^t$, the copy of the oracle $\Omega?$ used to detect the absence of leaders, resp. tokens. As explained in Remark 1, we only consider *strongly connected* graphs.

Whenever the oracle $\Omega?^l$ (resp. $\Omega?^t$) outputs 0, a black leader (resp. a black token) is created. The tokens keep moving through the network by swapping between two agents during an interaction. When a black token interacts with a white leader, the leader becomes a non-leader. When a white token interacts with a black leader, the leader becomes white. When a token interacts with a leader having the same color, then both the token and the leader turn into the opposite color.

Given an input assignment α for the Alg. 2, we denote by $\alpha.\Omega?^l_x$ (resp. $\alpha.\Omega?^t_x$) the value assigned by α to the (read-only) variable $\Omega?^l_x$ (resp. $\Omega?^t_x$). Similarly, given a configuration C, for every agent x, we denote by $C.leader_x$

---

[2] More formally, the behaviour $Self(\Omega?(2,1) \circ Beh(\mathcal{A}))$ implements the behaviour $\mathcal{ELE}$ (see Sec. 2).

Algorithm 2. Self-Stabilizing Leader Election

---

1 **variables** *agent x*
2    $\Omega?^l_x$: input (read-only) from the leader detector
3    $\Omega?^t_x$: input (read-only) from the token detector
4   *leader*$_x$: $\perp$ (non-leader), *white* or *black*
5   *token*$_x$: $\perp$ (no token), *white* or *black*
6 **protocol** *(initiator x, responder y)*
7   **if** $\Omega?^l_x = 0$ **then** *leader*$_x \leftarrow$ *black*
8   **if** $\Omega?^t_x = 0$ **then** *token*$_x \leftarrow$ *black*
9   **if** *token*$_x$ = *black* $\wedge$ *leader*$_y$ = *white* **then** *leader*$_y \leftarrow \perp$
10   **if** *token*$_x$ = *white* $\wedge$ *leader*$_y$ = *black* **then** *leader*$_y \leftarrow$ *white*
11   **if** *token*$_x$ = *leader*$_y$ = *black* **then** *token*$_x \leftarrow$ *leader*$_y \leftarrow$ *white*
12   **if** *token*$_x$ = *leader*$_y$ = *white* **then** *token*$_x \leftarrow$ *leader*$_y \leftarrow$ *black*
13   **if** *token*$_x \neq \perp \wedge$ *token*$_y \neq \perp$ **then** *token*$_x \leftarrow \perp$
14   *token*$_x \leftrightarrow$ *token*$_y$

(resp. $C.token_x$) the value of the variable *leader*$_x$ (resp. *token*$_x$) in the configuration C.

Given a configuration C, let $t(C)$ (resp. $l(C)$) be the total number of tokens (resp. leaders) in C. In C, if an agent x is a leader and an agent y holds a token (x and y not necessarily neighbours), we say that the leader at x and the token at y are *synchronized* if they have the same color. Then, we say that the configuration C contains a *synchronized pair of leader and token*. We consider an execution E of Alg. 2 and its infinite suffix SE (each couple (C, α) in SE occurs infinitely often in SE).

Lemma 3. *For every (C, α) in SE, there is a unique token in C and α assigns 1 to every variable $\Omega?^t_x$, i.e. $t(C) = 1$ and $\forall x, \alpha.\Omega?^t_x = 1$.*

Proof (Sketch). The oracle $\Omega?^t$ ensures that eventually there is at least one token. Since the number of tokens decreases only when two tokens merge, there is eventually always at least one token; whence eventually $\Omega?^t$ always outputs 1 everywhere. Finally, by global fairness, all the tokens eventually merge, and from that point there is exactly one (circulating) token[3].   □

Lemma 4. *Consider a configuration C that contains a synchronized pair of leader and token such that $l(C) \geq t(C) = 1$. Consider an input assignment α that assigns 1 to every variable $\Omega?^t_x$, i.e., for all x, $\alpha.\Omega?^t_x = 1$. Then for any configuration C' such that $(C, \alpha) \rightarrow C'$, C' contains a synchronized pair of leader and token and $l(C') \geq t(C') = 1$.*

Proof. The assumption on α ensures that no token is created during the step $(C, \alpha) \rightarrow C'$. If the unique token meets a leader with which it is synchronized, the leader remains a leader, and both flip their colors. Hence, C' still contains a unique token and some leader synchronized with this token.   □

---

[3] Note that this token may change its color.

Lemma 5. *There exists a configuration C in SE that contains a synchronized pair of leader and token such that $l(C) \geq t(C) = 1$.*

Proof (Sketch). We already know that every configuration in SE has a unique token. By contradiction, assume that no configuration in SE satisfies the condition. This means that in every configuration C in SE, every leader (if any) has a color opposite to the color of the unique token. Thanks to $\Omega?^l$, there is a configuration C in SE that has at least one leader, thus $l(C) \geq t(C) = 1$. If the token is white, all the leaders are black, and it is possible to move the token to whiten one of the leaders. The resulting configuration C' contains a synchronized pair of leader and token, and $l(C') \geq t(C') = 1$. By global fairness, C' occurs in SE. On the other hand, if the token is black, it is possible to turn all the white leaders into non-leaders and keep a black token. By global fairness, the resulting configuration C' occurs in SE. Since C' has no leader, thanks to the oracle $\Omega?^l$, a black leader is created at some point in SE. Hence, the corresponding configuration C'' has a synchronized pair of leader and token, and $l(C'') \geq t(C'') = 1$. $\square$

Lemma 6. *For every (C, α) in SE, C contains a synchronized pair of leader and token, $l(C) \geq t(C) = 1$ and for every agent x, $\alpha.\Omega?^l_x = \alpha.\Omega?^t_x = 1$.*

Proof (Sketch). The result follows from Lemmas 3, 4, 5 and the definition of $\Omega?^l$. $\square$

Theorem 3. *Alg. 2 is a self-stabilizing solution to $\mathcal{ELE}$ using $\Omega?(2, 1)$. Precisely, in any execution, there exists exactly one agent λ such that for every configuration C in SE, $C.leader_\lambda \neq \perp$ and for every agent $\mu \neq \lambda$, $C.leader_\mu = \perp$.*

Proof (Sketch). Thanks to Lem. 6, no leader is ever created during SE. In addition, in every configuration in SE, there is a leader synchronized with the token. On one hand, if the token is white, it can whiten all the black leaders, interact with one white leader, become black and turn all the white leaders into non-leaders. On the other hand, if the token is black, it can interact with a black leader (the leader with which it is synchronized) and become white; the next steps are the same as before. In both cases, the resulting configuration has exactly one leader. By global fairness, this configuration occurs in SE. Since no leader is created, there is actually a unique and permanent leader in SE. $\square$

7   Impossibility of Self-stabilizing Implementation of $\Omega?$ Using $\mathcal{ELE}$ under Global Fairness

We show that there is no self-stabilizing implementation of $\Omega?$ (i.e. $\Omega?(1, 1)$) using $\mathcal{ELE}$, even if we are allowed to use many copies of $\mathcal{ELE}$, under the global fairness assumption.

Theorem 4. *There is no non-initialized population protocol A such that, for some $k \geq 1$, the composition $B = (\mathcal{ELE} \otimes \cdots \otimes \mathcal{ELE}) \circ Beh(A)$, using k copies of $\mathcal{ELE}$, implements the behaviour $\Omega?$ over the family of all graphs under the global fairness assumption.*

Proof (Sketch). The result is proved by contradiction. Assume that such a protocol A exists. We consider a complete graph G of size $n \geq k+1$. We consider a run of the composition B on G, with a constant input trace $\alpha\alpha \ldots$ that assigns permanently 1 to a unique agent μ. In the corresponding execution E of A, at some point in SE, the output of the different $\mathcal{ELE}$ oracles have stabilized, and all the agents permanently output the value 1. However, by looking at the subgraph obtained by excluding μ, thanks to the assumption on A and the global fairness, there is a configuration in SE in which all the agents but μ output 0; whence a contradiction.   □

8 Discussion and Open Problems

Although an abundant literature has been devoted to leader election in the population protocol model, some problems remain open. One of the most challenging is maybe to decide whether or not an oracle is necessary for self-stabilizing solutions to $\mathcal{ELE}$ over rings. Angluin et al. [4], who raised first the issue, present non-uniform solutions (solutions depending on the size of the ring), but the question of an uniform solution has been open for several years. In [14], Fischer and Jiang tackle this issue, provided that the oracle Ω? is available.

The general framework we proposed allows to express several natural questions. We list some of them here.

In Sec. 3.2, we generalize Fischer and Jiang's oracle and present a lattice of oracles $\{\Omega?(k,d)\}_{k,d\geq 1}$ such that Ω? coincides with $\Omega?(1,1)$. Analyzing the relations among oracles, which are strong enough to solve leader election, is an interesting way to assess the hardness of this problem. For instance, in a previous work [6], the authors complement Fischer and Jiang's approach by showing that $\mathcal{ELE}$ is equivalent to Ω? over rings, for non-initialized protocols' behaviours, i.e., each problem is as hard as the other. It seems that the same technique as in [6] would show that all the oracles $\Omega?(k,d)$ are equivalent to $\mathcal{ELE}$ over rings, for non-initialized protocols' behaviours.

In addition, in this paper, we address the issue of comparing $\mathcal{ELE}$ with the oracles $\Omega?(k,d)$ over the family $\mathcal{F}_{all}$ of all graphs, for the family, denoted by PP_{NI}, of non-inialized protocols' behaviours. In Sec. 6, we show that $\mathcal{ELE} \preccurlyeq \Omega?(2,1)$, and in Sec. 7, we show that $\Omega? \not\preccurlyeq \mathcal{ELE}$. Since $\Omega? \preccurlyeq \Omega?(2,1)$, we have the strict relation $\mathcal{ELE} \prec \Omega?(2,1)$. In addition, it has been shown in [6] that $\Omega?(1,1)$ is sufficient to solve $\mathcal{ELE}$ over the family $BDeg(d)$ of d-bounded degree graphs (for any d), i.e. $\mathcal{ELE} \preccurlyeq \Omega?(1,1) \mod (BDeg(d), PP_{NI})$. It is an open problem to determine whether there exists a self-stabilizing implementation of $\mathcal{ELE}$ using $\Omega?(1,1)$ over $\mathcal{F}_{all}$ and if the relations $\Omega?(k,d) \preccurlyeq \Omega?(k',d') \mod (\mathcal{F}_{all}, PP_{NI})$ ($k \leq k'$ and $d \leq d'$) are strict when $k < k'$ or $d < d'$.

References

1. Angluin, D.: Local and global properties in networks of processors. In: 12th Symposium on the Theory of Computing, pp. 82–93. ACM (1980)

2. Angluin, D., Aspnes, J., Diamadi, Z., Fischer, M.J., Peralta, R.: Computation in networks of passively mobile finite-state sensors. In: PODC, pp. 290–299 (2004)
3. Angluin, D., Aspnes, J., Diamadi, Z., Fischer, M.J., Peralta, R.: Computation in networks of passively mobile finite-state sensors. Distributed Computing 18(4), 235–253 (2006)
4. Angluin, D., Aspnes, J., Fischer, M.J., Jiang, H.: Self-stabilizing population protocols. ACM Trans. Auton. Adapt. Syst. 3(4) (2008)
5. Beauquier, J., Blanchard, P., Burman, J.: Self-stabilizing leader election in population protocols over arbitrary communication graphs. Technical report, INRIA (2013), http://hal.archives-ouvertes.fr/hal-00867287
6. Beauquier, J., Blanchard, P., Burman, J., Denysyuk, O.: Oracles for self-stabilizing leader election in population protocols. Technical report, INRIA (2013), http://hal.archives-ouvertes.fr/hal-00839759
7. Boldi, P., Shammah, S., Vigna, S., Codenotti, B., Gemmell, P., Simon, J.: Symmetry breaking in anonymous networks: Characterizations. In: ISTCS, pp. 16–26 (1996)
8. Cai, S., Izumi, T., Wada, K.: How to prove impossibility under global fairness: On space complexity of self-stabilizing leader election on a population protocol model. Theory Comput. Syst. 50(3), 433–445 (2012)
9. Canepa, D., Potop-Butucaru, M.G.: Self-stabilizing tiny interaction protocols. In: WRAS, pp. 10:1–10:6 (2010)
10. Chandra, T.D., Hadzilacos, V., Toueg, S.: The weakest failure detector for solving consensus. J. ACM 43(4), 685–722 (1996)
11. Chandra, T.D., Toueg, S.: Unreliable failure detectors for reliable distributed systems. J. ACM 43(2), 225–267 (1996)
12. Charron-Bost, B., Hutle, M., Widder, J.: In search of lost time. Inf. Process. Lett. 110(21), 928–933 (2010)
13. Dijkstra, E.W.: Self-stabilizing systems in spite of distributed control. Commun. of the ACM 17(11), 643–644 (1974)
14. Fischer, M., Jiang, H.: Self-stabilizing leader election in networks of finite-state anonymous agents. In: OPODIS, pp. 395–409 (2006)
15. Fischer, M.H., Lynch, N.A., Paterson, M.S.: Impossibility of consensus with one faulty process. Journal of the ACM 32(2), 374–382 (1985)
16. Michail, O., Chatzigiannakis, I., Spirakis, P.G.: Mediated population protocols. Theor. Comput. Sci. 412(22), 2434–2450 (2011)
17. Michail, O., Chatzigiannakis, I., Spirakis, P.G.: Terminating population protocols via some minimal global knowledge assumptions. In: SSS, pp. 77–89 (2012)
18. Mizoguchi, R., Ono, H., Kijima, S., Yamashita, M.: On space complexity of self-stabilizing leader election in mediated population protocol. Distributed Computing 25(6), 451–460 (2012)

α-Register*

David Bonnin and Corentin Travers

LaBRI, University Bordeaux 1, France
name.surname@labri.fr

Abstract. It is well known that in an asynchronous message-passing system, one can emulate an atomic register providing that more than half of the processes are non-faulty. By contrast, when a majority of the processes may fail, simulating atomic register is not possible. This paper investigates weak variants of atomic registers that can be simulated tolerating a majority of processes failures. Specifically, the paper introduces a new class of registers, called α-register and shows how to emulate them.

For atomic registers, a read operation returns the last written value when there is no concurrent write operations. α-registers generalize atomic registers in the following sense: In any interval I, at most α values written before I are returned by the read operations in I. A simulation of an α-register tolerating f failures in a n-processes system is presented for $\alpha = 2M - 1$, where $M = \max(1, 2f - n + 2)$. The simulation is optimal up to a constant multiplicative factor: the paper establishes that α-registers cannot be simulated tolerating f failures if $\alpha \leq M$.

Keywords: Message passing, fault-tolerance, shared-memory simulation.

1 Introduction

Registers. A *register* is a basic shared object that allows processes to store and retrieve values. The state of a register consists in a value in some set $\mathcal{V}$; it supports two operation: WRITE(v), that changes its state to v and READ() that returns the value stored in the register. Several consistency conditions have been defined that specify correct responses for READ() operations overlapping concurrent WRITE() operations [22]. In their strongest form, registers are *atomic*: each operation appears to take place instantaneously at some point between its invocation and its response.

Twenty years ago, Attiya, Bar-Noy and Dolev showed that atomic registers can be emulated in asynchronous, crash prone message passing systems provided that a *majority* of the processes do not fail [5]. This fundamental result enables shared-memory algorithms to be automatically implemented in message passing environment. Furthermore, impossibility results and lower bounds established in the shared memory model can directly be translated to message passing. For example, the asynchronous computability theorem that characterizes tasks

---

* This work is supported in part by the ANR project DISPLEXITY.

R. Baldoni, N. Nisse, and M. van Steen (Eds.): OPODIS 2013, LNCS 8304, pp. 53–67, 2013.

wait-free solvable in shared memory [21] and its extensions to the t-resilient case [18] apply as well to the asynchronous message passing model with a majority of non-faulty processes.

Beyond the Majority Barrier. A key ingredient of the simulation of registers in message passing is a *quorum system*, that is a collection of sets of processes such that any two sets intersect. In Attiya, Bar-Noy and Dolev protocol (*ABD protocol* [5]), a quorum is any set of $n - f$ processes, where n is the total number of processes in the system and $f < \frac{n}{2}$ an upper bound on the number of failures. Intuitively, a WRITE(v) involves communicating v to a quorum while a READ() returns the most recent value in a quorum. By the intersection property, some process participate in both operations, allowing the READ() to return an up to date value. Quorums defined as set of $n - f$ processes are *live*, in the sense that any process can broadcast a request and eventually receives replies from $n - f$ processes. However, if less than a majority of the processes are non-faulty, i.e. $f \geq \frac{n}{2}$, contacting $n - f$ processes in a READ() operation may not ensure that the value returned by that operation is up to date. Indeed, simulating atomic registers while tolerating $f \geq \frac{n}{2}$ failures in asynchronous message passing is not possible [5].

A few approaches has been proposed to circumvent this impossibility. Probabilistic quorums systems allow two quorums to be non-intersecting with some small probability [1,16,23], leading to a small probability that READ() operations return stall values. Dynamic atomic storage systems, such as RAMBO [20] and DynaStore [3] emulate atomic registers in dynamic environments. They support a reconfiguration operation for adding or removing processes. Reconfiguration may thus be used to replace failed processes. However, failures are typically assumed to be limited when reconfigurations take place. The approaches [17,25] are also based on stronger model assumptions.

Another approach consists in relaxing the consistency guarantees of the implementation. *Eventual consistency* [15,24] essentially only requires that if finitely many WRITE() operations are performed, eventually every READ() operation returns the last written value. When availability is a primary concern, eventually consistent services has been implemented and deployed for large-scale, geo-replicated systems (e.g.,[10,13]). In this settings, network partitions may occur but operation must complete even in the case of such events.

The Question Addressed in the Paper. The paper investigates the following question:

> Given n and $\frac{n}{2} \leq f < n$, what type of (weak) register can be simulated in an n-processes asynchronous message passing system tolerating f failures?

By the ABD emulation, shared memory may be seen as an high-level language to design message passing algorithms tolerating a minority of failures. The question above thus amounts to finding an equivalent high level construct for the case in which a majority of the processes may fail.

Moreover, recently, non-trivial asynchronous algorithms for k-set agreement and k-parallel consensus[1] that tolerate $f(k) \geq \{\frac{n}{2}, k\}$ failures have been designed for message passing systems [8,9]. While the liveness of these algorithms depends on some additional assumption (such as, e.g., an eventual, non-faulty leader), the safety part relies solely on the bound $f(k)$ on the number of failures. As $f(k) \geq \frac{n}{2}$, the existence of those algorithms cannot be inferred from shared memory results. Identifying weak types of registers, that one can simulate when a majority of the processes could fail, might help understanding what can be computed in such systems.

Contributions of the Paper. The paper introduces a new type of registers, called *α-register* and shows an implementation in message passing systems tolerating a majority of faulty processes. Implementations of α-registers are required to be available, that is any WRITE() or READ() request must eventually complete, and partition-tolerant. Indeed, in an asynchronous system in which $f \geq \frac{n}{2}$ processes may fail, processes can be partitioned in two or more sets of at least $n - f$ processes, and messages exchanged between partitions may be arbitrarily delayed. Hence, according to the CAP theorem ([19], Corollary 1.1), it is unavoidable that some READ() operations return outdated values. The parameter α specifies how many distinct outdated values can be read in any interval I, that is values that have been written before I. When $\alpha = 1$, the definition boils down to atomic register.

In more detail, the contribution of the paper is threefold: (1) it introduces α-registers, a new type of register that generalizes atomic registers (Section 2); (2) for $f \geq \frac{n}{2}$ and $M = 2f - n + 2$, it presents a f-resilient message passing implementation of a single-writer multi-reader α-register with $\alpha = 2M - 1$ (Section 3); (3) finally, the paper establishes a lower bound linking f, n and α, namely there is no n-processes, f-resilient implementation of an α-register for $\alpha \leq M$ (Section 4). This lower bound implies that our α-register implementation is within an additive term of at most $\frac{\alpha}{4}$ of the maximal number of failures that can be tolerated.

2   Computational Model and Definition of α-Registers

Message Passing Asynchronous Distributed System. We consider a distributed system made of a set Π of n asynchronous processes $\{p_1, \ldots, p_n\}$, as described in e.g. [6,11]. Each process runs at its own speed, independently of the other processes.

Processes communicate by sending and receiving messages over a reliable but asynchronous network. Each pair of processes $\{p_i, p_j\}$ is connected by a bi-directional channel. Channels are reliable and asynchronous, meaning that each

---

[1] k-set agreement [12] and k-parallel consensus [2] generalize the consensus problem. In k-set agreement, at most k distinct values may be decided. k-parallel consensus consider k instances of consensus and requires each non-faulty process to decide in at least one of them.

message sent by p_i to p_j is received by p_j after some finite, but unknown, time; there is no global upper bound on message transfer delays. The algorithm in Section 3 assumes *FIFO* channels, that is for any pair of processes p_i, p_j, the order in which the messages sent by p_i to p_j are received is the same as the order in which they are sent.

The system is equipped with a global clock whose ticks range $\mathbb{T}$ is the set of positive integers. This clock is not available to the processes, it is used from an external point of view to state and prove properties about executions.

In a *step*, a process may send a message to some other process, performs arbitrary local computation and receives a message that has been previously sent to it but has not been already received. An *execution* is a possibly infinite sequence of steps. Processes may fail by *crashing*. A process that crashes prematurely halts and never recovers. In an execution, a process is *faulty* if it fails and *correct* otherwise. f denote an upper bound on the maximal number of processes that may fail.

Definition of α-Registers. As classical read/write registers, an α-Register supports two operations: WRITE(v), where v is a value taken from some set $\mathcal{V}$, and READ(). A WRITE(v) operation returns an acknowledgment ok and a READ() returns a value $u \in \mathcal{V} \cup \{\bot\}$ where u is the input of a WRITE() operation or the initial value $\bot$ of the α-Register. In an *admissible execution*, no process starts a WRITE(v) or READ() operation while its previous operation, if any, has not returned. The *execution interval* $I(op)$ of an operation instance op by process p begins when p calls WRITE() or READ() and ends when p returns from that call; if p never returns, $I(op)$ has no end. We sometimes simply say operation instead of operation instance. Two operations op_1 and op_2 are *concurrent* if $I(op_1) \cap I(op_2) \neq \emptyset$. A terminating operation op_1 *precedes* operation op_2 if $I(op_1) \cap I(op_2) = \emptyset$ and $I(op_1)$ ends before $I(op_2)$ begins. An operation op is *active* in an interval I if $I \cap I(op) \neq \emptyset$. To simplify the exposition, we assume without loss of generality that no two distinct WRITE() operations have the same input value[2].

In any admissible execution e, a α-Register satisfies the following properties.

1. *Termination.* Any READ() or WRITE(v) operation performed by a correct process terminates.
2. *Non-spurious value.* For any terminating READ() operation R that returns u, either $u = \bot$ or there exists a WRITE(u) operation that precedes or is concurrent with R.
3. *Chronological read.* Let R, R' be two terminating READ() operations performed by the same process in that order and let u, u' be the values returned. If $u \neq \bot$, then $u' \neq \bot$ and WRITE(u) precedes or is concurrent with WRITE(u').
4. *Non-triviality.* Let R be a READ() operation by process p and let u be the value returned by R. If there is a WRITE() operation by p that precedes R,

---

[2] This can be enforced by appending a sequence number and the id of the writer to each value to be written.

$u \neq \perp$. Moreover, if W is the last WRITE() operation by p that precedes R, WRITE(u) is either W or a WRITE() operation by another process that is concurrent with or is preceded by W.

5. *Propagation.* Let u be the input of a terminating WRITE() or the value returned by a READ() performed by a correct process. Eventually, for every terminating READ() operation R' with return value u' either $u = u'$ or WRITE(u) is concurrent with or precedes WRITE(u').

6. *α-Bounded reads.* In any interval I, the set of values that have been written by WRITE() operations that terminate before I and returned by the READ() operations whose execution interval is contained in I is of size at most α.

The termination property implies that an α-register is always available. In particular, any READ() operation by a non-faulty process always returns a value. The properties chronological read and non-triviality express consistency requirements in the context of a single process. Chronological read requires that successive READ() by the same process p_i do not return older values. Non-triviality intuitively requires that p_i "sees" its writes. After a WRITE(u) operation, every subsequent READ() by p_i returns a value as least as recent as u. The propagation properties implies that α-register are eventually consistent. If after some time no WRITE() operations are performed, eventually every READ() operation returns the last value written.

Since $f \geq \frac{n}{2}$, it can be shown by a partition argument that READ() may return arbitrary old values. Consider two sets Q_1, Q_2 of $n - f$ processes that do not intersect and suppose that every process not in $Q_1 \cup Q_2$ initially fails. As communication is asynchronous, messages exchanged between Q_1 and Q_2 may be delayed during an arbitrary long interval I. For some process $p_i \in Q_i, 1 \leq i \leq 2$, the operations by p_i may thus return after messages have been exchanged only with the processes in Q_i (this is indistinguishable for p_i from an execution in which every process not in Q_i fail before I.). Therefore, if p_1 performs WRITE() operations, the values it writes are not seen by p_2. Thus READ() operations by p_2 may return values that have written before any WRITE() operations by p_2.

Rather that bounding the staleness of values returned by READ() operation, which is impossible if asynchrony and a majority of failures have to be tolerated, the α-bounded read property imposes that not too many stale values, namely no more that α, are returned by READ() operations.

3   Single-Writer Multiple-Reader α-Register

This section presents a protocol (Algorithm 3.1) that implements a single-writer multiple-readers (SWMR) α-Register in an asynchronous system in which up to $f \leq n - 1$ processes may fail. The value of α depends on the number of failures the protocol tolerates, namely $\alpha = 2M - 1$, where $M = 2f - n + 2$ if $f \geq \frac{n}{2}$ and $M = 1$ otherwise. The algorithm assumes that channels are FIFO.

The algorithm is similar to the ABD protocol [5]. Each time a new value is written it is first associated with a unique timestamp (line 7). Timestamps are increasing so that more recent values get larger timestamps. As there is a single

Algorithm 3.1. SWMR α-Register (code for process p_i)

```
1:  INITIALIZATION
2:      seq_i ← 1; ⟨v_i, ts_i⟩ ← ⟨⊥, 0⟩; ⟨vr_i, tsr_i⟩ ← ⟨⊥, 0⟩;
3:      Qr_i ← ∅; Qe_i ← ∅; Qw_i ← ∅;
4:      Accept_i[1..n] ← [2, ..., 2]                         ▷ array of n integers initialized to 2
5:      for each p_j : 1 ≤ j ≤ n do send UPDATE(seq_i, ⟨v_i, ts_i⟩, 0)
6:  function WRITE(v)
7:      ⟨v_i, ts_i⟩ ← ⟨v, ts_i + 1⟩; seq_i ← seq_i + 1; Qw_i ← ∅;
8:      wait until |Qw_i| ≥ n − f;
9:      return ok
10: function READ( )
11:     n_iter ← 0;
12:     repeat
13:         ⟨vr_i, tsr_i⟩ ← ⟨v_i, ts_i⟩; seq_i ← seq_i + 1; Qr_i ← ∅; Qe_i ← ∅;
14:         wait until |Qr_i ∪ Qe_i| ≥ n − f;
15:         n_iter ← n_iter + 1
16:     until (|Qe_i| ≥ n − f) or (n_iter ≥ N)      ▷ N = 2(2f + 1)(⌊ n/(n−f) ⌋ + 1) + 1
17:     return vr_i
18: WHEN UPDATE(seq, ⟨v, ts⟩, old_seq) FROM PROCESS p_j IS RECEIVED
19:     if old_seq = seq_i then
20:         if ts = ts_i then Qw_i ← Qw_i ∪ {p_j}
21:         if ts > tsr_i then Qr_i ← Qr_i ∪ {p_j}
22:         if ts = tsr_i then Qe_i ← Qe_i ∪ {p_j}
23:     if ts > ts_i then
24:         if Accept_i[j] > 0 then Accept_i[j] ← Accept_i[j] − 1
25:         else ⟨v_i, ts_i⟩ ← ⟨v, ts⟩; Accept_i[1..n] ← [2, ..., 2]   ▷ Accept_i[j] = 0
26:     send UPDATE(seq_i, ⟨v_i, ts_i⟩, seq) to p_j
```

writer, no two values are associated with the same timestamp. Each process p_i maintains a pair of local variables $\langle v_i, ts_i \rangle$ which store the most recent value p_i knows of together with its timestamp. We say that p_i *accepts* a pair $\langle v, t \rangle$ when p_i changes $\langle v_i, ts_i \rangle$ to $\langle v, t \rangle$ (line 25).

Processes constantly exchange messages of type UPDATE that contain the most recent value and its timestamp known by the message's sender[3]. For any pair of processes p_i, p_j, UPDATE messages are exchanged between p_i and p_j following a "ping-pong" pattern. Initially, p_i and p_j send UPDATE messages to each other, and each time p_i (resp. p_j) receives UPDATE from p_i (resp. p_j), it replies with an UPDATE message. Each such message contains a triple $(sq, \langle v, ts \rangle, osq)$, where sq and osq are sequence numbers, and $\langle v, ts \rangle$ is the current value and timestamp of the process sending the message. p_i maintains a sequence number that is incremented each time p_i starts a new WRITE( ) operation or a new phase in a READ( ) operation

---

[3] The algorithm is not quiescent: processes keep sending and receiving messages even if no WRITE( ) or READ( ) operations are performed. The algorithm can be made quiescent at the price of an increasing complexity in the pseudo-code. We choose to ignore this issue to keep the pseudo-code simple.

(see below). If message $m = \text{UPDATE}(sq, \langle v, ts \rangle, osq)$ is sent by p_i in reply to a message $\text{UPDATE}(sq', \langle v', ts' \rangle, osq')$ from p_j (see line 18–line 26), then $osq = sq'$ and sq is the current sequence number of p_i. Thus, by comparing osq with its current sequence number, process p_j can determine whether m is related to its current operation or to a previous operation.

Write() Operations. The implementation of WRITE(v) is similar to the implementation in the ABD protocol. After a new timestamp t has been associated with v on line 7, the writer p_n changes its local variable $\langle v_n, ts_n \rangle$ to $\langle v, t \rangle$. It then waits until each process in a quorum of $(n - f)$ processes have accepted $\langle v, t \rangle$, and the operation then returns (line 8–line 9). In more detail, the local variable Qw_n, intended to contain a set of processes ids, is emptied at the beginning of the operation (line 7). Then, each time, a message UPDATE containing the pair $\langle v, t \rangle$ from a process p_j is received, p_j is added to Qw_n (line 20). The operation returns when $|Q_w| \geq n - f$.

The new pair $\langle v, t \rangle$ is disseminated by the UPDATE messages sent by the writer: once $\langle v_n, ts_n \rangle$ has been changed to $\langle v, t \rangle$, and until a new WRITE() operation is initiated, every UPDATE sent by p_n contains $\langle v, t \rangle$.

Value Dissemination. As newly written values are propagated asynchronously, at any point in time there might be pending UPDATE that have been sent to p_i by the processes in some set S, but not yet received by p_i. Each of these messages may contain a distinct pair $\langle \text{value}, \text{timestamp} \rangle$ from some set $\{\langle w_1, t_1 \rangle, \ldots, \langle w_m, t_m \rangle\}$. If $\langle v_i, ts_i \rangle$ changes each time p_i receives a newer value, the successive values of v_i might be $w_1, \ldots, w_m$. Furthermore, if the same happens at each process in set of size at least $n - f$, it could be the case that each value $w_1, \ldots, w_m$ is returned by a READ() operation. Instead, to avoid that READ() operations return many old values, we ensure that if $\langle v_i, ts_i \rangle$ changes from $\langle w, t \rangle$ to $\langle w', t' \rangle$, some process p_j stores $\langle w', t' \rangle$ after $\langle v_i, ts_i \rangle$ is set to $\langle w, t \rangle$ and before it is changed to $\langle w', t' \rangle$.

Due to the "ping-pong" pattern followed by messages exchanged between processes p_i and p_j, there are at most two messages that have been sent by p_j but have not yet been received by p_i at any point in time. Hence, if p_i receives three UPDATE from p_j in some interval I, the last one of these messages has been sent during I. The array $Accept_i$ is used to keep track of how many consecutive messages from the same process carrying new values have been received. Initially, $Accept_i[j] = 2$ and at any time, $Accept_i[j] \in \{0, 1, 2\}$ for any $j, 1 \leq j \leq n$. $Accept_i[j]$ is decremented each time p_i receives an UPDATE from p_j (line 24) carrying a value newer than p_i's current value. The array is also reset to $[2, \ldots, 2]$ when $\langle v_i, ts_i \rangle$ changes (line 25). Hence, $Accept_i[j] < 2$ means that p_i knows that its current value is outdated by p_j's current value (line 23 – line 24). If $\text{UPDATE}(*, \langle w, t \rangle, *)$ where $\langle w, t \rangle$ is newer than $\langle v_i, ts_i \rangle$ is received from p_j when $Accept_i[j] = 0$, $\langle v_i, ts_i \rangle$ is changed to $\langle w, t \rangle$ (line 25).

Suppose that at some point a set X of messages have not been yet received by p_i. Note that if, when some message $m \in X$ is received, $\langle v_i, ts_i \rangle$ changes, then no other message in X updates $\langle v_i, ts_i \rangle$. This is because $Accept_i$ is reset to

$[2, \ldots, 2]$ each time $\langle v_i, ts_i \rangle$ changes, messages are received in FIFO order and at any point time and for any process p_j, no more than two messages sent by p_j have not been received by p_i.

Read() Operations. A READ() operation (line 11–line 17) by process p_i consists in up to $N = O(\frac{fn}{n-f})$ iterations. Each iteration is identified by an increasing sequence number seq_i. At the beginning of iteration s, the pair $\langle v, ts \rangle$ currently hold by p_i is stored in $\langle vr_i, tsr_i \rangle$ and the two sets Qe_i and Qr_i, intended to contain processes that hold a pair equal to or more recent than, respectively, $\langle vr_i, tsr_i \rangle$ are emptied (line 13). An iteration terminates when p_i knows that at least $n - f$ processes store values as least as recent as vr_i, i.e., when $|Qr_i \cup Qe_i| \geq n - f$. The READ() operation terminates (1) immediately if $|Qe_i| \geq n - f$, i.e., for each process $p_j \in Qe_i$, there is a time at which $v_j = vr_i$ or (2) after N iterations have been performed. The value returned is then vr_i, the value of v_i at the beginning of the last iteration. We show in the proof that for every READ() operation op that returns a value w written before the operation starts, the operation terminates by condition (1) above. That is, each process p_j in a set Q of size $n - f$ stores v at some point during the interval of op. Intuitively, in each iteration for which condition (1) is not satisfied, p_i learns a newer value. This value is propagated to at least $n - f$ processes in the following iterations. Since the number of values that have been written before op starts and that can be learned and propagated is bounded by a function of f and n, every process knows the last value written before op starts after some constant number of iterations or new values are written concurrently with op.

Consider an interval I. Let w_ℓ be the value written by the last WRITE() that terminates before I. When WRITE(w_ℓ) returns, each process in a set Q_ℓ of size $n - f$ stores w_ℓ. Since a process can only replace its value with a newer one, the value stored by any process of Q_ℓ at any point in I is w_ℓ or a more recent value. Therefore, any value older than w_ℓ present in the system at the beginning of I is stored or contained in a message not yet received by a processes of $\Pi \setminus Q_\ell$. Let L, T be respectively the values stored by the processes of $\Pi \setminus Q_\ell$ and the values contained in the messages not yet received by the processes of $\Pi \setminus Q_\ell$.

In the worst case, $|L| = f$. The protocol ensures that if $v \in L$ is returned by a READ() performed during I, then at least $n - f$ processes stores v at some point during I. Since no process changes its value for an older one, the $n - f - 1$ oldest values in L cannot be returned by READ() operations. Hence, at most $f - (n - f - 1) = 2f - n + 1 = M - 1$ distinct values of L can be read in I.

As previously explained, for each process $p_i \in \Pi \setminus Q_\ell$, at most one message not yet received by p_i at the beginning of I may change the value v_i stored by p_i. Therefore, at most f values of T may be stored by the processes and thus be returned by READ() operations. As in the case of the values of L, the $n - f - 1$ oldest values cannot be returned by READ() operations. Therefore, at most $f - (n - f - 1) = 2f - n + 1 = M - 1$ distinct values of T can be read in I. In addition, w_ℓ may be read in I. It thus follows that the number of values written before I and returned by READ() operations during this interval is at most $2(M - 1) + 1 = 2M - 1 = \alpha$.

3.1 Proof of the Protocol

We consider an arbitrary infinite admissible execution α in which the unique writer is the process p_n. var_i denote the local variable var of process p_i and var_i^τ its value at time τ. Due to space constraints, some proofs are omitted. They can be found in [7].

Whenever the writer initiates a new WRITE() operation, it increases a counter whose value ts is assigned as a timestamp to the value v being written (line 7). This timestamp is unique and no other timestamp is ever associated to v. That is, for any process p_i, whenever the local variable v_i is changed to v, ts_i is changed accordingly to the timestamp ts associated with v (line 25). Values can thus be totally ordered according to their timestamp. In particular we say that value v is *newer* than or *more recent* than value v' is the timestamp ts assigned to v is larger than or equal to the timestamp ts' assigned to v' and we note $\langle v', ts' \rangle \preceq \langle v, ts \rangle$. Note that, for any process p_i, whenever the pair $\langle v_i, ts_i \rangle$ is modified (line 7 or line 25), it is replaced by a more recent value. That is,

Observation 1. *For every process p_i, and every times $\tau < \tau'$, $\langle v_i^\tau, ts_i^\tau \rangle \preceq \langle v_i^{\tau'}, ts_i^{\tau'} \rangle$.*

Each time a process p_i receives a message UPDATE from a process p_j, it sends back an UPDATE to process p_j (line 18 and line 26). Moreover, initially each process sends an UPDATE message to every process (line 5). It thus follows that messages UPDATE are perpetually exchanged between p_i and p_j if both processes are correct:

Lemma 1. *Let p_i, p_j be two correct processes. p_i receives infinitely many messages UPDATE from p_j.*

Proof. Initially, p_i sends an UPDATE message to p_j (initialization, line 5). By the code (line 18 and line 26), each time a correct process p receives a message UPDATE from a process q, p sends a message UPDATE to q. Since p_i and p_j are two correct processes, p_i receives infinitely many messages UPDATE from p_j. □

Next Lemma shows that whenever a correct process learns a new value, every other correct eventually learn that value or a more recent one. It forms the basis to show that READ() (Lemma 3) and WRITE() (Lemma 4) operations performed by correct processes terminate.

Lemma 2. *Let p_i, p_j be two correct processes. If at some time, $\langle v_i, ts_i \rangle = \langle v, ts \rangle \neq \langle \perp, 0 \rangle$, then eventually $\langle v, ts \rangle \preceq \langle v_j, ts_j \rangle$.*

Lemma 3. *Let p_i be a correct process. Every invocation of READ() by p_i returns.*

Lemma 4. *Assume that the writer p_n is a correct process. Every invocation of WRITE() by p_n returns.*

Proof of the Bounded Reads Property. Let $\mathcal{I}$ be an arbitrary interval. Let $\mathcal{R} = \{R_1, \ldots, R_m\}$ be a set of READ() operations whose execution intervals are contained in $\mathcal{I}$. Let w_i be the value returned by operation R_i and let $\mathcal{V}_R = \{w_1, \ldots, w_m\}$.

The remainder of this section is devoted to the proof of the following Lemma:

Lemma 5. $|\mathcal{V}_R \setminus \mathcal{V}_W| \leq 2M - 1$ *where* $M = \max(1, 2f - n + 2)$, *where* $\mathcal{V}_W$ *is the set of values written during* $\mathcal{I}$.

The values returned by each READ() of $\mathcal{R}$ is a value that is either written by one of the Write operation active during $\mathcal{I}$, or a value present in the system at the beginning of $\mathcal{I}$. A value v with timestamp ts is present in the system at the beginning of $\mathcal{I}$ if it is the value locally stored by a process, i.e., $\langle v, ts \rangle = \langle v_i, ts_i \rangle$ for some process p_i or $\langle v, ts \rangle$ is carried by a message UPDATE that has not yet been delivered. Let τ_b be the time at which $\mathcal{I}$ starts. We define:

- $\mathcal{V}_W$ the set of the values written during $\mathcal{I}$. That is, $w \in \mathcal{V}_W$ if $I(\text{WRITE}(w)) \cap \mathcal{I} \neq \emptyset$;
- $Inc_{j \to i}$ as the set of incoming messages UPDATE that have been sent by p_j but has not yet been received by p_i by time τ_b ;
- $\mathcal{V}_L = \{v : v \notin \mathcal{V}_W \text{ and } \exists p_i, \langle v_i, ts_i \rangle = \langle v, ts \rangle \text{ at time } \tau_b \}$;
- $\mathcal{V}_I = \{v : v \notin \mathcal{V}_L \cup \mathcal{V}_W \text{ and } \exists p_i, p_j, \text{UPDATE}(*, \langle v, ts \rangle, *) \in Inc_{j \to i}\}$;
- *vlast*, the value written by the last WRITE() operation that precedes $\mathcal{I}$.

That is, $\mathcal{V}_L$ is the set of values locally stored by the processes at the beginning of $\mathcal{I}$, while $\mathcal{V}_I$ is the set of values that are not locally stored, but part of the content of some messages still in transit. Note that $\mathcal{V}_R \setminus \mathcal{V}_W \subseteq \mathcal{V}_L \cup \mathcal{V}_I$.

We first observe that $|\mathcal{V}_L| \leq f + 1$ (Corollary 1). Essentially, this follows from the fact that a quorum $Qlast$ of at least $n - f$ processes must have accepted the value written by the last WRITE() operation $Wlast$ preceding $\mathcal{I}$ in order for that operation to return (Lemma 6). As a process replace the value it stores locally only with a more recent one (Observation 1), the value locally stored by each process $p_i \in Qlast$ is newer than or equal to *vlast*, at any time in $\mathcal{I}$. In other words, the value stored by p_i during $\mathcal{I}$ belongs to $\{vlast\} \cup \mathcal{V}_W$.

Lemma 6. *Let vlast be the value written by the last* WRITE() *operation preceding* $\mathcal{I}$ *and let tslast denote the timestamp associated with it. There is a set Qlast of at least* $n - f$ *processes such that at any time in* $\mathcal{I}$, $v_i \in \{vlast\} \cup \mathcal{V}_W$ *and* $ts_i \geq tslast$.

Proof. Let $Wlast$ be the last WRITE() that precedes $\mathcal{I}$. When this operation returns, p_n has received a message UPDATE$(*, \langle vlast, tslast \rangle, *)$ from each process in a set Q of size at least $n - f$ (line 8). This means that for each $p_i \in Q$, we have at some time $\langle v_i, ts_i \rangle = \langle vlast, tslast \rangle$ (line 26).

Moreover, by the code each value v written by the single writer p_n before *vlast* is associated with a timestamp strictly smaller than *tslast*. As each time the value stored locally (in v_i for process p_i) is modified, it is replaced by a more recent value, i.e., a value associated with a larger timestamp (Observation 1), it follows from the fact that $Wlast$ is the last write operation preceding $\mathcal{I}$ that for every process $p_j \in Q$, $ts_j \geq tslast$ and $v_j \in \{vlast\} \cup \mathcal{V}_W$ at any time in $\mathcal{I}$. $\square$

Corollary 1. $|\mathcal{V}_L| \leq f + 1$

Proof. By Lemma 6, at the beginning of $\mathcal{I}$, for each process $p_i \in Qlast$, $v_i \in \{vlast\} \cup \mathcal{V}_W$. As $\mathcal{V}_W \cap \mathcal{V}_L = \emptyset$ and as $|Qlast| \geq n - f$, $|\mathcal{V}_L| \leq f + 1$. $\qquad\square$

Consider two processes p_j and p_i. By the code p_j sends a message UPDATE to p_i each time it receives a message UPDATE from that process (line 18–line 26). Since initially both p_i and p_j send a message UPDATE to each other, it follows that at any time at most two messages UPDATE have been sent by p_j to p_i and has not yet been received by the latter:

Observation 2. *For every pair of processes* p_i, p_j, $|Inc_{j \to i}| \leq 2$.

Let $\mathcal{U} \subseteq \mathcal{V}_I$ be the set of values that are not locally stored by any process at the beginning of $\mathcal{I}$, but later stored by at least one process at some time in $\mathcal{I}$. That is,

$$u \in \mathcal{U} \iff u \in \mathcal{V}_I \text{ and at some time in } \mathcal{I}, v_i = u \text{ for some process } p_i$$

We upper-bound the size of $\mathcal{U}$ (Lemma 7) by f. This upper bound is a key ingredient in establishing that, for any value v read during $\mathcal{I}$, there is set of at least $(n - f)$ processes p_j that hold v at some point in $\mathcal{I}$, that is $v_j = v$ at some time in $\mathcal{I}$ (Lemma 8).

Lemma 7. $|\mathcal{U}| \leq f$

Proof. Let $u \in \mathcal{V}_I$ be a value and let ts denote the timestamp associated with it. Suppose that at some time τ in $\mathcal{I}$, $\langle v_x, ts_x \rangle = \langle u, ts \rangle$ for some process p_x. At the beginning of $\mathcal{I}$, no process stores locally u (for every process p_j, $v_j \neq u$) but a message UPDATE whose content contains u has been sent to some process but has not yet been received by that process.

Let p_i be the first process that, during $\mathcal{I}$ change its pair $\langle$local value, timestamp$\rangle$ to $\langle u, ts \rangle$ (at line 25). Since $u \notin \mathcal{V}_L$, this occurs when p_i receives a message $m = \text{UPDATE}(*, \langle u, ts \rangle, *)$ sent to it before $\mathcal{I}$, that is there exists a process p_j such that $m \in Inc_{j \to i}$.

Consider another value $u' \neq u$ that similarly to u (1) is contained in $\mathcal{V}_I$ and (2) at some time τ' in $\mathcal{I}$ is stored locally by some process $p_{x'}$ (i.e., at time τ', $v_{x'} = u'$). Let $p_{i'}$ be the first process that changes during $\mathcal{I}$ its local value $v_{i'}$ to u'. As explained above, this occurs when $p_{i'}$ receives a message $m' \in Inc_{j' \to i'}$.

Suppose for contradiction that $p_i = p_{i'}$. Assume without loss of generality that p_i first changes v_i to u and then later to u'. By the code, immediately after $\langle v_i, ts_i \rangle$ has been modified, the array $Accept_i$ is reset to $[2, \ldots, 2]$ (line 25). As the channels are FIFO, any message received from $p_{j'}$ by p_i during $\mathcal{I}$ and before m' is received are contained in $Inc_{j' \to i}$. Hence, after m has been received and before the reception of m', p_i has received at most $|Inc_{j' \to i}| - 1$ messages from $p_{j'}$. As $|Inc_{j' \to i}| \leq 2$ (Observation 2), it thus follows that $Accept_i[j'] > 0$ when m' is received by p_i (Recall that the counter $Accept_i[j']$ is decremented at most once each time a message from $p_{j'}$ is received line 23–line 24). Therefore (line 24–line 25), v_i remains unchanged when m' is received: a contradiction.

Finally, note that neither p_i nor $p_{i'}$ are contained in $Qlast$ since for each process p_j in this set, v_j is $vlast$ or a more recent value (Lemma 6). As $|Qlast| \geq n - f$, we conclude that $|\mathcal{U}| \leq f$. □

Lemma 8. *Let $R \in \mathcal{R}$ be a* READ*() operation and let v be the value returned by R. Either $v \in \mathcal{V}_W$, or there is a set Q_R of at least $n - f$ processes such that for each $p_i \in Q_R$, there is a time in $I(R)$ at which $v_i = v$.*

Finally, we bound the number of values that are read and, on one hand, stored by at least one process (Lemma 9), or, on the other hand, only contained in messages that have not yet been delivered at the beginning of $\mathcal{I}$ (Lemma 10). As any old value (i.e., a value not in $\mathcal{V}_W$) that is read during $\mathcal{I}$ is either stored locally by some process or contains in message not yet delivered at the beginning of $\mathcal{I}$, the bound on the number of values read during $\mathcal{I}$ follows.

Lemma 9. $|\mathcal{V}_L \cap \mathcal{V}_R| \leq M = 2f - n + 2$

Lemma 10. $|\mathcal{V}_I \cap \mathcal{V}_R| \leq M - 1 = 2f - n + 1$

Proof of Lemma 5. Any value that is returned by a READ() operation during $\mathcal{I}$ is either contained in $\mathcal{V}_L$ or $\mathcal{V}_I$ or $\mathcal{V}_W$. As $|\mathcal{V}_L \cap \mathcal{V}_R| \leq 2f - n + 2$ (Lemma 9), and $|\mathcal{V}_I \cap \mathcal{V}_R| \leq 2f - n + 1$ (Lemma 10), $|\mathcal{V}_R \setminus \mathcal{V}_W| = |\mathcal{V}_R \cap (\mathcal{V}_L \cup \mathcal{V}_I)| \leq (2f - n + 2) + (2f - n + 1) = 2M - 1$ □

Finally, the correctness of Algorithm 3.1 is implied by the following theorem.

Theorem 3. *Algorithm 3.1 implements a SWMR $(2M-1)$-register, where $M = 2f - n + 2$.*

Proof. Consider an admissible execution of Algorithm 3.1. The *termination* property of α-registers immediately follows from Lemma 3 and Lemma 4. For the *Non-spurious value*, the value returned by a READ() operation by process p_i is the value of the variable v_i at some time in the execution interval of the operation. At any point in the execution, v_i stores $\perp$ or a value that has been introduced by the writer.

For the *Chronological read* property, consider u, u' two values returned in that order by READ() operations performed by the same process and let t, t' be the timestamp associated with u, u' respectively. By Observation 1, $\langle u, t \rangle \preceq \langle u', t' \rangle$. Henceforth, WRITE(u) precedes WRITE(u') since there is a single writer. The *Non-triviality* property immediately follows from Observation 1: For the single writer, every READ() operation returns the input of its last preceding WRITE() or $\perp$ if there is no preceding WRITE().

To see why the *propagation* property is satisfied, let u be the input of a terminating WRITE() or the value returned by a READ() performed by a correct process p_i and let t denote its timestamp. By the code, at some point $\langle v_i, ts_i \rangle = \langle u, t \rangle$. Then, by Lemma 2, for every non-faulty process p_j, eventually $\langle u, t \rangle \preceq \langle v_j, ts_j \rangle$. Hence, eventually every value u' returned by READ() operations is either u or a value written after u. Finally, the α-*Bounded reads* property with $\alpha = 2M - 1$ follows immediately from Lemma 5. □

4   Lower bound

This section presents a lower bound on α for any implementation of an α-register. More precisely, it proves the following theorem:

Theorem 4. *Let n, f such that $f \geq \frac{n}{2}$. For any implementation of a SWMR α-register for n processes that tolerates f failures, $\alpha \geq M$.*

Proof. (Sketch) Without loss of generality, assume that A is a full information f-resilient protocol that implements a SWMR α-register. That is, the state of each process consists in its initial state and all its history and each time a process sends a message, it sends its entire state. The single writer is the process p_n.

We construct a family of executions of A. Each execution is parametrized by M integers $k_1, \ldots, k_M$. We show that for some values of $k_1, \ldots, k_M$, M distinct values are returned by READ() operations in an interval in which no WRITE() operation is active. Each execution is divided into two phases, each phase consisting in M sequential rounds.

Recall that $M = 2f - n + 2$. Let $\mathcal{V} = \{v_1, \ldots, v_M\}$ be a set of M distinct values. $k = (k_1, \ldots, k_M)$ is a M-tuple of positive integers. For $i, 1 \leq i \leq M$, let Q_i, Q'_i denote the sets of $n - f$ processes $Q_i = \{p_i\} \cup \{p_{f+2}, \ldots, p_n\}$ and $Q'_i = \{p_i\} \cup \{p_{M+1}, \ldots, p_{M+(n-f)-1}\}$, respectively. Observe that $Q_i \cap Q'_i = \{p_i\}$ since $M + (n - f) - 1 = f + 1$. Execution $\mathcal{E}_k$ is defined as follows:
First Phase. This phase consists in M rounds $r_1, \ldots, r_n$. For each $i, 1 \leq i < M$, round r_{i+1} begins after the end of round r_i. Only processes in Q_i take steps in round r_i. We first let every message that have been sent to processes in Q_i during the previous rounds (if any) to be received. Then, $k_i + 2$ operations on the α-register implemented by A are performed sequentially, in that order:

1. WRITE(v_i) is performed by process p_n;
2. Process p_n performs READ();
3. k_i READ() operations are performed by process p_i.

The messages sent to processes $p_j \notin Q_i$ are delayed until some time specified later. As $|Q_i| = n - f$, the execution is indistinguishable by processes in Q_i from an execution that is the same as $\mathcal{E}_k$ until the end of round r_{i-1} and in which the f processes $\notin Q_i$ fail at the beginning of r_i. As A tolerates f failures, every operation performed during round r_i terminates.

Note that the READ() operation performed by p_n return v_i by the non-triviality property of α-registers. Moreover, observe that if k_i is chosen large enough, the last READ() operation performed by p_i returns also v_i by the propagation property.
Second Phase. This phase consists also in M rounds $r'_1, \ldots r'_M$. In round r'_i, which begins after round r'_{i-1} has ended, only processes in Q'_i take steps. We first let the messages that have been sent to the processes in Q'_i during previous rounds $r'_j, 1 \leq j < i$ to be delivered. Then, process p_i performs a READ() operation. As in round r_i, the execution is indistinguishable to the processes in Q'_i from an

execution that is the same until the end of round r_{i-1} and in which the f processes $\notin Q_i$ fail at the beginning of r'_i. Since A tolerates f failures, the READ() operation terminates.

Assuming that k_i has been chosen large enough, the previous READ() operation performed by p_i (in round r_i) returns v_i. By the chronological read property of α-registers, the READ() by p_i in r'_i must return v_i or more recent value, that is a value v_j with $j > i$.

Consider the rounds $r_{i+1}, \ldots, r_M$. The set of processes that take steps in these rounds is $P = \{p_{i+1}, \ldots, p_M\} \cup \{p_{f+2}, \ldots, p_n\}$. On the other hand, the set of processes that takes steps during rounds $r'_1, \ldots, r'_i$ is $P' = \{p_1, \ldots, p_i\} \cup \{p_{M+1}, \ldots, p_{M+(n-f)-1}\}$. Note that $P \cap P' = \emptyset$. Moreover, by construction, for every pair of processes $p \in P, p' \in P'$, every message sent by p (if any) to p' during any round $r_{i+1}, \ldots, r_M$ has not been received by p' by the end of r'_i. Therefore, until the end of r'_i the execution is indistinguishable to the processes in P' from an execution $\mathcal{E}'$ that is the same except that rounds $r_{i+1}, \ldots, r_M$ do not occur in $\mathcal{E}'$. Therefore, the READ() performed by p_i in r'_i cannot return v_j, for any $j > i$. That is, this operation returns v_i.

In the second phase, no WRITE() operation is active. M distinct values are returned by READ() operation performed during this phase. Hence $\alpha \geq M$. $\square$

Remark. The lower bound can be slightly improved by a similar, though more involved, argument to yield $\alpha \geq M + 1$. See [7].

5   Conclusion

The paper has introduced α-registers. For n processes and at most $f \geq \frac{n}{2}$ failures, an implementation of a SWMR $(2M - 1)$-register is presented, where $M = 2f - n + 2$. The implementation is complemented by a lower bound stating that f-resilient simulation of an α-register for $\alpha < M + 1$ is impossible.

Many questions remain open for future research including closing the gap between the implementation and the lower bound, designing a multi-writer multi-reader implementation and understanding the computing power of α-registers. Another challenging direction is to generalize the bounded version of the ABD simulation [5]. Doing so may entail solving problems similar to the ones encountered in the design of fault-tolerant and self-stabilizing atomic registers [4,14].

References

1. Abraham, I., Malkhi, D.: Probabilistic quorums for dynamic systems. Distributed Computing 18(2), 113–124 (2005)
2. Afek, Y., Gafni, E., Rajsbaum, S., Raynal, M., Travers, C.: The k-simultaneous consensus problem. Distributed Computing 22(3), 185–195 (2010)
3. Aguilera, M.K., Keidar, I., Malkhi, D., Shraer, A.: Dynamic atomic storage without consensus. J. ACM 58(2), 7 (2011)

4. Alon, N., Attiya, H., Dolev, S., Dubois, S., Potop-Butucaru, M., Tixeuil, S.: Pragmatic self-stabilization of atomic memory in message-passing systems. In: Défago, X., Petit, F., Villain, V. (eds.) SSS 2011. LNCS, vol. 6976, pp. 19–31. Springer, Heidelberg (2011)

5. Attiya, H., Bar-Noy, A., Dolev, D.: Sharing memory robustly in message-passing systems. J. ACM 42(1), 124–142 (1995)

6. Attiya, H., Welch, J.: Distributed Computing. Wiley (2004)

7. Bonnin, D., Travers, C.: α-register. Technical report hal#00863060 (2013), http://hal.inria.fr/hal-00863060/PDF/

8. Bouzid, Z., Travers, C.: (anti-$\Omega^x \times \Sigma_z$)-based k-set agreement algorithms. In: Lu, C., Masuzawa, T., Mosbah, M. (eds.) OPODIS 2010. LNCS, vol. 6490, pp. 189–204. Springer, Heidelberg (2010)

9. Bouzid, Z., Travers, C.: Parallel consensus is harder than set agreement in message passing. In: ICDCS. IEEE Computer Society (2013)

10. Apache cassandra, http://cassandra.apache.org/

11. Chandra, T., Hadzilacos, V., Toueg, S.: The weakest failure detector for solving consensus. J. ACM 43(4), 685–722 (1996)

12. Chaudhuri, S.: More choices allow more faults: set consensus problems in totally asynchronous systems. Inf. Comput. 105(1), 132–158 (1993)

13. DeCandia, G., et al.: Dynamo: amazon's highly available key-value store. In: SOSP, pp. 205–220. ACM (2007)

14. Dolev, S., Dubois, S., Gradinariu Potop-Butucaru, M., Tixeuil, S.: Crash resilient and pseudo-stabilizing atomic registers. In: Baldoni, R., Flocchini, P., Binoy, R. (eds.) OPODIS 2012. LNCS, vol. 7702, pp. 135–150. Springer, Heidelberg (2012)

15. Fekete, A., Gupta, D., Luchangco, V., Lynch, N.A., Shvartsman, A.A.: Eventually-serializable data services. Theor. Comput. Sci. 220(1), 113–156 (1999)

16. Friedman, R., Kliot, G., Avin, C.: Probabilistic quorum systems in wireless ad hoc networks. ACM Trans. Comput. Syst. 28(3) (2010)

17. Friedman, R., Raynal, M., Travers, C.: Two abstractions for implementing atomic objects in dynamic systems. In: Anderson, J.H., Prencipe, G., Wattenhofer, R. (eds.) OPODIS 2005. LNCS, vol. 3974, pp. 73–87. Springer, Heidelberg (2006)

18. Gafni, E.: The extended bg-simulation and the characterization of t-resiliency. In: STOC, pp. 85–92. ACM (2009)

19. Gilbert, S., Lynch, N.A.: Brewer's conjecture and the feasibility of consistent, available, partition-tolerant web services. SIGACT News 33(2), 51–59 (2002)

20. Gilbert, S., Lynch, N.A., Shvartsman, A.A.: Rambo: a robust, reconfigurable atomic memory service for dynamic networks. Distributed Computing 23(4), 225–272 (2010)

21. Herlihy, M., Shavit, N.: The topological structure of asynchronous computability. J. ACM 46(6), 858–923 (1999)

22. Lamport, L.: On interprocess communication. Distributed Computing 1(2), 77–101 (1986)

23. Malkhi, D., Reiter, M.K., Wool, A., Wright, R.N.: Probabilistic quorum systems. Inf. Comput. 170(2), 184–206 (2001)

24. Terry, D.B., Demers, A.J., Petersen, K., Spreitzer, M., Theimer, M., Welch, B.B.: Session guarantees for weakly consistent replicated data. In: PDIS, pp. 140–149. IEEE Computer Society (1994)

25. Yu, H.: Overcoming the majority barrier in large-scale systems. In: Fich, F.E. (ed.) DISC 2003. LNCS, vol. 2848, pp. 352–366. Springer, Heidelberg (2003)

How (Not) to Shoot in Your Foot
with SDN Local Fast Failover
A Load-Connectivity Tradeoff

Michael Borokhovich[1],[*] and Stefan Schmid[2]

[1] Ben-Gurion University of the Negev, Israel
borokhom@cse.bgu.ac.il
[2] TU Berlin & T-Labs, Germany
stefan@net.t-labs.tu-berlin.de

Abstract. This paper studies the resilient routing and (in-band) fast failover mechanisms supported in Software-Defined Networks (SDN). We analyze the potential benefits and limitations of such failover mechanisms, and focus on two main metrics: (1) *correctness* (in terms of connectivity and loop-freeness) and (2) *load-balancing*. We make the following contributions. First, we show that in the *worst-case* (i.e., under adversarial link failures), the usefulness of local failover is rather limited: already a small number of failures will violate connectivity properties under *any* fast failover policy, even though the underlying substrate network remains highly connected. We then present randomized and deterministic algorithms to compute resilient forwarding sets; these algorithms achieve an almost optimal tradeoff. Our worst-case analysis is complemented with a simulation study.

1 Introduction

The *software-defined networking (SDN)* paradigm separates the control plane from the network data plane, and introduces a (software) *controller* that manages the *flows* in the network from a (logically) centralized perspective. This architecture has the potential to make the network management and operation more flexible and simpler, and to enable faster innovation also in the network core. For example, the controller may exploit application and network state information (including the switches under its control) to optimize the routing of the flows through the network, e.g., to implement isolation properties or improve performance.

However, the separation of the control from the data plane may have drawbacks. For example, a reactive flow control can introduce higher latencies due to the interaction of the switch with the remote controller. Moreover, the separation raises the question of what happens if the switches lose connectivity to the controller. One solution to mitigate these problems is to keep certain functionality closer to the switches or in the data plane [9].

An important tradeoff occurs in the context of network failures: Theoretically, e.g., a link failure, is best handled by the controller which has the logic to update

---

[*] Michael Borokhovich was supported in part by the Israel Science Foundation (grant 894/09).

R. Baldoni, N. Nisse, and M. van Steen (Eds.): OPODIS 2013, LNCS 8304, pp. 68–82, 2013.

forwarding rules according to the current network policies. However, as the indirection via the controller may take too long, modern network designs incorporate failover (or "backup") paths into the (switches' or routers') *forwarding tables*. For example, OpenFlow (since the 1.1 specification [6]), incorporates such a fast failover mechanism: it allows to predefine resilient and in-band failover routes which kick in upon a topological change. Only after the failover took place, the controller may learn about the new situation and install forwarding (and failover) rules accordingly.

Our Contributions. Given that the in-band failover tables need to be pre-computed and the corresponding rules are based on limited local network information only, we ask the question: "Can you shoot in your foot with local fast failover?" We formalize a simplified local failover problem, and first assume a conservative (or worst-case) perspective where link failures are chosen by an adversary who knows the entire network and all the pre-installed failover rules. For this setting, we provide a lower bound which shows that a safe fast failover can potentially come at a high network load, especially if the failover rules are destination-based only (Section 2). We then present randomized and deterministic algorithms to pre-compute resilient forwarding sets and show that the algorithms are (almost) optimal in the sense that they match the lower bound mentioned above (Section 3). Finally, we report on a simulation study (Section 4) which indicates that under random link failures, local fast failover performs better in general. In technical report [1], we give the formal specification of the two additional algorithms used in our simulations, and we extend the discussion to alternative adversary and traffic models.

Model and Terminology. We attend to the following model. We assume an SDN-network $G = (V, E)$ with n switches (or *nodes*) $V = \{v_1, \ldots, v_n\}$ (e.g., OpenFlow switches) connected by bidirectional links E. We assume that all nodes are directly connected, i.e., G forms a full mesh (a *clique*). This network serves an *all-to-one* communication pattern where any node $v_i \in V \setminus \{v_n\}$ communicates with a single destination v_n; in other words, we have $n - 1$ communicating (source-destination) pairs. Henceforth, by slightly abusing terminology, we will refer to the corresponding $n - 1$ communication paths as the *flows* $\mathcal{F} = \{f_1, \ldots, f_{n-1}\}$. The source-destination flows are unsplittable, i.e., each flow $f_i \ \forall i \in \{1, \ldots, k\}$ travels along a single path. For simplicity, we will assume that all flows f_i carry a constant amount of traffic $w = w(f_i)$, and that edge capacities $e \in E$ are infinite.

In order to ensure an efficient failover, each switch $v \in V$ can store the following kind of *failover rules*: Each rule $r \in R$ considers a specific local failure scenario, namely the set of failed incident links, and defines an alternative forwarding port for each source-destination pair. (This is slightly more general than what is provided e.g., by OpenFlow today: in OpenFlow, all paths need to resort to the same failover port, rending the connectivity-load tradeoff even worse.)

Formally, let $\Gamma(v) \ \forall v \in V$ denote the links (or equivalently: the *switch ports*) incident to node v in G, and let $\mathtt{FW}(v)$ define how the source-destination pairs (or flows) that are routed via node v (the "forwarding set"). A rule r is of the form: $r : \left(2^{\Gamma(v)}, \text{FW}(v)\right) \mapsto \text{FW}(v)$, that is, for each possible failure scenario $2^{\Gamma(v)}$ (i.e., the subset of ports which failed at v), the failover rule defines an alternative set of forwarding rules $\text{FW}(v)$ at v. Note that the number of rules can theoretically be large; however, as we will see, small rule tables are sufficient for the algorithms presented in this paper.

We study failover schemes that pursue two goals: (1) *Correctness:* Each source-destination pair is connected by a valid path; there are no forwarding loops. (2) *Performance:* The resulting flow allocations are well balanced. Formally, we want to minimize the load of the maximally loaded link in G after the failover: $\min \max_{e \in E} \lambda(e)$, where $\lambda(e)$ describes the number of flows f_i crossing edge e. Henceforth, let $\widehat{\lambda} = \max_{e \in E} \lambda(e)$ denote the maximum load.

For our randomized failover schemes, we will typically state our results *with high probability* (short: *w.h.p.*): this means that the corresponding claim holds with at least polynomial probability $1 - 1/n^c$ for an arbitrary constant c. Moreover, throughout this paper, log will refer to the *binary* logarithm.

2   You Must Shoot in Your Foot!

Let us first investigate the limitations of local failover mechanisms from a conservative worst-case perspective. Concretely, we will show that even in a fully meshed network (i.e., a *clique*), a small number of link failures can either quickly disrupt connectivity (i.e., the forwarding path of at least one source-destination pair is incorrect), or entail a high load. This is true even though the remaining physical network is still well connected: the minimum edge cut (short: *mincut*) is high, and there still exist many disjoint paths connecting each source-destination pair.

Theorem 1. *No local failover scheme can tolerate $n - 1$ or more link failures without disconnecting source-destination pairs, even though the remaining graph (i.e., after the link failures) is still $n/2$-connected.*

Proof. We consider a physical network that is fully meshed, and we assume a traffic matrix where all nodes communicate with a single destination v_n. To prove our claim, we will construct a set of links failures that creates a loop, for any local failover scheme. Consider a flow $v_1 \to v_n$ connecting the source-destination pair (v_1, v_n). The idea is that whenever the flow from v_1 would be directly forwarded to v_n in the absence of failures, we fail the corresponding physical link: that is, if v_1 would directly forward to v_n, we fail (v_1, v_n). Similarly, if v_1 forwards to (the backup) node v_i, and if v_i would send to v_n, we fail (v_i, v_n), etc. We do so until the number of intermediate (backup) nodes for the flow $v_1 \to v_n$ becomes $\lfloor \frac{n}{2} \rfloor$. This will require at most $\lfloor \frac{n}{2} \rfloor$ failures (of links to v_n) since every such failure adds at least one intermediate node.

In the following, let us assume that the last link on the path $v_1 \to v_n$ is (v_k, v_n). We simultaneously fail all the links (v_k, v_*), where v_* are all the nodes that are not the intermediate nodes on the path $v_1 \to v_n$, and not v_1. So, there are $n - \lfloor \frac{n}{2} \rfloor - 2$ nodes v_* (the minus 2 accounts for v_1 and v_k). By failing the links to v_*, we left v_k without a valid routing choice: All the remaining links from v_k point to nodes which are already on the path $v_1 \to v_n$, and a loop is inevitable.

In total, we have at most $\lfloor \frac{n}{2} \rfloor + n - \lfloor \frac{n}{2} \rfloor - 2 = n - 2$ failures. Notice, that the two nodes with the smallest degrees in the graph are the nodes v_n (with degree of at least $\frac{n}{2} - 1$) and v_k (with degree of at least $\frac{n}{2}$). The latter is true since the first $\lfloor \frac{n}{2} \rfloor$ failures were used to disconnect links to v_n, and another $n - \lfloor \frac{n}{2} \rfloor - 2$ failures were used to disconnect links from v_k. All the other nodes have a degree of $n - 2$.

The network is still $\frac{n}{2} - 1$ connected: the mincut of the network is at least $\frac{n}{2}$. Consider some cut with k nodes on the one side of the cut, and $n - k$ nodes on the other side. Obviously, one of the sets has a size of at most $\lfloor \frac{n}{2} \rfloor$; let us denote this smaller set by S. If S includes at least one of the nodes $V \setminus \{v_k, v_n\}$, then the number of outgoing edges form the set is at least $n - 2 - (|S| - 1)$, thus the mincut is at least $\frac{n}{2} - 1$. If S includes only both v_k and v_n, the mincut is at least $n - 1$ (the link (v_k, v_n) was failed). If only one of the nodes $\{v_k, v_n\}$ is in S, then the mincut is at least $\frac{n}{2} - 1$. $\qquad\square$

Regarding the maximal link load, we have the following lower bound.

Theorem 2. *For any local failover scheme tolerating φ link failures $(0 < \varphi < n)$ without disconnecting any source-destination pair, there exists a failure scenario which results in a link load of at least $\widehat{\lambda} \geq \sqrt{\varphi}$, although the minimum edge cut (mincut) of the network is still at least $n - \varphi - 1$.*

Proof. Let us first describe an adversarial strategy that induces a high load: Recall that in the absence of failures, each node v_i $(i \neq n)$ may use its direct link to v_n for forwarding. However, after some links failed, v_i may need to resort to the remaining (longer) paths from v_i to v_n. Since the failover scheme $\mathcal{S}$ tolerates φ failures and v_i remains connected to v_n, $\mathcal{S}$ will fail over to one of $\varphi + 1$ possible paths. To see this, let v_i^j $(j \in [1, \dots, \varphi])$ be one of the φ possible last hops on the path $(v_i \to \cdots \to v_i^j \to v_n)$, and let us consider the paths generated by $\mathcal{S}$:

$$(v_i \to v_n),$$
$$(v_i \to \cdots \to v_i^1 \to v_n),$$
$$(v_i \to \cdots \to v_i^1 \to \cdots \to v_i^2 \to v_n),$$
$$\cdots$$
$$(v_i \to \cdots \to v_i^1 \to \cdots \to v_i^2 \to \cdots \to v_i^\varphi \to v_n).$$

For example, the path $(v_i \to \cdots \to v_i^1 \to v_n)$ will be generated if the first failure is link (v_i, v_n), and the path $(v_i \to \cdots \to v_i^1 \to \cdots \to v_i^2 \to v_n)$ if the second failure is link (v_i^1, v_n) (see Fig. 1 for an illustration). Notice that the last hop v_i^j is unique for every path; otherwise, the loop-freeness property would be violated.

For each $i \in [1, \dots, n-1]$ (i.e., for each possible source) consider the set $A_i = \{v_i, v_i^1, \dots, v_i^{\sqrt{\varphi}}\}$, and accordingly, the multiset $\bigcup_i A_i$ is of size $|\bigcup_i A_i| = (n-1)(\sqrt{\varphi}+1)$ many nodes. Since we have $n - 1$ distinct nodes (we do not count

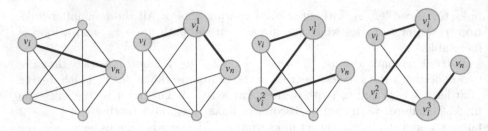

Fig. 1. *From left to right:* failover path $(v_i \to v_n)$ where each time the last hop to v_n is failed

$v_n)$, by a counting argument, there exists a node $w \in \bigcup_i A_i$ which appears in at least $\sqrt{\varphi}$ sets A_i.

If for each i such that $w \in A_i$, the adversary will cause v_i to route to v_n via w, then the load of the link (w, v_n) will be at least $\sqrt{\varphi}$. This can be achieved by failing at most $\sqrt{\varphi}$ links to v_n in each such set A_i. Thus, the adversary will fail $\sqrt{\varphi} \times \sqrt{\varphi} = \varphi$ links incident to v_n, while the maximum loaded link (w, v_n) will have a load of at least $\sqrt{\varphi}$.

It remains to prove that the network remains highly connected, despite these failures: The proof is simple. In a clique network without failures, the mincut is $n - 1$. In the worst case, each link failure will remove one link from some cut, and hence the mincut must eventually be at least $n - \varphi - 1$. By the same argument, there are at least $n - \varphi - 1$ many disjoint paths from each node v_i to the destination: initially, without failures, there are $n-1$ disjoint paths (a direct one and $n - 1$ indirect ones), and each failure affects at most one path. $\square$

Interestingly, it can be proved analogously that if a failover rule only depends on destination addresses, the situation is even worse.

Theorem 3. *Consider any local destination-based failover scheme in a clique graph. There exists a set of φ failures $(0 < \varphi < n)$, such that the remaining graph will have a mincut of $n - \varphi - 1$ and $\widehat{\lambda} \geq \varphi$.*

Proof. In order to construct a bad example, we first fail the direct link (v_1, v_n), and hence v_1 will need to reroute to some path with the last node before v_n being some node v_i. When we fail the link (v_i, v_n), v_i will have to reroute and some other node v_j will become the last hop on the path to v_n. We repeat this strategy to fail the links from the newly selected last hop and the destination v_n. This results in a routing path $v_1 \to \cdots \to v_i \to \cdots \to v_j \to \cdots \to w \to v_n$ with at least φ intermediate nodes. Since the algorithm is destination-based, i.e., forwarding rules depend only on the destination address of a packet, the load on the link (w, v_n) will be at least $\varphi + 1$: all the nodes on the path $v_1 \to v_n$ will send their packets via the same route. $\square$

3   How Not to Shoot in Your Foot!

We have seen that what can be achieved with local fast failover is rather limited. On the positive side, this section shows that there exist algorithms to pre-compute failover schemes which at least match the derived lower bounds: we present algorithms to pre-compute robust failover paths that *jointly optimize* the loop-freeness property and the load, i.e., find an almost optimal tradeoff.

Naturally, randomization can help to spread the communication load well, but we must ensure that paths remain loop-free. We first present such a randomized solution, discuss how to derandomize it, and finally look at deterministic failover algorithms.

We introduce a family of failover schemes S which can be represented in a generic *matrix form* $\delta_{i,j}$. Any failover scheme instance in this family will always forward a message directly to the destination if the corresponding link is available. Otherwise, if a given node v_i cannot reach the destination v_n via (v_i, v_n), it will resort to the sequence of alternatives represented as the row i in the matrix $\delta_{i,\cdot}$ (the "backup nodes" for v_i): v_i will first try to forward to node $\delta_{i,1}$, if this link is not available to node $\delta_{i,2}$, and so on. Similarly and more generally, starting from node $\delta_{i,j}$, if the link $(\delta_{i,j}, v_n)$ is not available, the failover scheme will try $\delta_{i,j+1}$, $\delta_{i,j+2}$, etc. In summary, the matrix representation can be depicted as follows:

$$\delta_{1,1}, \delta_{1,2}, \ldots, \delta_{1,n-2}$$

$$\cdots$$

$$\delta_{i,1}, \delta_{i,2}, \ldots, \delta_{i,n-2}$$

$$\cdots$$

$$\delta_{n-1,1}, \delta_{n-1,2}, \ldots, \delta_{n-1,n-2}$$

The following auxiliary claim characterizes the best adversarial strategy against the failover schemes S.

Claim 4. *For the family of failover schemes S, the highest load is induced if links towards the destination node v_n are failed.*

Proof. To achieve a load of φ on some link, the adversary first needs to bring at least φ flows to some node w. Consider a failover sequence $\delta_{i,\cdot}$ in which w is located at j's position, i.e., $\delta_{i,j} = w$. In order to bring the flow $v_i \to v_n$ to node w, the adversary needs to fail at least j links (every failure requires at most a single additional backup node). Thus, the adversary can remove the links to the destination from every node $\delta_{i,k}, k < j$ and from the source v_i. The optimality is due to the fact that once one of the *nodes* $\delta_{i,k}, k < j$ appears in other sequences, these failures are automatically reused: the links $(\delta_{i,k}, v_n)$ already failed. If the adversary would instead choose to fail other *links* (not towards the destination), e.g., $(\delta_{i,j}, \delta_{i,j+1})$, the failures can only be reused if the same *link* (and not only an endpoint) appears in other sequences before w. Therefore, we conclude that the strategy of failing the links to the destination is optimal: (1) it requires no more failures to bring a specific flow to w than any other strategy, and (2) link failures to the destination can strictly be reused more often than the failures of links to any other nodes.   □

3.1 Randomized Failover

What does a good failover matrix $\delta_{i,j}$ look like? Naively, one may choose the matrix entries (i.e., the "failover ports") uniformly at random from the set of next hops which are still available, and depending on the source and destination address, in order to balance the load. However, note that a random and *independent* choice will quickly introduce loops in the forwarding sequences: it is likely that a switch will forward traffic to a switch which was already visited before on the failover path.

Thus, our randomized failover scheme **RFS** will choose random *permutations*, i.e., for a source-destination pair (v_i, v_n), the sequence $\delta_{i,1}, \delta_{i,2}, \ldots, \delta_{i,n-2}$ (with $\delta_{i,j} \in V \setminus \{v_i, v_n\}$) is *always loop-free* (deterministically). Technically, **RFS** draws all $\delta_{i,j}$ uniformly at randomly from $V \setminus \{v_i, v_n\}$ but eliminates repetitions (e.g., by redrawing a repeated node). We can show that **RFS** is almost optimal, in the following sense.

Theorem 5. *Using the* **RFS** *scheme, in order to create a maximum load of* $\widehat{\lambda} = \sqrt{\varphi}$, *the adversary will have to fail at least* $\Omega \left(\frac{\varphi}{\log n} \right)$ *links w.h.p., where* $0 < \varphi < n$.

Proof. To create a link load of $\sqrt{\varphi}$ with the minimal number of link failures, the adversary must in particular be able to route at least $\sqrt{\varphi}$ flows to some node w. Given the $\sqrt{\varphi}$ load on the node, in the best case (for the adversary), the entire flow will be forwarded by w on a single outgoing link. (E.g., the link to the destination v_n.) We will show that w.h.p., it is impossible for the adversary to route more than $\sqrt{\varphi}$ flows to a single node.

The adversary can put a high load on some node w only if: 1) Node w is located close to the beginning of many sequences (i.e., is in a small "prefix" of the sequences); thus, a small number of failures is sufficient to redirect the flow to w. 2) Many nodes appearing before w in the sequence prefixes occur early in many other prefixes as well; thus, the adversary can "reuse" failed links to redirect also other source-destination pairs. Note that these two requirements may conflict, but to prove the lower bound on the number of required failures, we can assume that both conditions are satisfied: the set of $\sqrt{\varphi}$ sequences with the largest number of node repetitions in the w-prefixes also have the shortest w-prefixes.

With this intuition in mind, let us compute the probability that a node w appears more than approximately $\log n$ times at position j. Let Y_i^j be an indicator random variable that indicates whether w is located at position $j \in [1, \ldots, n-2]$ in sequence $i \in [1, \ldots, n-1]$. Let $Y^j = \sum_{i=1}^{n} Y_i^j$ be a random variable representing the number of times that w appears at position j. Since the failover sequences are random, $\Pr(Y_i^j = 1) = \frac{1}{n-2}$ (w is neither the source nor the destination) and thus, $\forall j, \mathbb{E}\left[Y^j\right] = \frac{n-1}{n-2}$. Applying the Chernoff bound on the sum of n *i.i.d.* Poisson trials, we obtain (for any $\delta > 0$):

$$\Pr\left(Y^j > (1+\delta)\mathbb{E}\left[Y^j\right]\right) \leq 2^{-\delta \mathbb{E}[Y^j]}$$

$$\Pr\left(Y^j > \frac{(1+3\log n)(n-1)}{n-2}\right) \leq 2^{-(3\log n)\times \frac{n-1}{n-2}}$$

$$\leq 2^{-3\log n} = 1/n^3.$$

Let us denote $z = \frac{(1+3\log n)(n-1)}{n-2}$ and rewrite:

$$\Pr\left(Y^j > z\right) \leq 1/n^3.$$

We can now apply a union bound argument[1] over all possible nodes w and over all possible positions j, which yields that with probability at least $1 - \frac{1}{n}$, any node will appear no more than z times at each position.

The adversary needs to select the $\sqrt{\varphi}$ sequences with the shortest w-prefixes. For a chosen sequence i, let us denote by k_i the prefix length for node w (the prefix length includes w itself). Since each node will appear no more than z times at each position (with probability of at least $1 - \frac{1}{n}$) the minimum length of a *total prefix* for any node w can be derived. Let us denote the minimum *total prefix* by k. Clearly, k is minimized for the shortest possible prefixes k_i. According to the analysis above, with high probability, there are no more than z prefixes of length 1, no more than z prefixes of length 2, and so on. Therefore:

$$k = \sum_{i=1}^{\sqrt{\varphi}} k_i \geq \sum_{i=1}^{z} 1 + \sum_{i=1}^{z} 2 + \cdots + \sum_{i=1}^{z} \frac{\sqrt{\varphi}}{z}$$

$$= z\left(1 + 2 + \cdots + \frac{\sqrt{\varphi}}{z}\right) = \frac{\varphi + \sqrt{\varphi}z}{2z} \geq \frac{\varphi}{2z}$$

$$\geq \frac{\varphi}{8\log n}. \tag{1}$$

Eq. 1 is true since for $n \geq 6$, $\frac{(1+3\log n)(n-1)}{n-2} \leq 8\log n$.

In conclusion, we know that in order to achieve a load of $\sqrt{\varphi}$, the adversary has to fail the entire total prefix of w that consists of at least $\frac{\varphi}{8\log n}$ nodes. However, the nodes in the prefixes are not necessarily all distinct, and the number of links the adversary needs to fail only depends on the *distinct* nodes in the *total prefix* of the node w. The latter is true due to the fact that the best adversarial strategy is to fail only the links to the destination since in this case every such failure is reused once the same node appears again in the *total prefix* of w (see Claim 4). Hence, we next compute the minimum number of distinct nodes D in any set of k random nodes. As we are interested in lower bounding D, we can choose k minimal, i.e., $k = \frac{\varphi}{8\log n}$. The analysis follows from a *balls-and-bins* argument

---

[1] The union bound argument says that the probability of the union of the events is no greater than the sum of the probabilities of the individual events.

where bins represent node IDs and balls are the k positions that should be failed by the adversary. Thus, D is a number of occupied bins (i.e., bins that contain at least one ball). Let D_i be a binary random variable indicating that the i-th ball falls into an empty bin (i.e., $D = \sum_{i=1}^{k} D_i$). So, $\Pr(D_i = 1) \geq \frac{n-1-k}{n-1}$. Since $k = \frac{\varphi}{8 \log n}$ and $\varphi < n$, we obtain that:

$$\Pr(D_i = 1) \geq \frac{n-1-k}{n-1} \geq \frac{8 \log n - 1}{8 \log n} \geq 0.8.$$

Thus, $\mathbb{E}[D] = k\mathbb{E}[D_i] \geq 0.8k$. Now we can apply the Chernoff bound (for any $\delta \in (0, 1]$):

$$\Pr(D \leq (1 - \delta)0.8k) \leq \Pr(D \leq (1 - \delta)\mathbb{E}[D])$$
$$\leq e^{-\mathbb{E}[D]\delta^2/2} \leq e^{-0.8k\delta^2/2}.$$

By taking $\delta = 0.5$ we obtain $\Pr(D \leq 0.4k) \leq e^{-0.1k}$.

It remains to prove that this bound still holds under the union bound for all $\binom{n-1}{\sqrt{\varphi}}$ possible sets of sequences that the adversary can choose. In other words, we have to ensure that $\binom{n}{\sqrt{\varphi}}e^{-0.1k} \leq \frac{1}{n}$ (we took a larger number, since: $\binom{n}{\sqrt{\varphi}} \geq \binom{n-1}{\sqrt{\varphi}}$).

$$\binom{n}{\sqrt{\varphi}}e^{-0.1k} \leq n^{\sqrt{\varphi}}e^{-0.1k} = n^{\sqrt{\varphi}}e^{-\frac{\varphi}{80 \log n}} \tag{2}$$
$$= e^{\sqrt{\varphi}\ln n - \frac{\varphi}{80 \log n}} = e^{\varphi\left(\frac{\ln n}{\sqrt{\varphi}} - \frac{1}{80 \log n}\right)}$$
$$\leq e^{\varphi\left(\frac{\log n}{\sqrt{\varphi}} - \frac{1}{80 \log n}\right)}.$$

For $\varphi \geq 82^2 \log^4 n$, we have $\left(\frac{\log n}{\sqrt{\varphi}} - \frac{1}{80 \log n}\right) \leq \frac{-2}{82^2 \log n}$, and hence $\binom{n}{\sqrt{\varphi}}e^{-0.1k} \leq e^{\frac{-2\varphi}{82^2 \log n}} \leq e^{-2\log^3 n} \leq \frac{1}{n^2}$. Since $\binom{n}{\sqrt{\varphi}}\Pr(D \leq 0.4k) = \binom{n}{\sqrt{\varphi}}\Pr(D \leq \frac{0.4\varphi}{8 \log n}) \leq \frac{1}{n^2}$, w.h.p., any set of $\sqrt{\varphi}$ sequences (i.e., w-prefixes) will require $\Omega(\frac{\varphi}{\log n})$ failures. $\square$

3.2   Deterministic Failover

Theoretically, the result of Theorem 5 can be derandomized, i.e., the **RFS** scheme can *deterministically* ensure low loads. The idea is that we could *verify* whether an (improbable) situation occurred and the random sequences generated by **RFS** actually yield a *high* load (we just need to check all possible loads at any w); if so, another set of random permutations is generated. However, this verification is computationally expensive.

We hence now initiate the discussion of efficient deterministic schemes. In particular, we propose an optimal failover scheme (which matches our lower bound in Section 2), at least for small φ. Similar to **RFS**, the deterministic failover scheme **DFS** is defined by a *failover matrix* $\delta_{i,j}$; however, here $\delta_{i,j}$ will simply refer to a node's *index* (and not the node itself): We define the index of any node v_ℓ to be $\ell - 1$, i.e., the nodes $\{v_1, v_2, \ldots, v_n\}$ are mapped to the indices $\{0, 1, \ldots, n - 1\}$. Given a destination node v_n, **DFS** is defined by the following index matrix:

$$1, 2, 4, 8, \ldots, \left(0 + 2^{\lfloor \log n \rfloor}\right) \quad \mathrm{mod}\ n$$

$$2, 3, 5, 9, \ldots, \left(1 + 2^{\lfloor \log n \rfloor}\right) \quad \mathrm{mod}\ n$$

$$3, 4, 6, 10 \ldots, \left(2 + 2^{\lfloor \log n \rfloor}\right) \quad \mathrm{mod}\ n$$

$$\cdots$$

In general, the index in sequence $i \in [1, \ldots, n-1]$ at position $j \in [1, \ldots, \lfloor \log n \rfloor]$ is $\delta_{i,j} = (i-1) + 2^{j-1} \mod n$. For example, if the link (v_1, v_n) fails, v_1 will reroute via the node with index 1, i.e., via v_2; and so on. We can show the following result.

Theorem 6. *The* **DFS** *scheme achieves a maximum load of* $\widehat{\lambda} = O(\sqrt{\varphi})$ *in any scenario with* $\varphi < \lfloor \log n \rfloor$ *failures.*

Proof. We will prove something even stronger: the adversary cannot choose link failures such that any *node* w forwards more than $\sqrt{\varphi}$ flows. Clearly, an upper bound on the node load is an upper bound on the (incident) links: in the worst case, w will forward all traffic to the same link. To create a high load at some node w, the adversary needs to find failover sequences in the matrix $\delta_{i,j}$ where the node w appears close to the *beginning* of the sequence, and fail *all* the links (v_i, v_n), where v_i is a node preceding w in a sequence: i.e., the adversary fails the *total prefix* of w. Note that failing the links to the destination is the best strategy for the adversary as failures are automatically reusable in other sequences (see Claim 4).

The following two claims will help us to show that the adversary wastes its entire failure budget in order to achieve a maximum load of $\sqrt{\varphi}$.

Claim 7. *Every node index participates in only* $\lfloor \log n \rfloor$ *sequences.*

Proof. The **DFS** failover matrix is defined as $\delta_{i,j} = (i-1) + 2^{j-1} \mod n$, where $i \in [1, \ldots, n-1]$ and $j \in [1, \ldots, \lfloor \log n \rfloor]$. From this construction, it follows that there are no index repetitions in the matrix columns. Since there are $\lfloor \log n \rfloor$ columns, the claim follows. $\square$

Claim 8. *For any node index* ℓ, *all* ℓ-*prefixes (sets of indices preceding* ℓ *in the sequences) are disjoint.*

Proof. Let us define $m = i - 1$ and $k = \ell - 1$. The index in sequence $m \in [0, \ldots, n - 2]$ at position $k \in [0, \ldots, \lfloor \log n \rfloor - 1]$ is $m + 2^k \mod n$. Consider a sequence m' where the index w appears at position k' and a sequence m'' where the index ℓ appears at position k''. Without loss of generality, assume that $k'' > k'$. Let $m' + 2^{k^*}$ mod n and $m'' + 2^{k^{**}}$ mod n represent the indices in the prefixes of ℓ in sequences m' and m'' accordingly. Assume by contradiction that these indices are the same. We have that

$$m' + 2^{k'} = m'' + 2^{k''} \quad \text{mod } n$$
$$m' + 2^{k^*} = m'' + 2^{k^{**}} \quad \text{mod } n \text{ (assumption)}$$

and hence

$$m' - m'' = 2^{k''} - 2^{k'} + n \cdot C_1$$
$$m' - m'' = 2^{k^{**}} - 2^{k^*} + n \cdot C_2$$

Therefore

$$2^{k^{**}} - 2^{k''} + 2^{k'} - 2^{k^*} = n \cdot C_3$$

where C_1, C_2 and C_3 are some integer constants.

Notice that $\max(2^{k^{**}}, 2^{k''}, 2^{k'}, 2^{k^*}) < n$, so the only possible values for C_3 are: $\{-1, 0, 1\}$. Moreover, $(2^{k^{**}} - 2^{k''}) < 0$, while $(2^{k'} - 2^{k^*}) > 0$, and since the absolute value of these differences is bounded by $2^{\lfloor \log n \rfloor - 1} \leq 0.5n$, we can write:

$$-0.5n < 2^{k^{**}} - 2^{k''} + 2^{k'} - 2^{k^*} < 0.5n.$$

Thus, 0 remains the only possible value for C_3. The values $\{2^{k^{**}}, 2^{k''}, 2^{k'}, 2^{k^*}\}$ are distinct since there are no repetitions in the columns of the sequence matrix. Since $2^{k''} > 2^{k^{**}} + 2^{k'} + 2^{k^*}$, due to a geometric series argument (the largest element is greater than the sum of all previous elements), we can state that

$$2^{k^{**}} - 2^{k''} + 2^{k'} - 2^{k^*} < 0.$$

We conclude that there is no integer constant C_3 satisfying our assumption $m' + 2^{k^*} = m'' + 2^{k^{**}}$ mod n (i.e., there are two identical indices in the ℓ-prefixes). $\qquad\square$

Armed with these claims, we are ready to continue with the proof. Since all prefixes are disjoint, the adversary cannot reuse failures of one flow for another. Thus, the adversary will be able to route *one* flow to w using a single failure (by finding a sequence in which w appears at the first position); to add another flow, the adversary takes a sequence δ_i in which w is located at position 2 and will fail the links (v_i, v_n), and $(v_{(\delta_{i,1})+1}, v_n)$. And so on. Thus, the number of used failures can be represented as

$$1 + 2 + 3 + 4 + \cdots + L \leq \varphi$$

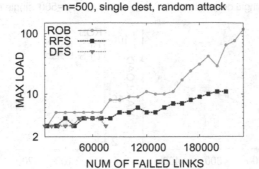

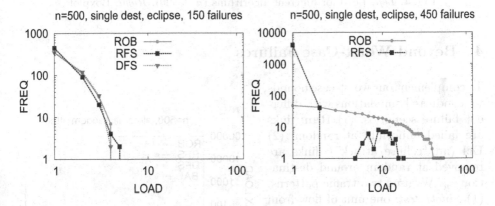

Fig. 3. Load distribution over links

where L is the number of flows passing through w on the way to the destination v_n. So:

$$1 + 2 + 3 + 4 + \cdots + L \leq \varphi$$
$$\frac{L(L+1)}{2} \leq \varphi$$
$$L < \sqrt{2\varphi}.$$

Note that the index of the destination node ($n - 1$ in our case) can appear inside the failover sequences. In this case, the index will be skipped since the link to it from the source already failed. By skipping one index, we shorten the failover sequence by 1, and since every sequence has length $\lfloor \log n \rfloor$, our failover scheme holds for any $\varphi < \lfloor \log n \rfloor$. $\qquad\square$

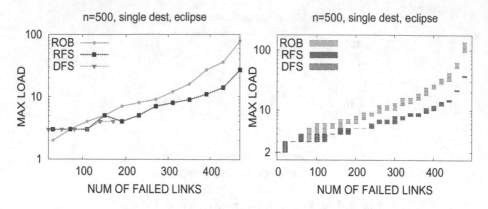

Fig. 4. *Left:* Load for different algorithms ($n = 500$). *Right:* Boxplot

4   Beyond Worst-Case Failures

To complement our worst-case bounds, we conducted simulations with different failure scenarios. (1) **Ran**: links are failed uniformly at random; (2) **Ecl** (an "eclipse attack"): links are removed at random around destination v_n. We used two traffic patterns. (1) *Single dest:* one unit of flow from each node to v_n; (2) *all-to-all:* one unit of flow from each node to every other node. In addition to our failover schemes **RFS** and **DFS**, we also simulate the following naive strategies. (1) **Bal** ("balanced"): If the destination cannot be reached directly, forward to an available port chosen uniformly at random (depending on the destina-

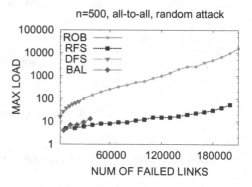

Fig. 5. Max load in all-to-all communication

tion and the set of failed links). This strategy seeks to balance traffic but does not ensure loop-freeness. (2) **Rob** ("robust"): If the destination cannot be reached, forward to the available neighboring switch which has the lowest identifier (in a modulo manner, and assuming that switches have unique identifiers). We start with the *single dest* traffic pattern. Figure 4 (*left*) plots the load as a function of the number of failed links under **Ecl**. (Note the logarithmic scale on the y-axis.) We observe that compared to Theorem 5 which deals with worst-case failures, **RFS** performs significantly better: while our conservative bound suggests that to create a load of $10 = \sqrt{\varphi}$, $\varphi/\log n = 100/\log 500 = 11.15$ failures are needed, more than 300 are necessary in our experiment. Moreover, we observe that **RFS** yields a much lower load than the naive approach **Rob**; for the single-destination scenario, **Bal** is similar to **Rob** and is not shown explicitly here. The **DFS** algorithm also gives a low load; however, it is only defined up to a certain φ (see Theorem 6). The variance of these experiments is typically small, see the boxplot in Figure 4 (*right*).

As expected, under **Ran** failures, the load is generally lower, and our scheme can tolerate more failures without creating loops (see Figure 2).

As the maximal link load reveals a partial picture only, Figure 3 studies the load distribution over multiple links (under **Ecl**), once for 150 failures (*left*) and once for 450 failures (*right*). Obviously, most links hardly contain more than one or two flows under **RFS**; again, under the naive **Rob** strategy (and similarly for **Bal**), the situation is worse.

Let us have a look at alternative traffic matrices. Figure 5 shows the results for an all-to-all communication pattern (under **Ran**). Interestingly, for **RFS**, the load is not much higher than in the single-destination scenario; this confirms the good load-balancing properties of **RFS**. However, we also see that **DFS** performs poorly and needs to be generalized for the multi-destination scenario. Finally, we note that in this scenario, we can exploit **Bal**'s flexibility and in contrast to the single destination case, the algorithm significantly outperforms **Rob** (in terms of load).

5 Related Work

This work is motivated by the trend towards Software-Defined Networking and in particularly the fast failover mechanism which supports the in-band masking of failures (see Section 5.8 of the OpenFlow 1.1 specification). However, as the convergence time of routing algorithms is often relatively high compared to packet forwarding speeds, ranging from 10s of milliseconds to seconds depending on the network [2], many networks today incorporate some robustness already in the forwarding tables of a router or switch: Thus, robust routing concepts and link protection schemes have been studied intensively for many years, also outside SDN.

For example, robust Multiprotocol Label Switching (MPLS) supports local and global path protection to compute shortest backup paths around an outage area [7,11], where "shortest" is often meant in terms of congestion [8,12]. Related to our connectivity and load-balancing tradeoff is also the work by Suchara et al. [10] who analyze how to jointly optimize traffic engineering and failure recovery from pre-installed MPLS backup paths. However, in contrast to our paper, their solution is path-based and not local, and the focus is on robust optimization.

Alternative solutions to make routing more resilient rely on special header bits (e.g., to determine when to switch from primary to backup paths, as in *MPLS Fast Reroute* [7], or to encode failure information to make failure-aware forwarding decisions [3,5]), or on fly table modifications [4]. Recently, Feigenbaum et al. [2] made an interesting first step towards a better theoretical understanding of resilient SDN tables. The authors prove that routing tables can provide guaranteed resilience (i.e., loop-freeness) against a *single* failure, when the network remains connected.

6 Conclusion

So, will or won't you shoot in your foot with fast failover? Our results show that there exists an interesting tradeoff between a "safe" and "efficient" failover. The usefulness of the local failover depends on whether link failures are rather adversarial or random, and on how flexibly the failover rules can be specified. In particular, we have seen that the possibilities of destination-based failover schemes are very limited. But also more expressive failover schemes where flows can be forwarded depending on arbitrary local matching rules (this is more general than today's OpenFlow specification), can lead to high network loads in the worst case. On the positive side, relatively simple algorithms exist which match these lower bounds.

Acknowledgments. We would like to thank Chen Avin for valuable discussions and advice.

References

1. Borokhovich, M., Schmid, S.: How (not) to shoot in your foot with sdn local fast failover: A load-connectivity tradeoff. Technical Report, arXiv:1309.3150 (2013)
2. Ba, J.F., et al.: On the resilience of routing tables. In: Proc. ACM Symposium on Principles of Distributed Computing (PODC), pp. 237–238 (2012)
3. Lakshminarayanan, K., Caesar, M., Rangan, M., Anderson, T., Shenker, S., Stoica, I.: Achieving convergence-free routing using failure-carrying packets. In: Proc. SIGCOMM, pp. 241–252 (2007)
4. Liu, J., Yan, B., Shenker, S., Schapira, M.: Data-driven network connectivity. In: Proc. HotNets, pp. 8:1–8:6 (2011)
5. Lor, S.S., Landa, R., Rio, M.: Packet re-cycling: eliminating packet losses due to network failures. In: Proc. HotNets, pp. 2:1–2:6 (2010)
6. OpenFlow Spec (2013),
   http://www.openflow.org/documents/openflow-spec-v1.1.0.pdf,
   openflow.org
7. Pan, P., Swallow, G., Atlas, A.: Fast reroute extensions to RSVP-TE for LSP tunnels. In: RFC 4090 (2005)
8. Saito, H., Yoshida, M.: An optimal recovery LSP assignment scheme for MPLS fast reroute. In: Proc. NETWORKS (2002)
9. Schmid, S., Suomela, J.: Exploiting locality in distributed sdn control. In: Proc. SIGCOMM Workshop on Hot Topics in Software Defined Networking, HotSDN (2013)
10. Suchara, M., Xu, D., Doverspike, R., Johnson, D., Rexford, J.: Network architecture for joint failure recovery and traffic engineering. In: Proc. ACM SIGMETRICS, pp. 97–108 (2011)
11. Vasseur, J.-P., Pickavet, M., Demeester, P.: Network Recovery: Protection and Restoration of Optical, SONET-SDH, IP, and MPLS. Morgan Kaufmann Publishers Inc. (2004)
12. Wang, D., Li, G.: Efficient distributed bandwidth management for MPLS fast reroute. IEEE/ACM Trans. Netw. (2008)

Message Passing or Shared Memory: Evaluating the Delegation Abstraction for Multicores

Irina Calciu[1], Dave Dice[2], Tim Harris[2], Maurice Herlihy[1,2], Alex Kogan[2],
Virendra Marathe[2], and Mark Moir[2]

[1] Brown University
{irina,mph}@cs.brown.edu
[2] Oracle Labs
{dave.dice,timothy.l.harris,alex.kogan,
virendra.marathe,mark.moir}@oracle.com

Abstract. Even for small multi-core systems, it has become harder and harder to support a simple shared memory abstraction: processors access some memory regions more quickly than others, a phenomenon called *non-uniform memory access* (NUMA). These trends have prompted researchers to investigate alternative programming abstractions based on message passing rather than cache-coherent shared memory. To advance a pragmatic understanding of these models' strengths and weaknesses, we have explored a range of different message passing and shared memory designs, for a variety of concurrent data structures, running on different multicore architectures. Our goal was to evaluate which combinations perform best, and where simple software or hardware optimizations might have the most impact. We observe that different approaches perform best in different circumstances, and that the communication overhead of message passing can often outweigh its benefits. Nonetheless, we discuss ways in which this balance may shift in the future. Overall, we conclude that, by emphasizing high-level shared data abstractions, software should be designed to be largely independent of the choice of low-level communication mechanism.

Keywords: NUMA, message passing, shared memory, delegation, locks, concurrent data structures.

1 Introduction

As modern processor architectures evolve, programming abstractions are straining to keep up. The transition from single-core to increasingly multi-core architectures means that scalability, that is, the ability to exploit parallelism and manage concurrency, has become a central concern for software system design.

Even for small multi-core systems, it has become harder and harder to support a simple shared-memory abstraction. This abstraction is already starting to fail with respect to performance: processors observe that some memory regions can be accessed more quickly than others, a phenomenon called *non-uniform memory access* (NUMA). Once a concern primarily for large-scale, high-performance workloads, NUMA effects are increasingly visible to smaller,

R. Baldoni, N. Nisse, and M. van Steen (Eds.): OPODIS 2013, LNCS 8304, pp. 83–97, 2013.

"everyday" programs. In the long term, some researchers have even suggested that cache coherence will no longer be feasible across a single multi-core chip, or that individual cores may perform better in the absence of coherence.

In reaction to these trends, researchers have investigated alternative programming abstractions in which—even within a shared-memory system—coordination is based on message passing rather than via direct access to shared-memory data structures. A key example is a design pattern we call *delegation*, in which one thread requests that another thread perform an operation on its behalf, and the request and response (if any) are sent by message passing. For instance, Barrelfish [1] runs a separate kernel on each core, and cores communicate only via a message passing interface, itself implemented in shared memory.

Advocates for delegation appeal to its simplicity: it promises to support application designs that span NUMA architectures, heterogeneous architectures, and even architectures that lack global coherence. Moving platform-specific engineering concerns—such as cache line sizes or idiosyncratic coherence protocols—out of the application and into the message passing substrate could ease porting applications from one platform to another, or from one platform to its successor.

Many of these proposals (surveyed in Section 7), however, are *ad-hoc* in nature, focusing on a specific implementation of a specific data structure, yielding little insight into where the message passing abstraction performs better than the shared-memory abstraction. Our contribution is to explore a range of message passing and shared-memory designs, on various benchmarks running on different multicore architectures, and to evaluate which combinations perform best.

This paper does not take sides in the ongoing debate about the relative merits of shared-memory versus message-passing abstractions [9]. In contrast, our contribution is to advance a pragmatic understanding of these models' strengths and weaknesses. In particular, such debates often present a false dichotomy: that we must choose between these models, and that one is superior. Instead, by emphasizing high-level abstractions, software can be designed largely independently of the choice of low-level communication mechanism. The choice itself should be based on pragmatic performance evaluations.

We will use the following terminology. Many modern large-scale multiprocessor architectures are composed of multiple *sockets* or *nodes*, each encompassing multiple *cores*. Each core has access to a local cache hierarchy and to dynamic random-access memory (DRAM), along with multiple *threads*. Such systems typically utilize a cache-coherence protocol, which creates the illusion that threads share a common memory. Nevertheless, as noted, cache coherence protocols do not hide NUMA effects in the form of differences in the times needed to communicate between local and remote memories. We use the term *NUMA domain* to indicate a set of threads with identical memory access times.

Section 2 explains our notion of delegation, and Section 3 describes alternative ways of implementing message passing on a shared-memory multicore. Section 4 describes the range of benchmarks used, and Section 5 describes the experimental results, which are discussed further in Section 6. Section 7 surveys related work, and Section 8 presents conclusions.

2   Delegation

In *delegation*, access to a data structure is mediated by one or more *server threads*, which are the only threads allowed to manipulate the data directly. Even though all threads share a common (NUMA) memory, client and server threads communicate by a message passing protocol whose implementation is optimized to take advantage of the underlying shared memory.

When a client thread needs to apply an operation to a data structure, it *delegates* that operation by sending a request message to the server thread. When the server thread receives the message, it carries out the operation directly on the data structure, and stores the result value, if any, in a client-allocated buffer.

Delegation is attractive for several reasons. First, the server thread can operate directly on the data structure—without synchronizing its accesses with other threads, in which case programmers need not worry about synchronization. Furthermore, a server thread may encounter fewer cache misses and generate less coherence traffic than threads operating on the data structure directly. Advocates of delegation often suggest that delegation can produce more robust software designs: to be cost-effective, applications must be designed to work over a wide range of parallel platforms, making it difficult to optimize shared data structures for any specific platform. Delegation introduces an abstraction layer, allowing implementations to be optimized for different platforms with no changes to applications. This abstraction layer allows applications to be more easily scaled out from multicores to multiple machines by replacing the shared-memory communication protocol with one that operates over a distributed system.

Nevertheless, delegation also has its pitfalls. From the point of view of an individual operation on a data structure, a central question is how the time spent operating directly on a shared data structure compares with the cost of (i) sending and receiving messages, (ii) queuing time of a message at a server thread, and (iii) execution time at the server thread. Server threads are statically assigned to cores, so they may be idle some of the time. If there are enough cores, however, this static assignment removes the need for complex mechanisms to enable server threads to be quickly identified and dispatched (as with active messages [17]). Server threads may become a bottleneck, so the underlying data structure may need to be partitioned and delegated to multiple servers, a problem similar to moving from coarse to fine-grained locking. Finally, delegation imposes some inconvenience on programmers, as operation requests and responses must be marshalled into messages before being sent and unmarshalled upon receipt.

In short, while delegation has some attractive properties, it does not follow that delegation-based data structures are inherently preferable.

3   Communication

Message Passing. Clients communicate with servers via a message-passing protocol implemented in shared memory. Although we consider several different mechanisms for communication, the messages themselves are similar across schemes.

Each message contains an *opcode* identifying the requested operation (e.g., add a key-value pair to a table), one or more arguments (the values to add), a pointer to a buffer where the call's result is to be stored (e.g., the value returned by a get() call), and a ready flag that the server sets when the result is ready. We follow the convention that the client manages the memory occupied by messages (most messages are allocated on the client's stack).

In our experiments, when a client issues a request, it blocks until the response is available. Straightfoward alternative approaches could allow clients to issue requests for multiple operations to be performed in parallel.

We evaluate three communication mechanisms: MPSCChannel, InletQueue, and DNCInletQueue, each making different synchronization trade-offs.

The MPSCChannel (multiple producer, single consumer), based on the "Multilane" structure of Dice and Otenko [6], uses an array of request slots. A shared variable PutCursor indicates the next available slot. A client uses a *compare-and-swap* (CAS) instruction to increment PutCursor atomically, and uses the previous value (modulo the size of the array) to choose a slot. Because that slot might still be in use, the client repeatedly calls CAS to swap null with a pointer to its request message. The server uses a private variable TakeCursor to cycle around slots, waiting for each one to contain a non-null pointer to a request. It then reads the opcode and arguments from the request, resets the array slot to null (making it available to other clients), performs the operation, stores the result in the buffer provided, and finally sets the ready flag.

In NUMA architectures, memory accesses not satisfied by local cache are substantially slower when applied to remote memory than to local memory. A disadvantage of MPSCChannel is that it requires threads to repeatedly apply CAS to remote PutCursor locations, and, more rarely, to remote slots.

The InletQueue channel provides one slot per NUMA domain. Each client uses CAS to attempt to replace the slot's null value with a pointer to its message. When the server reads the request, it resets the slot to null to make it available again, performs the operation, copies the result (if any) to the client's buffer, and sets the message's ready flag.

The DNCInletQueue channel ("direct, no CAS") uses only load and store operations to access remotely share variables, and a lock for synchronization among threads on a single node. In this channel, the node's slot contains the message itself, not just a pointer to the message. When a client thread acquires the lock for its node's slot, it copies the request message into the slot, including a pointer to the client buffer where the result is to be stored.

The motivation for DNCInletQueue is to ensure that the mechanism used for actual inter-socket communication is as simple as possible (simple stores by the client and simple loads by the server): synchronization such as acquiring the lock that protects the slot is performed only among threads on the same node. (With InletQueue, although only clients on the same node attempt to modify the slot, slots are still shared remotely with the server reading them.) We believe this approach creates the best opportunity for potential future hardware optimizations to reduce communication overhead. Even without such optimizations, the "direct" aspect of DNCInletQueue ensures that, when a server reads a slot written by a client, it already knows the operation to perform. In contrast, methods that send a pointer to the message require the server to initiate another round of inter-socket communication to fetch the message contents.

Shared Memory. For shared-memory mechanisms, we consider lock-based structures employing the following kinds of locks: a simple spin lock, the MCS lock [11], and fair and unfair versions of a NUMA-aware "cohort" lock C−TKT−MCS [5] that uses MCS for synchronization between threads on the same socket, and a global ticket lock to explicitly manage when the lock is handed off to a thread on another socket. Handing off the lock preferentially within a socket can reduce lock handoff time, and increase cache locality for data accessed in the critical section. However, doing so blindly can result in "gross unfairness", in which high throughput is achieved, but some threads are essentially starved. Perhaps surprisingly, depending on the architecture, this phenomenon can occur even with simple locks that do not explicitly seek to keep the lock within a socket. Thus, it is important to manage such pitfalls. We therefore include "fair" and "unfair" variants of C−TKT−MCS (denoted as C−TKT−MCS-fair and C−TKT−MCS-unfair, respectively). The fair version imposes a limit on how many times the lock can be handed off within a socket, avoiding grossly unfair behavior.

4 Benchmarks

In this paper, we restrict our analysis to two representative cases among the data structures we explored. Both implement a *map* interface, storing key-value pairs with standard insert (), remove() and get() operations.

4.1 Concurrent Hash Maps

The hash map is partitioned into multiple pieces; with delegation, each is managed by a server thread. Each partition has a preconfigured number of *buckets*, where each bucket is a linked list of *chunks*. Each chunk is a fixed-size, cache-line-aligned structure that holds a set of key-value pairs whose keys lie within a fixed range. Chunk size is a multiple of 64 bytes (the unit of cache coherence). To speed searches, adjacent chunks' key ranges do not overlap, and each chunk records the maximum key that it stores. Chunks in a bucket are sorted by their maximum stored keys, but key-value pairs within a chunk are unordered.

Each bucket is a linked list of cache-aligned chunks, instead of the more traditional list of key-value pairs, because loading each chunk brings in multiple key-value pairs, reducing cache coherence traffic. This structure should benefit both shared-memory and delegation-based methods. However, it is likely to favor shared-memory more because delegation ensures that all accesses to this data are from the same NUMA node, resulting in more effective use of lower level caches, and more ability to place data in memory near where it will be accessed.

To store a key-value pair in a partition, the key is hashed to identify the bucket where the pair will be stored. The bucket's list of chunks is then scanned to identify which chunk should contain the given key, skipping chunks whose maximum key is smaller than it. The target chunk is then scanned linearly for the given key. If found, the value is updated. If not, but the chunk is full, the chunk is split and half of its elements are moved to a new chunk, making space for the new pair. Chunk size is subject to a trade-off: smaller chunks are better for cache locality, but larger chunks reduce the frequency of splitting.

Although many other possibilities exist, we have chosen shared-memory and delegation-based implementations that each exploit a key advantage they have over the other. For delegation, by having a single server thread manage each partition, it can do so without additional synchronization. For shared memory, we have chosen an example in which fine-grained locking is straightforward: a fixed-size hash map implemented using a single lock for each bucket, allowing one thread per bucket to access the hash map concurrently.

While multiple server threads could also use this technique to collectively manage a partition, this would impose overhead on each operation, introduce issues such as how clients balance requests over these multiple servers, and require additional hardware threads to be reserved for the additional severs. In contrast, in the lock-based hash map, as long as the total number of buckets (and therefore locks) remains the same, the actual number of partitions has almost no effect on performance or on the number of hardware threads required.

4.2   Concurrent Linked Lists

The concurrent linked list is a degenerate hash map, where each partition consists of a single bucket. In particular, each bucket is a linked list of chunks, as explained in Section 4.1. One important point is that the whole partition/bucket is protected by a single lock, so the concurrency achievable in the lock-based linked list is bounded by the number of partitions, much as with delegation.

4.3   Workloads

We used both *small* and *large* workloads. The small (large) hash map has 500 (50,000) buckets per partition. In the small (large) workload, the hash map is initialized by storing a key-value pair with a randomly-chosen key 1000 (100,000) times. For the linked list, the small (large) workload initializes the list by storing a key-value pair with a randomly-chosen key 1000 (100,000) times. Thus, the small workload has better cache locality than the large workload. After some experimentation, we sized chunks to accommodate 64 key-value pairs.

We experimented with three *mixes* of operations: *read-only*, consisting entirely of get() calls, *write-only*, consisting of 50% insert () and 50% remove() calls, and *read-write*, a mixture of 50% get(), 25% insert (), and 25% remove() calls. There are too many combinations of data structures, architectures, and workloads to present them all, so we focus here on the most interesting cases.

Results were qualitatively similar for the three operation mixes (read-only, write-only, and read-write); for brevity, we present only the read-write results.

5   Performance Results

The experiments were conducted on two systems with different architectures. The first is an 8-socket Nehalem system [13] ("X4800"), each socket containing a Xeon X7560 processor chip with 8 hyperthreaded cores running at a 2.26Ghz clock frequency, with a total of 128 hardware threads. The second system is an Oracle T4-4 [14] ("SPARC T4-4"), which consists of 4 T4 SPARC sockets, each socket containing 8 cores, and each core containing 8 hardware thread contexts, for a total of 256 hardware thread contexts, running at a 3 GHz clock frequency.

For the delegation-based implementations, server threads were placed uniformly among the sockets (see Section 5.4 for additional details). Placement of client threads was controlled by the OS in all cases. In each experiment, each thread repeatedly chooses at random whether to insert or delete an item, and performs the operation. No "external" work is performed between operations. We measure the total number of operations completed by all threads over a measurement period of ten seconds, and report throughput as the number of operations performed by all client threads per millisecond. Each experiment was repeated 6 times, and the average throughput for each configuration is reported.

5.1   Hash Map

The first set of experiments was conducted on the concurrent hash map data structure of Section 4.1. We experimented with both small (Figures 1(a) and 1(b)), and large (Figure 1(c) and 1(d)) workloads. Unless stated otherwise, the number of partitions (as well as the number of server threads in the case of delegation) is constant at 8 (which is equal to or a small multiple of the number of sockets), and we use the read-write operation mix.

Figure 1 shows that, for the hash map benchmark, shared-memory mechanisms with any of locks performs much better than delegation for any channel type. This difference is because the fine-grained locking employed in our hash map implementation allows many threads to manipulate the shared hash map data structure concurrently. On the other hand, concurrency is limited by the number of servers in the case of delegation. Indeed, the performance of delegation scales only up to 32 threads (which is more than 8, the number of servers, because clients perform work such as choosing a random key and determining which server thread will perform the operation before sending the request).

For small hash maps (Figures 1 (a) and (b)), all locks eventually stop scaling (and most perform worse) as the number of threads increases. This is due to contention on the (relatively) small number of buckets/locks. The performance of delegation, though worse than that of locking, is less sensitive to this contention because it is limited primarily by the sequential server threads, whose performance is largely insensitive to contention on message queues. (Nonetheless, MPSCChannel's centralized PutCursor makes it more sensitive to contention than the other message queues.)

The simple MCS lock is typically the best-performing lock at low contention. However, MCS's performance degrades under heavy contention. By contrast,

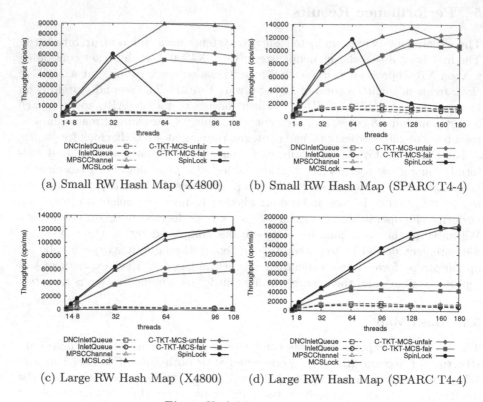

(a) Small RW Hash Map (X4800)   (b) Small RW Hash Map (SPARC T4-4)

(c) Large RW Hash Map (X4800)   (d) Large RW Hash Map (SPARC T4-4)

Fig. 1. Hash Map experiment

NUMA-aware locks perform better under high contention because there is an increased likelihood that locks can be handed off to threads on the same socket. The unfair C–TKT–MCS variant provides better high-contention performance than the fair variant because it permits more consecutive, local hand-offs. We return to this point in Section 5.3.

5.2   Linked List

Figure 2 summarizes results for the linked list benchmark. As noted in Section 4.2, each partition contains just one bucket protected by a lock. Furthermore, each operation performs more memory accesses with the linked list than with the hash map, as all key-value pairs of a partition are stored in one bucket. A larger number of memory accesses per operation favors delegation if better server cache locality outweighs the cost of client-server communication. Indeed, the delegation methods performed considerably better than all locking schemes for large linked lists (Figures 2(c) and 2(d)), where operations access a large number of memory locations during list traversals. For small linked lists

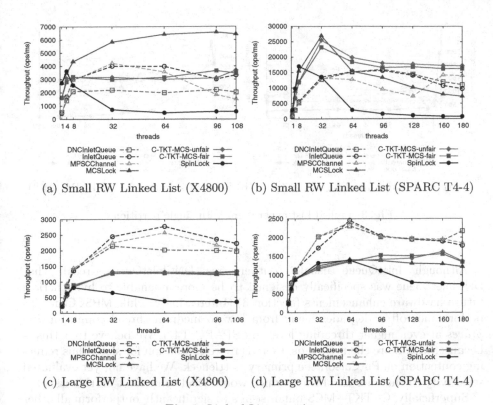

(a) Small RW Linked List (X4800)   (b) Small RW Linked List (SPARC T4-4)

(c) Large RW Linked List (X4800)   (d) Large RW Linked List (SPARC T4-4)

Fig. 2. Linked List experiment

(Figures 2(a) and 2(b)), delegation provided competitive performance, losing only to MCS on X4800, and to the C−TKT−MCS variants on SPARC T4-4.

Although the simple MCS algorithm [11] provides superior performance in many cases, its performance degrades severely in some cases. There are two reasons for this. First, when contention increases, MCS has no facility to encourage consecutive lock handoffs within the same socket. As a result, the `Tail` variable that is modified by every lock acquisition "bounces" around the system frequently. This in turn causes data accessed in the critical section to similarly bounce around the system. NUMA-aware locks are able to avoid this effect and thus outperform MCS in this case (Figure 2(b)).

To evaluate these mechanisms in less balanced workloads, we repeated the experiment using only one partition, representing a partition that receives a disproportionate fraction of the requests, or alternatively a configuration in which there are not enough partitions, so all partitions may be overloaded.

Results are shown in Figure 3. (We omit results for large linked lists for this case, as sequential execution of operations dominates performance. Thus, the synchronization mechanism used has little bearing on performance.)

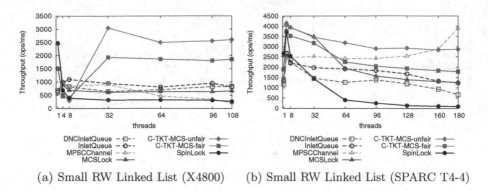

(a) Small RW Linked List (X4800)   (b) Small RW Linked List (SPARC T4-4)

Fig. 3. Linked List experiment with single partition

Although InletQueue usually outperforms DNCInletQueue, recall that DNCInletQueue was specifically designed to be more amenable to hypothetical future hardware enhancements (Section 3). Interestingly, while MPSCChannel's performance often degrades going from low to medium thread counts, it *improves* at even higher threading levels on SPARC T4-4. We believe that this is because, with more client threads, there is more contention for slots, thus reducing contention on PutCursor, the primary bottleneck. We have not yet evaluated sensitivity to the number of slots, which would shed some light on this issue.

Superficially, C−TKT−MCS-unfair seems to significantly outperform all other methods and—to a lesser extent—both C−TKT−MCS-fair and MPSCChannel also stand out. However, some caution is needed in interpreting these results. As discussed in Section 3, some methods provide deceptively high throughput by "gross unfairness": they provide high throughput to some threads, while other threads receive much lower throughput or even starve completely. If this issue is overlooked, it is easy to conclude that a method that would be unacceptable in practice delivers the best results. We discuss this issue in more detail next.

5.3   Fairness

As a crude indicator of unfairness, we use *spread*, defined as the maximum per-thread count divided by the minimum per-thread count (plus one to avoid divide by zero). If the throughput of all threads is approximately equal, the spread will be close to 1. Methods that are grossly unfair—particularly those that starve some threads completely—exhibit very high spread.

In Figure 3(b), C−TKT−MCS-unfair consistently delivers the highest or nearly the highest throughput, but exhibits a spread value of over 560,000 at 32 threads. Its fair counterpart typically exhibits a spread value close to 1 (we occasionally see values of up to 4.5), but delivers significantly lower throughput in most cases. Similarly, on X4800, C−TKT−MCS-unfair exhibits spread over 1,000,000 in the highest contention case (single partition, 108 threads), while C−TKT−MCS-fair almost always yields spread very close to 1 (with rare outliers not exceeding 40).

The delegation methods also exhibited high spread values (for example, on SPARC T4-4, up to 2,100 for InletQueue, 670,000 for DNCInletQueue and 360,000 for MPSCChannel; the situation is not as bad on X4800, but still we occasionally see spread values up to 1,200).

Next we describe a preliminary exploration of how the fairness of the delegation methods might be improved. MPSCChannel suffers from CAS contention on the remotely-shared PutCursor variable. InletQueue applies CAS on the message slot to swap in a pointer to the message, and DNCInletQueue uses a simple spinlock to acquire ownership of the message slot. When threads compete in this manner, unfairness can result because a thread that releases the message slot has the corresponding synchronization variable in cache and is therefore likely to be able to acquire the slot again before another thread can.

To address this issue, we experimented with simple backoff mechanisms whereby, if a thread experiences too many consecutive CAS failures, it sets a flag causing all threads accessing that channel to pause before retrying, conditional on a function of their thread IDs and the number of times the slot lock has been acquired. This reduces contention and gives "priority" to different threads over time. This eliminated the gross unfairness on SPARC T4-4 without impacting throughput, but we still observed spread values of up to 2000 for DNCInletQueue and 1150 for InletQueue, indicating that there is still considerable room for improvement. We found that parameters controlling the threshold and backoff could be tuned to different points in a tradeoff between spread and throughput. We are still experimenting to improve our results here.

Unlike SPARC T4-4, X4800 yielded spread values at worst in the low hundreds even before these optimizations, which were less effective on X4800, although we have not yet tuned them for this platform.

5.4 Hardware-Related Details

In earlier experiments, InletQueue and DNCInletQueue degraded significantly at higher thread counts. After some investigation, we hypothesized that this was due to "sibling rivalry": client threads executing on the same core as a server thread would compete with the server thread for resources, thus indirectly slowing client threads making requests to that server. To address this issue, when placing a server thread on a core, we reserve all other hardware threads on that core so that they are not used by clients. This resulted in a significant improvement, allowing the delegation methods to outperform all others across the threading range for large linked lists on both platforms, for example Figures 2(c) and 2(d). Although this dedicates more hardware threads to delegation, these threads could potentially be used to benefit the server, rather than interfering with it. We leave investigation of this direction for future work.

This experience highlights one potential downside of delegation. Apart from using cores that might otherwise be used by additional application threads, reserving sibling threads requires server threads to be "pinned" to a specific hardware thread, which can be a mixed blessing. First, overriding the operating system's thread placement policy prevents it from choosing the best placement based on the current workload. This is clearly demonstrated in Figure 2(b): at low thread counts, the lock-based methods have a significant advantage because the operating system is able to place all threads on the same socket.

On the other hand, a fixed relationship between data and the hardware threads that access it can be exploited in some contexts. To illustrate this point we performed an experiment (not shown), in which we controlled the placement of these structures so that each delegation message queue was allocated on the same NUMA node as the corresponding server thread. In contrast, these structures are usually allocated by a single thread at initialization and are thus all allocated in physical memory of the same NUMA node.

This simple placement optimization substantially improved the performance of delegation on X4800, especially for InletQueue and DNCInletQueue; the latter improved by more than 2x in most cases. This may be counterintuitive given that these structures are likely to remain in cache. However, on X4800, each memory access requires communication with the location's "home node" (see [4] for a detailed explanation). Thus, locating each communication structure near the server thread that accesses it most often improves performance.

The substantial performance gains achieved by even this modest optimization reinforces our belief that significantly more could be achieved if hardware were explicitly optimized for such communication patterns.

Reducing coherence traffic between nodes can reduce consumption of inter-socket bandwidth, which may in turn avoid a system-wide bottleneck that may indirectly reduce performance [3]. The delegation methods we have presented were in large part motivated by similar concerns. Using hardware performance counters, we have found that the delegation methods typically generate a small number of remote cache misses per operation (typically around 4-5, although we sometimes observe significantly higher rates in high-contention cases). Software techniques—such as discussed in [2,8], and hardware optimizations tailored for these communication patterns could both significantly reduce this number.

However, recent progress in building NUMA-aware locks [5] has changed the landscape. By limiting how often locks (and therefore associated data) migrate between sockets—while avoiding gross unfairness exhibited by locks that do so "accidentally", such locks can reduce the per-operation remote cache miss rate almost to zero by performing large numbers of operations protected by a lock on one socket before allowing the lock to migrate to another. This depends on sufficient demand for a lock within a socket, suggesting that such techniques are excellent for avoiding performance disasters due to lock contention, but may not be as effective in scalable applications with little lock contention.

6   Discussion

Our results show that delegation can sometimes outperform direct shared-memory approaches, particularly when operations access enough data to ensure that the benefits of delegation outweigh its communication costs. Nonetheless, the best shared-memory mechanisms often performed about the same as or substantially better than the delegation mechanisms. Synchronization granularity is a key issue, for both locking and delegation. For easily partitionable data structures, like those considered in this paper, fine-grained locking is straightforward. For delegation, it is similarly straightforward to partition the data structure, allowing multiple server threads to service requests from client threads in parallel, but finer granularity requires additional hardware threads to be used.

While granularity affects both approaches in similar ways, there are interesting differences. Suppose, for example, that we want to make our hash map resizeable. Resizing is straightforward in the case of delegation, because operations need not synchronize with each other. In contrast, resizing a hash map implemented with per-bucket locks is more challenging, as the resizing must be coordinated with threads accessing the partition using these locks.

Different challenges and opportunities exist when workloads face contention. NUMA-aware locks such as the C−TKT−MCS variants can help limit the performance degradation of lock-based approaches, although these locks impose overhead in the hopefully more common case in which there is no lock contention.

With delegation, an overloaded server thread can become a bottleneck. Client-side techniques that combine multiple requests into one equivalent one, thus reducing the communication costs and the demand on the server thread, may improve performance. Elimination [8] can be used to complete operations without communicating with the server at all [2].

Server-side techniques may help too. For example, a server thread experiencing high demand could repartition its own partition and create an additional server thread to manage it. This may be effective if the execution of operations is the bottleneck. If, however, the communication channel for requests is the bottleneck, simple repartitioning will not help, and more ambitious techniques would be required in which client threads also become aware of the repartitioning.

7 Related Work

Lozi et al. [10] propose structuring a client-server system so that one or more cores are dedicated to server threads that execute critical sections on behalf of client threads. Client and server threads communicate through an array of contexts, one per client. A client's context includes the lock address, the critical section's private variables, and a function that encapsulates the critical section's code. (Note that some effort is required to encapsulate critical sections in this way.) Clients and servers use atomic operations on shared variables to signal when a request starts and completes. The authors observe that their scheme improves lock access contention and cache locality, but do not explore alternative signaling or communication structures.

Suleman et al. [16] consider an asymmetric multicore architecture encompassing a small number of high-performance cores and many smaller, less powerful cores. The paper examines architectural support for delegating critical sections to the high-performance cores; evaluation is via an in-house simulator.

Metreveli *et al.* [12] describe *CPHASH*, a concurrent hash map that uses a form of delegation to enhance cache locality. They show that delegation can outperform locking for one data structure on one platform configuration. Our goal, in contrast, is to characterize the relative merits of delegation and direct shared-memory mechanisms in a range of data structures, communication mechanisms, workloads, and platform configurations.

Hendler *et al.* [7] and Oyama *et al.* [15] propose mechanisms in which threads execute operations on behalf of others while holding a lock. (Again, critical sections must be encapsulated as self-contained functions.) This approach resembles delegation: a single thread serially executes multiple operations. But that thread is determined dynamically, not statically as in delegation schemes.

8   Conclusions

Delegation works well when the data structure can be partitioned so that it fits in the servers' collective caches. Delegation also works well when critical sections encompass many memory accesses, as in the case of the linked lists, because the communication overhead is outweighed by the savings in cache misses and coherence traffic. These savings are more substantial when the cost of remote memory access is high, allowing delegation to beat efficient NUMA-aware locks.

However, delegation is often outperformed by the best locking implementations. In particular, when critical sections are short, and especially in "small" workloads in which data accessed in the critical section is likely to be cached, locking approaches require little or no remote communication, while delegation still pays in communication overhead but delivers less benefit.

Nevertheless, as the number of sockets in multicore machines grows, so will the cost of remote memory access. Furthermore, techniques not explored in this paper (such as elimination and combining), as well as potential hardware improvements, may make delegation more attractive in the future.

Our experience has shown that low-level hardware details can make a considerable difference to the behavior of synchronization algorithms. Thus, we conclude that multicore applications should be designed around high-level data abstractions, hiding the low-level communication details, so that one mechanism can be replaced by another as workloads and platforms change.

Acknowledgments. We are grateful to Nir Shavit for useful input and feedback and to Bill Bridge and Garret Swart who suggested this research direction.

References

1. Baumann, A., Barham, P., Dagand, P.-E., Harris, T., Isaacs, R., Peter, S., Roscoe, T., Schüpbach, A., Singhania, A.: The multikernel: a new OS architecture for scalable multicore systems. In: Proc. ACM SIGOPS Symposium on Operating Systems Principles (SOSP), pp. 29–44 (2009)

2. Calciu, I., Gottschlich, J.E., Herlihy, M.: Using elimination and delegation to implement a scalable NUMA-friendly stack. In: Proc. Usenix Workshop on Hot Topics in Parallelism (HotPar) (2013)
3. Dashti, M., Fedorova, F., Funston, J., Gaud, F., Lachaize, R., Lachaize, B., Quema, V., Quema, M.: Traffic management: a holistic approach to memory placement on NUMA systems. In: Proc. Conf. on Arch. Support for Prog. Lang. and Op. Systems (ASPLOS), pp. 381–394 (2013)
4. Dice, D.: NUMA-aware placement of communication variables (November 2012), blogs.oracle.com/dave/entry/numa_aware_placement_of_communication1
5. Dice, D., Marathe, V.J., Shavit, N.: Lock cohorting: a general technique for designing NUMA locks. In: Proc. ACM Symp. on Principles and Practice of Parallel Programming (PPoPP), pp. 247–256 (2012)
6. Dice, D., Otenko, O.: Brief announcement: multilane - a concurrent blocking multiset. In: Proc. ACM SPAA, pp. 313–314 (2011)
7. Hendler, D., Incze, I., Shavit, N., Tzafrir, M.: Flat-combining and the synchronization parallelism tradeoff. In: Proceedings of the Twenty Third ACM Symposium on Parallelism in Algorithms and Architectures (SPAA), pp. 355–364 (June 2010)
8. Hendler, D., Shavit, N., Yerushalmi, L.: A scalable lock-free stack algorithm. In: Proc. ACM Symposium on Parallelism in Algorithms and Architectures (SPAA), pp. 206–215 (2004)
9. Lauer, H.C., Needham, R.M.: On the duality of operating system structures. SIGOPS Oper. Syst. Rev. 13(2), 3–19 (1979)
10. Lozi, J.-P., David, F., Thomas, G., Lawall, J., Muller, G.: Remote core locking: Migrating critical-section execution to improve the performance of multithreaded applications. In: Proc. USENIX Annual Technical Conference, ATC (2012)
11. Mellor-Crummey, J.M., Scott, M.L.: Algorithms for scalable synchronization on shared-memory multiprocessors. ACM Trans. Comput. Syst. 9(1), 21–65 (1991)
12. Metreveli, Z., Zeldovich, N., Kaashoek, M.F.: Cphash: a cache-partitioned hash table. In: Proc. ACM SIGPLAN Symposium on Principles and Practice of Parallel Programming, PPoPP 2012, pp. 319–320. ACM, New York (2012)
13. Oracle Corporation. Oracle's Sun Fire X4800 Server Architecture (2010), www.oracle.com/technetwork/articles/systems-hardware-architecture/sf4800g5-architecture-163848.pdf
14. Oracle Corporation. Oracle's SPARC T4-1, SPARC T4-2, SPARC T4-4, and SPARC T4-1B Server Architecture (2012), www.oracle.com/technetwork/server-storage/sun-sparc-enterprise/documentation/o11-090-sparc-t4-arch-496245.pdf
15. Oyama, Y., Taura, K., Yonezawa, A.: Executing parallel programs with synchronization bottlenecks efficiently. In: Proc. Int. Workshop on Parallel and Distributed Computing for Symbolic and Irregular Applications, PDSIA (1999)
16. Suleman, M.A., Mutlu, O., Qureshi, M.K., Patt, Y.N.: Accelerating critical section execution with asymmetric multi-core architectures. In: Proc. Conf. on Arch. Support for Prog. Lang. and Op. Systems (ASPLOS), pp. 253–264 (2009)
17. von Eicken, T., Culler, D.E., Goldstein, S.C., Schauser, K.E.: Active messages: a mechanism for integrated communication and computation. In: Proc. Int. Symposium on Computer Architecture (ISCA), pp. 256–266 (1992)

Reputation-Based Mechanisms for Evolutionary Master-Worker Computing

Evgenia Christoforou[1,2], Antonio Fernández Anta[1], Chryssis Georgiou[3],
Miguel A. Mosteiro[4], and Angel (Anxo) Sánchez[2,5]

[1] Institute IMDEA Networks, Madrid, Spain
[2] Universidad Carlos III de Madrid, Madrid, Spain
[3] University of Cyprus, Nicosia, Cyprus
[4] Kean University, Union, NJ, USA & Univ. Rey Juan Carlos, Madrid, Spain
[5] BIFI Institute, Zaragoza, Spain

Abstract. We consider Internet-based Master-Worker task computing systems, such as SETI@home, where a master sends tasks to potentially unreliable workers, and the workers execute and report back the result. We model such computations using evolutionary dynamics and consider three type of workers: *altruistic*, *malicious* and *rational*. Altruistic workers always compute and return the correct result, malicious workers always return an incorrect result, and rational (selfish) workers decide to be truthful or to cheat, based on the strategy that increases their benefit. The goal of the master is to reach eventual correctness, that is, reach a state of the computation that always receives the correct results. To this respect, we propose a mechanism that uses *reinforcement learning* to induce a correct behavior to rational workers; to cope with malice we employ *reputation schemes*. We analyze our reputation-based mechanism modeling it as a Markov chain and we give provable guarantees under which truthful behavior can be ensured. Simulation results, obtained using parameter values that are likely to occur in practice, reveal interesting trade-offs between various metrics, parameters and reputation types, affecting cost, time of convergence to a truthful behavior and tolerance to cheaters.

Keywords: Volunteer computing, evolutionary game theory, reinforcement learning, reputation.

1 Introduction

Motivation and Prior Work. The need for high-performance computing and the growing use of personal computers and their capabilities (i.e. CPU and GPU), and the wide access to the Internet, have led to the development of Internet-based computing. At present, Internet-based computing is mostly embraced by the scientific community in the form of volunteer computing; where computing resources are volunteered by the public to help solve scientific problems. Among the most popular volunteering projects is SETI@home [22] running on the BOINC [4] platform. A profit-seeking computation platform has also been developed by Amazon, called Mechanical Turk [3]. Although the potential is great, the use of Internet-based computing is limited by the untrustworthy nature of the platform's components [4, 18].

R. Baldoni, N. Nisse, and M. van Steen (Eds.): OPODIS 2013, LNCS 8304, pp. 98–113, 2013.
© Springer International Publishing Switzerland 2013

In Internet-based Master-Worker task computing systems a master process sends tasks, across the Internet, to worker processes, that execute and report back the result. However, these workers are not trustworthy, and hence might report incorrect results [4, 5, 20]. Prior work has considered different approaches in tackling the problem. A classical Distributing Computing approach is to model the malfunctioning (due to a hardware or a software error) or cheating (intentional wrongdoer) as *malicious* workers that wish to hamper the computation and thus always return an incorrect result. The non-faulty workers are viewed as *altruistic* ones [8] that always return the correct result. Under this view, malicious-tolerant protocols have been considered, e.g., [14, 21, 24], where the master decides on the correct result based on majority voting. A Game-theoretic approach is to assume that workers are *rational* [1, 17, 25], that is, a worker decides whether to truthfully compute and return the correct result or return a bogus result, based on the strategy that best serves its self-interest (increases its benefit). Under this view, incentive-based algorithmic mechanisms have been devised, e.g., [15, 30], that employ reward/punish schemes to "enforce" rational workers to act correctly.

In prior work [16], all three types were considered, and both approaches were combined in order to produce an algorithmic mechanism that provides incentives to rational workers to act correctly, while alleviating the malicious workers' actions. All the solutions described are *one-shot (or stateless)* in the sense that the master decides about the outcome of an interaction with the workers involving a specific task, without using any knowledge gained by prior interactions. In [10], we took advantage of the repeated interactions between the master and the workers, assuming the presence of only rational workers. For this purpose, we studied the *dynamics of evolution* [23] of such master-worker computations through *reinforcement learning* [27] where both the master and the workers adjust their strategies based on their prior interaction. The objective of the master is to reach a state in the computation after which it always obtains the correct results, while the workers attempt to increase their benefit. Hence, prior work either considered all three types of workers in one-shot computation, or multi-round interactions assuming only rational workers.

In volunteer computing workers join projects to support a scientific goal and/or to gain prestige [5], while in non-volunteer computing workers expect payment. Whatever the reason, Internet-based computing can not be considered a reliable platform [4, 5, 18, 20]. Thus provable guarantees must be given that the designed mechanism provides a reliable platform, especially in commercial platforms where one can not consider altruistic workers. The existence of all three types of workers must be assumed since workers can have a predefined behavior (malicious or altruistic) or not (rational) as we observe from the survey conducted by BOINC [8] and the behavior of its users [7]. A mechanism must be designed that benefits from the repeated interaction with the workers and thus detaches the knowledge of the distribution over the type of workers from the assumptions (in comparison with [16]).

Our contributions

- We design such an algorithmic mechanism that uses reinforcement learning to induce a correct behavior to rational workers while coping with malice using *reputation*. We consider a centralized reputation scheme controlled by the master that may use three different reputation metrics to calculate each worker's reputation. The first is adopted from [26], the second, which we introduce, allows for a more drastic change of reputation and the third is inspired by BOINC's reputation scheme [6].

- We analyze our reputation-based mechanism modeling it as a Markov chain and we identify conditions under which truthful behavior can be ensured. We analytically prove that by using the second reputation type (introduced in this work for the first time) reliable computation is eventually achieved.
- Simulation results, obtained using parameter values that are likely to occur in practice, reveal interesting trade-offs between various metrics and parameters, such as cost, time of convergence to a truthful behavior, tolerance to cheaters and the type of reputation metric employed. Simulations also reveal better performance of our reputation type (second type) in several realistic cases.

Background and related work. As part of our mechanism we use reinforcement learning to induce the correct behavior of rational workers. Reinforcement learning [27] models how system entities, or *learners*, interact with the environment to decide upon a strategy, and use their experience to select or avoid actions according to the consequences observed. Positive payoffs increase the probability of the strategy just chosen, and negative payoffs reduce this probability. Payoffs are seen as parameterizations of players' responses to their experiences. There are several models of reinforcement learning. A well-known model is that of Bush and Mosteller [9]; this is an aspiration-based reinforcement learning model where negative effects on the probability distribution over strategies are possible, and learning does not fade with time. The learners adapt by comparing their experience with an *aspiration* level. In our work we adapt this reinforcement learning model and we consider a simple aspiration scheme where aspiration is fixed by the workers and does not change during the evolutionary process.

The master reinforces its strategy as a function of the reputation calculated for each worker. Reputation has been widely considered in on-line communities that deal with untrustworthy entities, such as online auctions (e.g., eBay) or P2P file sharing sites (e.g., BitTorrent); it provides a mean of evaluating the degree of trust of an entity [19]. Reputation measures can be characterized in many ways, for example, as objective or subjective, centralized or decentralized. An objective measure comes from an objective assessment process while a subjective measure comes from the subjective belief that each evaluating entity has. In a centralized reputation scheme a central authority evaluates the entities by calculating the ratings received from each participating entity. In a decentralized system entities share their experience with other entities in a distributed manner. In our work, we use the master as a central authority that objectively calculates the reputation of each worker, based on its interaction with it; this centralized approach is also used by BOINC.

The BOINC system itself uses a form of reputation [6] for an optional policy called adaptive replication. This policy avoids replication in the event that a job has been sent to a highly reliable worker. The philosophy of this reputation scheme is to require a long time for the worker to gain a good reputation but a short time to lose it. Our proposed mechanism differs significantly from the one that is used in BOINC. One important difference is that we use auditing to check the validity of the worker's answers while BOINC uses only replication; in this respect, we have a more generic mechanism that also guarantees reliability of the system. Notwithstanding inspired by the way BOINC handles reputation we have designed a BOINC-like reputation type in our mechanism (called type three).

Sonnek et al. [26] use an adaptive reputation-based technique for task scheduling in volunteer setting (i.e., projects running BOINC). Reputation is used as a mechanism to reduce the degree of redundancy while keeping it possible for the master to verify the results by allocating more reliable nodes. In our work we do not focus on scheduling tasks to more reliable workers to increase reliability but rather we design a mechanism that forces the system to evolve to a reliable state. We also demonstrate several tradeoff between reaching a reliable state fast and master's cost. We have created a reputation function (called reputation type 1) that is analogous to the reputation function used in [26] to evaluate this function's performance in our setting.

Aiyer et al. [2] introduced the BAR model to reason about systems with Byzantine (malicious), Altruistic, and Rational participants. They introduced the notion of *BAR-tolerant* protocols, i.e., protocols that are resilient to both Byzantine faults and rational manipulation. As an application, they designed a cooperative backup service for P2P systems, based on a BAR-tolerant replicated state machine. More recent works have considered other problems in the BAR model (e.g., data transfer [29]). Although the objectives and the model considered are different, our reputation-based mechanism can be considered, in some sense, to be BAR-tolerant.

2 Model

In this section we characterize our model and we present the concepts of auditing, payoffs, rewards and aspiration. We also give a formal definition of the three reputation types used by our mechanism.

Master-Worker Framework. We consider a master and a set W of n workers. The computation is broken into *rounds*, and in each round the master sends a task to the workers to compute and return the result. Based on the workers' replies, the master must decide which is the value most likely to be the correct result for this round. We assume that tasks have a unique solution; although such limitation reduces the scope of application of the presented mechanism [28], there are plenty of computations where the correct solution is unique: e.g., any mathematical function.

Eventual Correctness. The goal of the master is to eventually obtain a reliable computational platform: After some finite number of rounds, the system must guarantee that the master (with minimal cost) obtains the correct task results in every round with probability 1. We call such property *eventual correctness*.

Worker Types. We consider three type of workers: *rational, altruistic* and *malicious*. Rational workers are selfish in a game-theoretic sense and their aim is to maximize their utility (benefit). In the context of this paper, a worker is *honest* in a round, when it truthfully computes and returns the correct result, and it *cheats* when it returns some incorrect value. Altruistic and malicious workers have a predefined behavior, to always be honest or cheat, respectively. Instead, a rational worker decides to be honest or cheat depending on which strategy maximizes its utility. We denote by $p_{Ci}(r)$ the probability of a rational worker i cheating in round r. This probability is not fixed and the worker adjusts it over the course of the computation. The master is not aware of the worker types, neither of a distribution of types (our mechanism does not rely on any statistical information).

While workers make their decision individually and with no coordination, following [24] and [14], we assume that all the workers that cheat in a round return the same incorrect value; this yields a worst case scenario (and hence analysis) for the master with respect to obtaining the correct result using mechanisms where the result is the outcome of voting. It subsumes models where cheaters do not necessarily return the same answer. (This can be seen as a weak form of collusion.)

For simplicity, unless otherwise stated, we assume that workers do not change their type over time. Observe that in practice it is possible that changes occur. For example, a rational worker might become malicious due to a bug, or a malicious worker (e.g., a worker under the influence of a virus) become altruistic (e.g., if an antivirus software reinstates it). If this may happen, then all our results still apply for long enough periods between two changes.

Auditing, Payoffs, Rewards and Aspiration. To induce the rational workers to be honest, the master employs, when necessary, *auditing* and *reward/punish* schemes. The master, in a round, might decide to audit the response of the workers, at a cost. In this work, auditing means that the master computes the task by itself, and checks which workers have been honest. We denote by $p_{\mathcal{A}}(r)$ the probability of the master auditing the responses of the workers in round r. The master can change this auditing probability over the course of the computation, but restricted to a minimum value $p_{\mathcal{A}}^{min} > 0$. When the master audits, it can accurately reward and punish workers. When the master does not audit, it rewards only those in the weighted majority (see below) of the replies received and punishes no one.

In this work we consider three worker payoff parameters: (a) $WP_{\mathcal{C}}$: worker's punishment for being caught cheating, (b) $WC_{\mathcal{T}}$: worker's cost for computing a task, and (c) $WB_{\mathcal{Y}}$: worker's benefit (typically payment) from the master's reward. Also, following [9], we assume that, in every round, a worker i has an *aspiration* a_i: the minimum benefit it expects to obtain in a round. In order to motivate the worker to participate in the computation, the master usually ensures that $WB_{\mathcal{Y}} \geq a_i$; in other words, the worker has the potential of its aspiration to be covered. We assume that the master knows the aspirations. Finally, we assume that the master has the freedom of choosing $WB_{\mathcal{Y}}$ and $WP_{\mathcal{C}}$ with goal of eventual correctness.

Reputation. The reputation of each worker is measured by the master; a centralized reputation mechanism is used. In fact, the workers are unaware that a reputation scheme is in place, and their interaction with the master does not reveal any information about reputation; i.e., the payoffs do not depend on a worker's reputation.

In this work, we consider three reputation metrics. The first one, called *type 1* is analogous to a reputation metric used in [26] and the third one, called *type 3* is inspired by BOINC. We also define our own type called *type 2* that is not influenced by any other reputation type, and as we show in Section 4 it possesses beneficial properties. In all types, the reputation of a worker is determined based on the number of times it was found truthful. Hence, the master may update the reputation of the workers only when it audits. We denote by $aud(r)$ the number of rounds the master audited up to round r, and by $v_i(r)$ we refer to the number of auditing rounds in which worker i was found truthful up to round r. We let $\rho_i(r)$ denote the *reputation* of worker i after round r, and for a given set of workers $Y \subseteq W$ we let $\rho_Y(r) = \sum_{i \in Y} \rho_i(r)$ be the aggregated reputation of the workers in Y, by aggregating we refer to summing the reputation values. Then, the three reputation types we consider are the following:

Type 1: $\rho_i(r) = (v_i(r) + 1)/(aud(r) + 2)$.

Type 2: $\rho_i(r) = \varepsilon^{aud(r)-v_i(r)}$, for $\varepsilon \in (0, 1)$, when $aud(r) > 0$, and $\rho_i(r) = 1/2$, otherwise.

Type 3: Here we define $\beta_i(r)$ as the error rate of worker i at round r and by $A = 0.05$ the error bound. Reputation for this type is calculated as follows:

Step 1:

$\beta_i(r) \leftarrow 0.1$

if *worker truthful* **then**

$\quad \beta_i(r) \leftarrow \beta_i(r) \cdot 0.95 \ \backslash\backslash$ *calculating error rate*

else $\beta_i(r) \leftarrow \beta_i(r) + 0.1$

Step 2:

if $\beta_i(r) > A$ **then**

$\quad \rho_i(r) \leftarrow 0.001 \ \backslash\backslash$ *calculating reputation*

else $\rho_i(r) \leftarrow 1 - \sqrt{\frac{\beta_i(r)}{A}}$

In each round, when the master *does not audit*, the result is obtained from the *weighted majority* as follows. Consider a round r. Let $F(r)$ denote the subset of workers that returned an incorrect result, i.e., the rational workers who chose to cheat plus the malicious ones; recall that we assume as a worst case that all cheaters return the same value. Then, $W \setminus F(r)$ is the subset of workers that returned the correct value, i.e., the rational workers who chose to be truthful plus the altruistic ones. Then, if $\rho_{W \setminus F(r)}(r) > \rho_{F(r)}(r)$, the master will accept the correct value, otherwise it will accept an incorrect value. The mechanism, presented in the next section, employs auditing and appropriate incentives so that rational workers become truthful with high reputation, while malicious workers (alternatively altruistic workers) end up having very low (altr. very high) reputation after a few auditing rounds.

3   Reputation-Based Mechanism

We now present our reputation-based mechanism. The mechanism is composed by an algorithm run by the master and an algorithm run by each worker.

Master's Algorithm. The algorithm begins by choosing the initial probability of auditing and the initial reputation (same for all workers). The initial probability of auditing will be set according to the information the master has about the environment (e.g., workers' initial p_C). For example, if it has no information about the environment, a possibly safe approach is to initially set $p_A = 0.5$. The master also chooses the reputation type to use (e.g., type 1, 2 or 3).

After that, at each round, the master sends a task to all workers and, when all answers are received, the master audits the answers with probability p_A. In the case the answers are not audited, the master accepts the value returned by the weighed majority, and continues to the next round with the same probability of auditing and the same reputation values for each worker. In the case the answers are audited, the value p_A of the next round is reinforced (i.e., modified according to the accumulated reputation of the cheaters) and the reputations of the workers are updated based on their responses. Then, the master rewards/penalizes the workers appropriately. Specifically, if the master audits and a worker i is a cheater (i.e., $i \in F$), then $\Pi_i = -WP_C$; if i is honest, then $\Pi_i = WB_y$. If the master does not audit, and i returns the value of the weighted majority (i.e., $i \in W_m$), then $\Pi_i = WB_y$, otherwise $\Pi_i = 0$.

We include a threshold, denoted by τ, that represents the master's *tolerance* to cheating (typically, we will assume $\tau = 1/2$ in our simulations). If the ratio of the aggregated reputation of cheaters with respect to the total is larger than τ, p_A is increased, and decreased otherwise. The amount by which p_A changes depends on the difference

Algorithm 1. Master's Algorithm

$p_A \leftarrow x$, where $x \in [p_A^{min}, 1]$
$aud = 0$
// initially all workers have the same reputation
$\forall i \in W : v_i = 0; \rho_i = 0.5$
for $r \leftarrow 1$ **to** ∞ **do**
  send a task T to all workers in W
  upon receiving all answers **do**
    audit the answers with probability p_A
    if the answers were not audited **then**
      // weighted majority, coin flip in case of a tie
      accept the value returned by workers in $W_m \subseteq W$,
              where $\rho_{W_m} > \rho_{W \setminus W_m}$
    else   // the master audits
      $aud \leftarrow aud + 1$
      Let $F \subseteq W$ be the set of workers that cheated.
      $\forall i \in W :$
        if $i \notin F$ **then** $v_i \leftarrow v_i + 1$ // honest workers
        update reputation ρ_i of worker i
      $p_A \leftarrow \min\{1, \max\{p_A^{min}, p_A + \alpha_m(\frac{\rho_F}{\rho_W} - \tau)\}\}$
    $\forall i \in W :$ **return** payoff Π_i to worker i

Algorithm 2. Algorithm for Rational Worker i

$p_{Ci} \leftarrow y$, where $y \in [0, 1]$
for $r \leftarrow 1$ **to** ∞ **do**
  receive a task T from the master
  $S_i \leftarrow -1$ with probability p_{Ci},
  and $S_i \leftarrow 1$ otherwise
  if $S_i = 1$ **then**
    $\sigma \leftarrow compute(T)$,
  else
    $\sigma \leftarrow arbitrary \ solution$
  send response σ to the master
  get payoff Π_i
  if $S_i = 1$ **then**
    $\Pi_i \leftarrow \Pi_i - WC_T$
  $p_{Ci} \leftarrow \max\{0, \min\{1, p_{Ci} - \alpha_w(\Pi_i - a_i)S_i\}\}$

between these values, modulated by a *learning rate* α_m. This latter value determines to what extent the newly acquired information will override the old information. (For example, if $\alpha_m = 0$ the master will never adjust p_A.) A pseudocode of the algorithm described is given as Algorithm 1.

Workers' Algorithm. This algorithm is run only by rational workers (recall that altruistic and malicious workers have a predefined behavior).[1] The execution of the algorithm begins with each rational worker i deciding an initial probability of cheating p_{Ci}. In each round, each worker receives a task from the master and, with probability $1 - p_{Ci}$ computes the task and replies to the master with the correct answer. Otherwise, it fabricates an answer, and sends the incorrect response to the master. We use a flag S_i to model the stochastic decision of a worker i to cheat or not. After receiving its payoff, each worker i changes its p_{Ci} according to payoff Π_i, the chosen strategy S_i, and its aspiration a_i.

The workers have a *learning rate* α_w. In this work, we assume that all workers have the same learning rate, that is, they learn in the same manner (see also the discussion in [27]; the learning rate is called step-size there); note that our analysis can be adjusted to accommodate also workers with different learning rates. A pseudocode of the algorithm is given as Algorithm 2.

4   Analysis

We now analyze the reputation-based mechanism. We model the evolution of the mechanism as a Markov Chain, and then discuss the necessary and sufficient conditions for achieving eventual correctness. Modeling a reputation-based mechanism as a Markov Chain is more involved than previous models that do not consider reputation (e.g. [10]).

---

[1] Since the workers are not aware that a reputation scheme is used, this algorithm is the one considered in [10]; we describe it here for self-containment.

The Markov Chain. Let the state of the Markov chain be given by a vector s. The components of s are: for the master, the probability of auditing p_A and the number of audits before state s, denoted as aud; and for each *rational* worker i, the probability of cheating p_{Ci} and the number of *validations* (i.e., the worker was honest when the master audited) before state s, denoted as v_i. To refer to any component x of vector s we use $x(s)$. Then, $s = \langle p_A(s), aud(s), p_{C1}(s), p_{C2}(s), \ldots, p_{Cn}(s), v_1(s), v_2(s), \ldots, v_n(s) \rangle$.

In order to specify the transition function, we consider the execution of the protocol divided in rounds. In each round, probabilities and *counts* (i.e. numbers of validations and audits) are updated by the mechanism as defined in Algorithms 1 and 2. The state at the end of round r is denoted as s_r. Abusing the notation, we will use $x(r)$ instead of $x(s_r)$ to denote component x of vector s_r. The workers' decisions, the number of cheaters, and the payoffs of each round $r > 0$ are the stochastic outcome of the probabilities and counts at the end of round $r - 1$. We specify the transition from s_{r-1} to s_r by the actions taken by the master and the workers during round r.

In the definition of the transition function that follows, the probabilities are limited to $p_A(s) \in [p_A^{min}, 1]$ and for each rational worker i to $p_{Ci}(s) \in [0, 1]$, for any state s. The initial state s_0 is arbitrary but restricted to the same limitations. Let $P_F(r)$ be the probability that the set of cheaters in round r is exactly $F \subseteq W$. (That is, $P_F(r) = \prod_{j \in F} p_{Cj}(r-1) \prod_{k \notin F}(1 - p_{Ck}(r-1))$.) Then, the transition from state s_{r-1} to s_r is as follows.

- Malicious workers always have $p_C = 1$ and altruistic workers always have $p_C = 0$.
- With probability $p_A(r-1) \cdot P_F(r)$, the master audits when the set of cheaters is F. Then, according to Algorithms 1 and 2, the new state is as follows.

   For the master: $p_A(r) = p_A(r-1) + \alpha_m (\rho_F(r)/\rho_W(r) - \tau)$ and $aud(r) = aud(r-1) + 1$.

   (1) For each worker $i \in F$: $v_i(r) = v_i(r-1)$ and, if i is rational, then $p_{Ci}(r) = p_{Ci}(r-1) - \alpha_w(a_i + WP_C)$.

   (2) For each worker $i \notin F$: $v_i(r) = v_i(r-1) + 1$ and, if i is rational, then $p_{Ci}(r) = p_{Ci}(r-1) + \alpha_w(a_i - (WB_y - WC_T))$.

- With probability $(1 - p_A(r-1))P_F(r)$, the master does not audit when the set of cheaters is F. Then, according to Algorithms 1 and 2, the following updates are carried out.

   For the master: $p_A(r) = p_A(r-1)$ and $aud(r) = aud(r-1)$.

   For each worker $i \in W$: $v_i(r) = v_i(r-1)$.

   For each rational worker $i \in F$,

   (3) if $\rho_F(r) > \rho_{W \setminus F}(r)$ then $p_{Ci}(r) = p_{Ci}(r-1) + \alpha_w(WB_y - a_i)$,

   (4) if $\rho_F(r) < \rho_{W \setminus F}(r)$ then $p_{Ci}(r) = p_{Ci}(r-1) - \alpha_w \cdot a_i$,

   For each rational worker $i \notin F$,

   (5) if $\rho_F(r) > \rho_{W \setminus F}(r)$ then $p_{Ci}(r) = p_{Ci}(r-1) + \alpha_w(a_i + WC_T)$,

   (6) if $\rho_F(r) < \rho_{W \setminus F}(r)$ then $p_{Ci}(r) = p_{Ci}(r-1) + \alpha_w(a_i - (WB_y - WC_T))$.

Recall that, in case of a tie in the weighted majority, the master flips a coin to choose one of the answers, and assigns payoffs accordingly. If that is the case, transitions (3)–(6) apply according to that outcome.

Conditions for Eventual Correctness. We show now the conditions under which the system can guarantee eventual correctness. The analysis is carried out for a universal class of reputation functions characterized by the following properties.

Property 1: For any $X \subset W$ and $Y \subset W$, if the Markov chain evolves in such a way that $\forall i \in X, \lim_{r\to\infty}(v_i(r)/aud(r)) = 1$ and $\forall j \in Y, \lim_{r\to\infty}(v_j(r)/aud(r)) = 0$, then there is some r^* such that $\forall r > r^*, \rho_X(r) > \rho_Y(r)$.

Property 2: For any $X \subset W$ and $Y \subset W$, if $aud(r+1) = aud(r)+1$ and $\forall j \in X \cup Y$ it is $v_j(r+1) = v_j(r)+1$ then $\rho_X(r) > \rho_Y(r) \Rightarrow \rho_X(r+1) > \rho_Y(r+1)$.

Observe that all reputation functions (type 1, type 2 and type 3) we consider (cf. Section 2), satisfy Property 1. However, regarding Property 2, while reputation type 2 satisfies it, reputation type 1 and 3 do not. As we show below, this *makes a difference* with respect to guaranteeing eventual correctness.

The following terminology will be used throughout. For any given state s, a set X of workers is called a *reputable set* if $\rho_X(r) > \rho_{W \setminus X}(r)$. In any given state s, let a worker i be called an *honest worker* if $p_{Ci}(s) = 0$. Let a state s be called a *truthful state* if the set of honest workers in state s is reputable. Let a *truthful set* be any set of truthful states. Let a worker be called a *covered worker* if the payoff of returning the correct answer is at least its aspiration plus the computing cost. I.e., for a covered worker i, it is $WB_y \geq a_i + WC_T$. We refer to the opposite cases as *uncovered worker* ($WB_y < a_i + WC_T$), *cheater worker* ($p_{Ci}(s) = 1$), *untruthful state* (the set of cheaters in that state is reputable), and *untruthful set*, respectively. Let a set of states S be called *closed* if, once the chain is in any state $s \in S$, it will not move to any state $s' \notin S$. (A singleton closed set is called an *absorbing* state.) For any given set of states S, we say that the chain *reaches* (resp. *leaves*) the set S if the chain reaches some state $s \in S$ (resp. reaches some state $s \notin S$).

In the master's algorithm, a non-zero probability of auditing is always guaranteed. This is a necessary condition. Otherwise, unless the altruistic workers outnumber the rest, a closed untruthful set is reachable, as we show in [11].

Eventual correctness follows if we can show that the Markov chain always ends in a closed truthful set. We prove first that having at least one worker that is altruistic or covered rational is necessary for a closed truthful set to exist. Then we prove that it is also sufficient.

Lemma 1. *If all workers are malicious or uncovered rationals, no truthful set S is closed, if the reputation type satisfies Property 2.*

Proof. Let us consider some state s of a truthful set S. Let Z be the set of honest workers in s. Since s is truthful, then Z is reputable. Since there are no altruistic workers, the workers in Z must be uncovered rational. Let us assume that being in state s the master audits in round r. From Property 2, since all nodes in Z are honest in r, Z is reputable after r. From transition (2), after round r, each worker $i \in Z$ has $p_{Ci}(r) > 0$. Hence, the new state is not truthful, and S is not closed.

Lemma 2. *If at least one worker is altruistic or covered rational, a truthful set S is reachable from any initial state, if the reputation type satisfies Properties 1 and 2.*

Proof (Proof Sketch). Let C be the set of workers that are altruistic or covered rational. From any initial state, there is a non-zero probability that the master audits in all subsequent rounds. Then, from transition (2) and Properties 1 and 2, there is a non-zero probability of reaching a truthful state s^* in which (a) all workers in C are honest and (b) C is reputable. Once s^* is reached, all subsequent states satisfy these two properties (which define the set S), independently of whether the master audits (from transition (2) and (6), and Property 2).

Now, putting together Lemmas 1 and 2 we obtain the following theorem.

Theorem 1. *Having at least one worker altruistic or covered rational is necessary and sufficient to eventually reach a truthful set S from any initial state, and hence to guarantee eventual correctness, if the reputation type satisfies Properties 1 and 2.*

Observe that if there is no knowledge on the distribution of the workers among the three types (altruistic, malicious and rationals), the only strategy to make sure eventual correctness is achieved, if possible, is to cover all workers. Of course, if all workers are malicious (an unlikely situation, as shown in [8, 12, 13]) there is no possibility of reaching eventual correctness.

5   Simulations

This section complements our analytical results with illustrative simulations. The graphical representation of the data obtained captures the tradeoffs between reliability and cost and among all three reputation types, concepts not visible through the analysis. Here we present simulations for a variety of parameter combinations likely to occur in practice (extracted from [12,13]) and similar to our earlier work [10]. We have designed our own simulation setup by implementing our mechanism (the master's and the workers' algorithms, including the three types of reputation discussed above) using C++. The simulations were contacted on a dual-core AMD Opteron 2.5GHz processor, with 2GB RAM running CentOS version 5.3. We consider a total of 9 workers (that will be rational, altruistic or malicious in the different experiments). The figures represent the average over 10 executions of the implementation, unless otherwise stated (when we show the behavior of typical, individual realizations). The chosen parameters are indicated in the figures. Note that, for simplicity, we consider that all workers have the same aspiration level $a_i = 0.1$, although we have checked that with random values the results are similar to those presented here, provided their variance is not very large. We consider the same learning rate for the master and the workers, i.e., $\alpha = \alpha_m = \alpha_w = 0.1$. Note that the learning rate, as discussed for example in [27] (called step-size there), is generally set to a small constant value for practical reasons. Finally we set $\tau = 0.5$ (see [10]), $p_{\mathcal{A}}^{min} = 0.01$ and $\epsilon = 0.5$ in reputation type 2.

The contents of this section can be summarized as follows: In the next paragraph we present results considering only rational workers and, subsequently, results involving all three type of workers. We continue with a discussion on the number of workers that must be covered and how the choice of reputation affects this. Finally, we briefly show that our mechanism is robust even in the event of having workers changing their behavior (e.g. rational workers becoming malicious due to software or hardware error). An exhaustive account of simulation results is presented in [11].

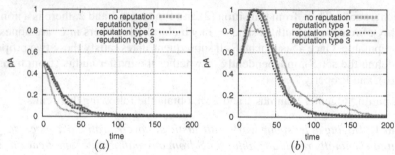

Fig. 1. Rational workers. Auditing probability of the master as a function of time (number of rounds) for parameters $p_\mathcal{A} = 0.5$, $\alpha = 0.1$, $a_i = 0.1$, $WB_\mathcal{Y} = 1$, $WP_\mathcal{C} = 0$ and $WC_\mathcal{T} = 0.1$. (a) initial $p_C = 0.5$ (b) initial $p_C = 1$.

Rational Workers. While the main reason for introducing reputation was to cope with malicious workers, as a first step we checked whether reputation improves the algorithm performance for rational workers only. In this case as we see in Figure 1, the first two reputation types give similar results as in the case of no reputation. Reputation type 3, on the other hand, seems to perform better in the case that the initial $p_C = 0.5$, while in the case of $p_C = 1$ the system has a slower convergence rate, but the auditing probability at the first 50 rounds is lower. This has to do with the fact that in type 3 the reputation of a worker is reinforced indirectly, what is directly reinforced depending on the workers honesty is the error rate. Our observations in Figure 1 reveals an interesting tradeoff: depending on whether the master has information on the workers' initial behavior or on the auditing that is willing to perform it will choose to use or not reputation type 3. From Figure 1 we can also see that the mechanism of [10] is enough to bring rational workers to produce the correct output, precisely because of their rationality. Although Figure 1 depicts the $p_\mathcal{A}$ of the master and not the p_C of the workers we have observed (see [11]) that for all the initial p_C studied, by the time the master's auditing probability reaches $p_\mathcal{A}^{min}$, the system had already reached eventual correctness.

Figure 2 allows to compare the behavior of the three reputation types, with reputation ratio defined as $\sum_{i \in W} \rho_i S_i / |W|$. Reputation type 1 leads rational workers to reputation values close to 1 (at a rate that depends on the value of the initial p_C). However, when type 2 is applied reputation takes values between (0,0.3). This happens because when the master catches a worker cheating, its reputation decreases exponentially, never increasing again. Reputation type 3, on the other hand, allows for dramatic increases and decreases of reputation. This is a result of the indirect way we calculate reputation type 3, as we mentioned above.

Different Types of Workers. We now move to our main case of interest and include different types of workers in our experiments. Figure 3 shows results for the extreme case, with malicious workers, no altruistic workers, and rational workers that initially cheat with probability $p_C = 1$. We observe that if the master does not use reputation and a majority of malicious workers exist, then the master is enforced by the mechanism to audit in every round. Even with a majority of rational workers, it takes a long time for the master to reach $p_\mathcal{A}^{min}$, if reputation is not used. Introducing reputation can indeed cope with the challenge of having a majority of malicious workers. For type 1, the larger the

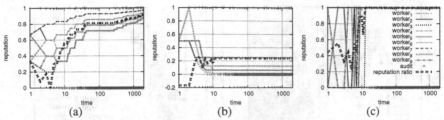

Fig. 2. Rational workers, for an individual realization with initially $p_C = 0.5$, $p_A = 0.5$, $WC_T = 0.1$, $WP_C = 0$, $\alpha = 0.1$ and $a_i = 0.1$. Left, reputation type 1. Middle, reputation type 2. Right, reputation type 3.

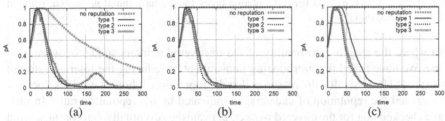

Fig. 3. Master's auditing probability as a function of time in the presence of rational and malicious workers. Parameters in all plots, rationals' initial $p_C = 1$, master's initial $p_A = 0.5$, $WC_T = 0.1$, $WP_C = 0$ and $\alpha = 0.1$, $a_i = 0.1$. In (a) 4 malicious and 5 rationals, (b) 5 malicious and 4 rationals , (c) 8 malicious and 1 rational.

number of malicious workers, the slower the master reaches p_A^{min}. On the contrary, the time to convergence to the p_A^{min} is independent of the number of malicious workers for reputation type 2. This is due to the different dynamical behavior of the two reputations discussed above. For reputation type 3, if a majority of rationals exists then convergence is slower. This is counter-intuitive, but as we mentioned before it is linked to the way reputation and error rate are calculated. On the other hand, with type 3, p_A is slightly lower in the first rounds. We thus conclude that reputation type 2 gives better results, as long as at least one rational worker exists and the master is willing to audit slightly more in the first rounds. We have checked that if rational workers are replaced by altruistic ones, the performance of the two reputation schemes improves, as expected.

Covering only a Subset of Rational Workers. In the previous paragraphs we considered only cases where the master was covering all workers, that is, $WB_y > a + WC_T$ for all workers. For the case with malicious workers, as explained in Section 4, this is unavoidable. But for the case with rational workers, as was argued in the same section, we may avoid covering all workers, a scenario which we now explore. In Figure 4(a) the extreme case of only one covered worker is presented. We see that with reputation type 1 our system does not converge, *which is consistent with the results of* Section 4. The master tolerates a significant percentage of cheaters (since $\tau = 0.5$), creating a very unstable system, where the probability of the master receiving the correct answer varies greatly. This occurs because by tolerating more cheaters, the master creates a

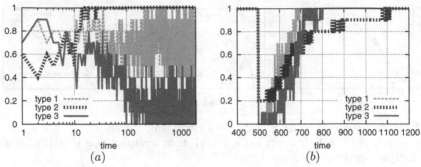

Fig. 4. Correct reply rate as a function of time. Parameters are initial $p_\mathcal{A} = 0.5$ $WC_\mathcal{T} = 0.1$, $WP_\mathcal{C} = 0$ and $\alpha = 0.1$, $a_i = 0.1$. (a) Reputation types 1 and 3 have initial $p_\mathcal{C} = 0.5$, while in type 2, $p_\mathcal{C} = 1$. (b) initial $p_\mathcal{C} = 1$.

system where the cheating probability of the uncovered workers spikes between zero and one. We have found that introducing punishment or reducing the tolerance do not fix this (see [11]). Basically, the reason is that the reputation of honest workers does not always exceed the reputation of cheaters, as indicated by the reputation ratio. In fact, we have checked that for the covered worker, $p_\mathcal{C}$ vanishes eventually, but for uncovered workers this is not the case: their $p_\mathcal{C}$ takes values usually below 0.6 but close to that value. Given that uncovered workers' $p_\mathcal{C}$ is greater than zero, it may occur that the master does not audit and a number of uncovered workers, with reputation higher than the rest, cheat. Because of this, even if the master maintains a high auditing probability, eventual correctness is not guaranteed.

For reputation type 3 our system does not converge, *which is consistent with the results of* Section 4 since reputation type 3 does not satisfy Property 2. Type 3 gives even worse results than type 1, since the correct reply ratio is always lower compared to type 1. Finally, Figure 4(a) shows that our system always converges using reputation type 2, as expected by the analysis in Section 4. A collection of elaborative simulation figures (see [11]) show that, the exponential dynamics of this reputation type works for all the parameters considered, and the master always reaches $p_\mathcal{A} = p_\mathcal{A}^{min}$ with eventual correctness. In addition, we see that the master also decreases its auditing cost, unlike the case of reputation type 1 where $p_\mathcal{A}$ goes to values close to one in order for the master to receive the correct reply with a high probability. We have also verified that when a majority (5 out of 9) of workers is covered, the system converges independently of the reputation type used.

Finally, for the sake of experimentation we checked that our mechanism reaches eventual correctness (with reputation type 2) by covering only 1 out of the 5 rational workers when the other 4 are malicious ones. The performance of the system to reach eventual correctness is similar to the analogous case where all workers are covered. Reputation type 1 and 3 have the same problems as before, whereas the fact that reputation becomes constant with type 2 allows rational covered workers to form a reputable set by itself and achieve fast eventual correctness.

Dynamic Change of Roles. As a further check of the stability of our procedure, we now study the case when after convergence is reached some workers change their type, possibly due to a software or hardware error. We simulate a situation in which 5 out of 9 rational workers suddenly change their behavior to malicious at time 500, a worst-case scenario. Figure 4 shows that after the rational behavior of 5 workers turns to malicious, convergence is reached again after a few hundred rounds and eventual correctness resumes. Notice that it takes more time for reputation type 2 to deal with the changes in the workers' behavior because this reputation can never increase, and hence the system will reach eventual correctness only when the reputation of the workers that turned malicious becomes less than the reputation of the workers that stayed rational. It also takes more time for reputation type 3 to deal with the changes in the worker' behavior. In the case of reputation type 1 not only the reputation of the workers that turned malicious decreases but also the reputation of the workers that stayed rational increases. Therefore, reputation type 1 exhibits better performance in dealing with dynamic changes of behavior than reputation types 2 and 3.

6   Conclusions and Future work

In this work we study a malicious-tolerant generic mechanism that uses reputation. We consider three reputation types, and give provable guarantees that only reputation type 2 (first presented here) provides eventual correctness in the case of covering only one altruistic or rational worker, something that is confirmed by our simulations. We show that reputation type 2 has more potential in commercial platforms where high reliability together with low auditing cost, rewarding few workers and fast convergence are required. This will help in developing reliable commercial Internet-based Master-Worker Computing services. From our simulations we make one more interesting observation: in the case when only rational workers exist and reputation type 3 (BOINC-like) is used, although the system takes more time to converge, in every round auditing is lower. Thus, reputation type 3 may fit better in volunteering setting where workers are most probably altruistic or rational and fast convergence can be sacrificed for lower auditing. In particular, our simulations reveal interesting tradeoffs between our reputation types and parameters and show that our mechanism is a generic one that can be adjusted to various settings. In a follow-up work we plan to investigate what happens if workers are connected to each other, forming a network (i.e, a social network through which they can communicate) or if malicious workers develop a more intelligent strategy against the system. Also the degree of trust among the players has to be considered and modeled in this scenario.

Acknowledgments.   This work is supported by the Cyprus Research Promotion Foundation grant ΤΠΕ/ΠΛΗΡΟ/0609(ΒΕ)/05, the National Science Foundation (CCF-0937829, CCF-1114930), Kean University UFRI grant, Comunidad de Madrid grants S2009TIC-1692 and MODELICO-CM, and MICINN grants TEC2011-29688-C02-01 and PRODIEVO, and National Natural Science Foundation of China grant 61020106002.

References

1. Abraham, I., Dolev, D., Goden, R., Halpern, J.Y.: Distributed computing meets game theory: Robust mechanisms for rational secret sharing and multiparty computation. In: Proc. of PODC 2006, pp. 53–62 (2006)

2. Aiyer, A.S., Alvisi, L., Clement, A., Dahlin, M., Martin, J., Porth, C.: BAR fault tolerance for cooperative services. In: Proc. of SOSP 2005, pp. 45–58 (2005)
3. Amazon's Mechanical Turk, https://www.mturk.com
4. Anderson, D.: BOINC: A system for public-resource computing and storage. In: GRID (2004)
5. Anderson, D.: Volunteer computing: the ultimate cloud. Crossroads 16(3), 7–10 (2010)
6. BOINC reputation,
   http://boinc.berkeley.edu/trac/wiki/AdaptiveReplication
7. BOINC stats, http://boincstats.com/en/forum/10/4597
8. BOINC user survey, http://boinc.berkeley.edu/poll_results.php
9. Bush, R.R., Mosteller, F.: Stochastic Models for Learning. Wiley (1955)
10. Christoforou, E., Fernández Anta, A., Georgiou, C., Mosteiro, M., Sánchez, A.: Applying the dynamics of evolution to achieve reliability in master-worker computing. Concurrency and Computation: Practice and Experience (2013); A preliminary version appears in Euro-Par (2012)
11. Christoforou, E., Fernandez Anta, A., Georgiou, C., Mosteiro, M.A., Sánchez, A.: Reputation-based Mechanisms for Evolutionary Master-Worker Computing. ArXiv (2013)
12. The Einstein@home project, http://einstein.phys.uwm.edu
13. Estrada, T., Taufer, M., Anderson, D.P.: Performance prediction and analysis of BOINC projects: An empirical study with EmBOINC. J. of Grid Computing 7(4), 537–554 (2009)
14. Fernández, A., Georgiou, C., Lopez, L., Santos, A.: Reliable Internet-based computing in the presence of malicious workers. Parallel Processing Letters 22(1) (2012)
15. Fernández Anta, A., Georgiou, C., Mosteiro, M.A.: Designing mechanisms for reliable Internet-based computing. In: Proc. of NCA 2008, pp. 315–324 (2008)
16. Fernández Anta, A., Georgiou, C., Mosteiro, M.A.: Algorithmic Mechanisms for Internet-based Master-Worker Computing with Untrusted and Selfish Workers. In: IPDPS 2010 (2010)
17. Golle, P., Mironov, I.: Uncheatable distributed computations. In: Naccache, D. (ed.) CT-RSA 2001. LNCS, vol. 2020, pp. 425–440. Springer, Heidelberg (2001)
18. Heien, E.M., Anderson, D.P., Hagihara, K.: Computing low latency batches with unreliable workers in volunteer computing environments. J. of Grid Computing (2009)
19. Josang, A., Ismail, R., Boyd, C.: A Survey of Trust and Reputation Systems for Online Service Provision. Decision Support Systems Journal 43(2), 618–644 (2007)
20. Kondo, D., Araujo, F., Malecot, P., Domingues, P., Silva, L.M., Fedak, G., Cappello, F.: Characterizing result errors in internet desktop grids. In: Kermarrec, A.-M., Bougé, L., Priol, T. (eds.) Euro-Par 2007. LNCS, vol. 4641, pp. 361–371. Springer, Heidelberg (2007)
21. Konwar, K.M., Rajasekaran, S., Shvartsman, M.M.A.A.: Robust network supercomputing with malicious processes. In: Dolev, S. (ed.) DISC 2006. LNCS, vol. 4167, pp. 474–488. Springer, Heidelberg (2006)
22. Korpela, E., Werthimer, D., Anderson, D., Cobb, J., Lebofsky, M.: SETI@home: Massively distributed computing for SETI. Computing in Science and Engineering (2001)
23. Maynard-Smith, J.: Evolution and the Theory of Games. Cambridge University Press (1982)
24. Sarmenta, L.: Sabotage-tolerance mechanisms for volunteer computing systems. Future Generation Computer Systems 18(4), 561–572 (2002)
25. Shneidman, J., Parkes, D.C.: Rationality and self-interest in P2P networks. In: Kaashoek, M.F., Stoica, I. (eds.) IPTPS 2003. LNCS, vol. 2735, pp. 139–148. Springer, Heidelberg (2003)

26. Sonnek, J., Chandra, A., Weissman, J.B.: Adaptive Reputation-Based Scheduling on Unreliable Distributed Infrastructures. IEEE TPDS 18(11) (2007)
27. Szepesvári, C.: Algorithms for Reinforcement Learning. Synthesis Lectures on Artificial Intelligence and Machine Learning. Morgan & Claypool Publishers (2010)
28. Taufer, M., Anderson, D., Cicotti, P., Brooks, C.L.: Homogeneous redundancy: a technique to ensure integrity of molecular simulation results using public computing. In: IPDPS (2005)
29. Vilaça, X., Denysyuk, O., Rodrigues, L.: Asynchrony and Collusion in the N-party BAR Transfer Problem. In: Even, G., Halldórsson, M.M. (eds.) SIROCCO 2012. LNCS, vol. 7355, pp. 183–194. Springer, Heidelberg (2012)
30. Yurkewych, M., Levine, B.N., Rosenberg, A.L.: On the cost-ineffectiveness of redundancy in commercial P2P computing. In: Proc. of CCS 2005, pp. 280–288 (2005)

State-Driven Testing of Distributed Systems

Domenico Cotroneo, Roberto Natella, Stefano Russo, and Fabio Scippacercola

Università degli Studi di Napoli Federico II
{cotroneo,roberto.natella,sterusso,fabio.scippacercola}@unina.it

Abstract. In distributed systems, failures are often caused by software faults that manifest themselves only when the system enters a particular, rarely occurring system state. It thus becomes important to identify these failure-prone states during testing. We propose a state-driven testing approach for distributed systems, able to execute tests in hard-to-reach states in a repeatable and accurate way. Moreover, we present the implementation and experimental evaluation of the approach in the context of a fault-tolerant flight data processing system. Experimental results confirm the feasibility of the approach, and the accuracy and reproducibility of tests.

Keywords: Experimental Dependability Assessment, Fault Tolerance, Fault Injection, Workload, Genetic Algorithms, State-based Testing.

1 Introduction

Distributed computing systems are today adopted in many business- and safety-critical domains, such as air traffic control, healthcare, and e-banking systems. In these contexts, it is mandatory to perform rigorous verification and validation activities to assure that distributed systems are highly dependable.

As a matter of fact, distributed systems tend to fail in subtle ways. These failures can be caused by software faults that manifest themselves only when the system enters a particular, rarely occurring system state [1,2,3]. Failure-prone states often evade testing since they only occur for specific sets of events and inputs (*workload*), as showed in several studies on testing of distributed systems, including filesystems [4,2], DBMSs [5], and multicast and group membership protocols [2,6,7]. Thus, identifying these states during testing is a challenging problem. This problem is exacerbated by the non-determinism of distributed systems, the need for minimal instrumentation of the system, and the presence of Off-The-Shelf (OTS) components whose internals are unknown. Past studies have mainly focused on exercising the system using synthetically generated workloads [8,9], or using workloads derived from performance benchmarks [5,10,11]. Nevertheless, these approaches are not effective at covering rare (*hard-to-reach*) states. Other approaches generate a workload from stochastic or non-deterministic models of the system, but do not scale well for complex systems [12,13].

In our previous paper [14], we proposed a workload generation technique that automatically drives the system's state towards the hard-to-reach states. In this

R. Baldoni, N. Nisse, and M. van Steen (Eds.): OPODIS 2013, LNCS 8304, pp. 114–128, 2013.

paper, we integrate this technique into a state-driven testing approach able to execute tests in hard-to-reach states, in a repeatable and accurate way. It does not rely on a detailed model of the system in terms of probabilities or time, and is suitable for testing the *actual implementation* of complex, OTS-based, distributed systems. Moreover, we present the implementation and experimental evaluation of the approach in the context of a fault-tolerant flight data processing system. The evaluation shows the feasibility of the approach and its ability to perform accurate and reproducible tests in the correct global state.

The paper is organized as follows. Section 2 presents past work on state-driven testing of distributed systems. Section 3 provides basic concepts and assumptions, and Section 4 describes our approach. Section 5 and 6 presents the experimental evaluation. Section 7 closes the paper.

2   Related Work

Studies on the verification of distributed systems can be classified into two broad classes: analytical-simulation studies and experimental ones. Experimental studies, in which our work is included, assess the actual implementation of a system by executing it. They include, for instance, fault injection methods, which assess fault tolerance mechanisms and algorithms through the deliberate injection of faults in the actual system or in a prototype [6].

In experimental studies, *model-based testing* (MBT) approaches are commonly adopted for generating test cases from a formal description of the system [15]. For instance, *conformance testing* and *FSM-based testing* approaches generate test cases aimed at covering the states of the model and at assuring that the system evolves as described by the model. Early approaches adopted graph-searching techniques to identify inputs able to drive the system along a given path in the state model [16]. Later approaches [12,13] extended these approaches to drive the system state in spite of random factors that change the system state in unpredictable ways. Nevertheless, the application of these approaches in complex systems is limited by scalability issues due to state space explosion, the need for a detailed model of the system, and by restrictive assumptions they implicitly make about the system: for instance, they only consider "stable" states, in which the system waits for inputs or events [17].

For these reasons, fault injection approaches do not rely on a system model to generate a workload. Some of them assess dependability by adopting a workload *representative* of the real workload that will be experienced during operation [5,10,11], in a similar way to performance benchmarks. In other cases, synthetic workloads are randomly generated, according to a high-level workload specification provided by the tester [18], for instance in terms of input probability distributions [8,19]. Moreover, most fault injection studies randomly inject faults during an experiment, repeating this process several times [20,21,7]. In these studies, the tested system states are limited to those exercised by the considered workload, and testers must manually tune the workload in order to bring the system in "hard-to-reach" states, from which they can perform tests.

Moreover, random injection can require a significant number of experiments to "hit" the system at hard-to-reach states.

More advanced fault injection approaches trigger the injection when a specific events occur in the system [22,23,2], and perform an a-posteriori state-based sampling of experiments to compute dependability measures [24,2]. For instance, Loki [2] considers the global state of a distributed system for triggering fault injection: to assure that a fault has been injected in a desired state, it performs an off-line analysis of execution traces and repeats the experiment if the injection has been triggered in a wrong state. These approaches still rely on a workload provided by the tester, either hand-written or using a representative workload, which does not assure that all important states are covered during testing. Compared to these works, our approach actively tunes the workload in order to cover a specific state specified by the tester, thus complementing experimental assessment approaches such as Loki. In our preliminary work [14], we discussed the issues behind state-driven workload generation in distributed systems, and first proposed the use of genetic algorithms to this aim. In this study, we integrate this technique in a comprehensive approach for fault injection testing.

3   Basic Concepts and Problem Statement

In the design of our approach for state-driven testing, we make practical assumptions about the architecture of the distributed system (DS) under evaluation. We consider DSs in which a set of *services* is exported by a *frontend* process, masking the complexity of the system to its users (Fig. 1). A client sends requests to the frontend process by means of one or more messages, the frontend interacts with the other processes of the DS and, once the computation has finished, replies to the client. In this context, a *workload W* consists of a set of service requests generated during an execution. This view of DSs applies to several systems, including orchestrated web services and three-tier web applications.

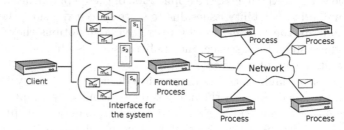

Fig. 1. The distributed system under test

In state-based testing of DSs, the workload is adopted to bring the system in a global state defined by the tester (*target state*), in order to let him to perform a test right after the DS has reached the target state, for instance by submitting a set of inputs or by injecting a fault while the system is in the target state. The aim of *State-Driven Workload Generation* (SDWG) is to search for a workload

$\overline{W}$ such that the likelihood that the system under test (SUT) spends at least a period τ in the target state is high enough to allow the accurate and reproducible test execution in the desired state. The τ period includes the time for allowing a *Test Executor* (TE) program to notice (by collecting event logs) that the DS has entered the target state during the experiment, and to perform the test after the state has been reached: for instance, in a fault injection experiment, the TE (e.g., a fault injection tool) will require a small amount of time to corrupt a message or to kill a process [22,2].

The state of an individual process in the DS is referred to as *local state*, whereas the *global state* of the DS, denoted with $s \in \mathcal{S}$, is the union of all the local states. The target state or, more in general, the *set of target states* $\mathcal{S}_G$, is a subset of the global states in which the tester aims to perform a state-driven test. Local and global states of the DS, and target states, should be defined by the tester before generating state-driven tests. We refer as the system model to a high-level specification (using a formalism such as Finite State Machines (FSM) or Petri Nets (PN)) by which the tester describes the set of global states, including the target states. The tester should define the system model on the basis of system requirements and its high-level design. The model should account for the state of local resources and the state of computation at each process, in addition to the testing goals. The model can be specified using well-known formalisms such as Finite State Machines (FSM) or Petri Nets (PN). Using the system model, the tester can focus workload generation on those target states that are important for testing. For instance, to test the effectiveness of a deadlock detection mechanism in a distributed DBMS, the system model should reflect the contents of the lock table and distributed transactions. The target state can be expressed in terms of markings of a PN, e.g., in terms of number of tokens in places that represent the ongoing execution of a transaction. More detailed examples of high-level system models adopted for fault injection testing of a distributed filesystem and a group membership protocol are provided respectively in [4,25].

For SDWG, we only require a relatively simple model that reflects the software under development at a high-level of abstraction, which should not necessarily provide details about low-level hardware and software layers of the system (e.g., OS, middleware). In particular, we do not require the system model to characterize the *time* and the *probability* of events in the system, but only the *relationship* between events and states: time, including communication and computation delays, can be unfeasible to characterize even in a probabilistic way, especially for complex distributed systems with third-party and OTS components, whose internals are unknown. Since the time of events are unknown, transitions in the system model are not timed, and only express the relationship between events and the state of the DS (according to [15], it is an *untimed, non-deterministic* and *operational* model). In our approach, the system model is used *after* the execution of the workload to obtain, from raw event logs of an execution, the sequence of states that the system has followed during the execution, and refine the workload based on this feedback.

4 A State-Driven Testing Approach

We are proposing an MBT approach composed by two phases, namely the *workload search phase* and the *testing phase*. The workload search phase is only briefly summarized in the following because it has been the focus of our previous work [14], while the testing phase is the main topic of this section. Fig. 2 shows the overall approach: the tester first *searches a workload*, using the *Workload Generator* (WG) [14], then performs the actual *testing* of the SUT using the workload $\overline{W}$ found in the previous phase. To find $\overline{W}$, the WG applies "candidate" workloads to the system, and evaluates whether such workloads bring the system in the target state. The WG determines if the system reached the target state collecting the system events during the execution, e.g., messages and outputs produced by processes, and analyzing them after the events have been "translated" into the history of global states traversed by the system.

In the testing phase, the tester links his module, the *Test Executor*, to the WG: the TE is responsible for executing tests, e.g., it may be deputed to inject a fault and to observe its effects on the SUT. The WG supplies again the workload $\overline{W}$ to the system, but here, during runtime, it triggers the Text Executor when it notices the occurrence of *test triggering conditions* (e.g., a specific sequence of messages sent within the system). These conditions are defined by the WG such that the likelihood that the test is performed in the target global state is maximized. This likelihood represents an evaluation of the accuracy and reproducibility of the test when using a given workload. Test reproducibility allows their re-execution after applying a fix, given that the fix does not impact the system model or the execution of $\overline{W}$. The likelihood is evaluated by the WG during the workload search phase, so the search can be stopped when it is high enough. Moreover, after the execution of a test, the WG framework checks whether the experiment has been conducted in the correct global state, in order to assure the correctness of results. The test is repeated in the unlikely case that the state of the test was not the desired one.

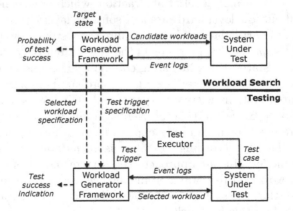

Fig. 2. Workflow of a test using the proposed approach

4.1   The Workload Search Phase

In the workload search phase (described in detail in [14]), the WG interacts with the DS under test in a closed-loop configuration. It exercises the DS with a workload, analyzes its behavior, and modifies the workload until a specified target state is reached. In this loop, the WG alternates an *on-line* phase, in which the DS is executed, and an *off-line* phase, in which the behavior of the DS is analyzed. The distinction between the off-line and on-line phases allows to reduce the intrusiveness of the WG, since only minimal information is collected during the on-line phase, and most of the processing for analyzing the system evolution and computing the workload occurs off-line. In the on-line phase, the WG executes for a fixed time period the DS with a candidate workload W. Then, in the off-line phase, the WG analyzes the behavior of the system through *event logs* collected during execution, and evaluates whether the target state has been reached. The off-line phase adopts a *Petri Net system model* (Section 3) to obtain, from raw event logs of an execution, the sequence of states that the system reached during the execution, and how much time has been spent at each state. Candidate workloads are iteratively generated and executed until the target state is reached with a given probability and for a given sojourn time.

Local events are collected at each process of the DS, timestamping them using local clocks. When the experiment is over, an *off-line synchronization algorithm* is executed to align the events of an execution on a single *global timeline* [2,26]. Off-line synchronization was preferred over on-line synchronization approaches, such as NTP, since on-line synchronization protocols exchange packets during the execution of the system and can thus interfere with its evolution. For each event, the algorithm estimates a *lower* and an *upper bound* of its timestamp, representing the *uncertainty interval* of the event. We showed in [27] that, when a PN system model is adopted, the global state of the system can be exactly identified in those periods where uncertainty intervals do not overlap.

A workload W leads the SUT to traverse one or more global states $s_k \in \mathcal{S}$, and sojourns in each of them for a finite time $d_k > 0$. The behavior of the SUT under a workload W is evaluated from a set of executions. The sequence of all the states traversed by the system in an execution under the workload W forms an *execution report* $r_i \in \mathcal{R}_W$, where each state traversal is denoted with (s_k, d_k).

The WG adopts a *workload configuration* $w_c \in W_c$ to represent workloads; w_c is a vector of parameters, representing the frequency and the type of requests to be sent, i.e., the workload to be generated. The tester should specify, for each parameter, a discrete set of allowed values (e.g., values uniformly distributed within a range). The WG explores, with the *WL Navigator* component, several combinations of such parameters to find a combination able to reach the target state. The parameters represent the periodicity of the messages exchanged with the DS and other customizable factors, such as the *delays* to introduce in the processes for increasing the likelihood of sojourning in the target state for long enough. The WL Navigator makes use of a *Genetic Algorithm* (GA) to search for a suitable w_c. It starts from a random configuration, and then generates new candidate workloads by randomly mutating and combining candidate solutions from a previous iteration, by replacing a parameter value of an existing solution with (i) another value from the set of allowed values, or (ii) a value of the same parameter taken from another workload. The quality (*fitness*) of w_c is evaluated using a *fitness function*, which takes into account the "distance" between the tentative solution and the target states, and the "continuity" of periods spent in the target state. Based on w_c, the WG generates the actual workload W, by acting as a client of the SUT using a *WL Feeder* component.

4.2   The Testing Phase

After the workload search, in which a state-driven workload has been found, the system is actually tested using the selected workload. During a test (Fig. 2), events logs are still collected, and they are analyzed at run-time to trigger the Test Executor when the WG notices that the SUT has reached the target state. However, due to delays in the transmission of events and to the lack of clock synchronization, the test could be triggered in a global state that is different than the desired target state. In order to avoid incorrect experiments, we perform off-line synchronization after the test, analyze the execution report, and evaluate whether the test has been triggered in the correct global state. If the execution report points out that the test was triggered in an incorrect or undetermined state, the experiment is discarded and must be repeated. This "optimistic" approach is based on the observation that *if the system sojourns in the target state for long enough, it is likely that the test will triggered in a correct global state*, which is also assumed by other testing and fault injection tools for distributed systems such as the Loki [2]. Therefore, we can expect that tests will be correctly triggered most of times, and that only a few experiments will be discarded, as the WG seeks for a workload that maximizes the sojourn time in the target state. In any case, the off-line analysis assures that incorrect experiments are discarded and do not affect the evaluation of the system.

An important issue that we noticed in a preliminary implementation of our approach is that, even if the workload brings the SUT in the target state for a long enough time, it often happens that, during the same execution, the DS enters the target state only for short periods: in such cases, the test would be incorrectly triggered since the system leaves the target state during the execution of the test. In other terms, during an execution, there can be many state traversals shorter than the required τ, and only a few traversals longer than τ, where τ is the time required for the execution of the test (see Section 3). To mitigate this issue and to improve the likelihood of triggering the test in the desired global state, we adopt the following test triggering mechanism: we avoid (incorrect) triggering when a target state is traversed only for a short time, by raising the trigger only when *a triggering-delay θ has been elapsed since the system entered in the target state*. Figure 3 shows an example of the whole process, based on a hypothetical system model with two places and two transitions. The test trigger specification consists of the following conditions: (i) the place p_2 should have at least one token, and (ii) the first condition should hold for at least a delay θ^*. During the testing phase, the WG collects events and updates its internal representation of the global state. When an event happens, a message is sent to the WG, which updates the system state and checks if the target state (e.g., the marking $\langle p_1 = 0, p_2 = 1 \rangle$) has been reached. If so, the θ-delay starts, and the test is triggered only after θ is elapsed. Therefore, traversals of the target states shorter than θ^* will be filtered out. The delay θ is selected by the WG as follows, by maximizing the probability of correct test execution. A test is correct (i.e., triggered in the correct global state) if, after the SUT has reached the target state and remains in that state for a period θ, it does not change state for an additional period τ to allow the Test Executor to perform the test. We estimate the *probability of test success* $pts_{S_{G},\tau}(\theta)$ by:

$$pts_{S_{G},\tau}(\theta) = \Pr\left(d_k \geq \theta + \tau \mid d_k \geq \theta \wedge s_k \in S_G\right) \cdot$$
$$\cdot \Pr\left(\exists e_k = (s_k, d_k) \in r_w : d_k \geq \theta \wedge s_k \in S_G\right) \quad (1)$$

where the first factor of the product represents the probability to stay in the target state s_k for $\theta + \tau$ given that the triggering delay has been elapsed, and the second factor represents the probability that the workload will bring the system in target state for a long enough period at least once during the experiment. These probabilities can be empirically estimated from the execution reports collected during the workload search. The value of the triggering-delay θ^* is selected by the WG by maximizing the *pts*:

$$\theta^* = \arg \max_{\theta \in [0;\theta_{max}]} pts_{S_{G},\tau}(\theta) \quad (2)$$

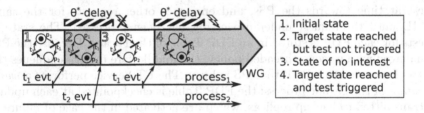

	1. Initial state
	2. Target state reached but test not triggered
	3. State of no interest
	4. Target state reached and test triggered

Fig. 3. Test triggering based on event logs and on a triggering delay

5 Case Study

We implemented and evaluated our approach within the Flight Data Processing System (FDPS) described in [14], and here summarized. FDPS is a distributed software developed in C++ which uses CARDAMOM, a fault-tolerant CORBA-compliant middleware. It is a part of an Air Traffic Control (ATC) system, in charge of managing and keeping up-to-date Flight Data Plans (FDPs).

The architecture of FDPS (Fig. 4) is composed by a Façade component, which acts as the frontend of the system and is replicated by the CARDAMOM Fault-Tolerance (FT) Service, and by a set of three Processing Servers (PSs), managed by the Load-Balancing (LB) Service. Service requests are delivered to the Façade by the middleware: the Façade forwards requests to a specific PS according to a round robin scheduler; once the requests are completed, they are sent back to the Façade, which disseminates the updated FDP through a Data Distribution Service (DDS) and replies to the clients.

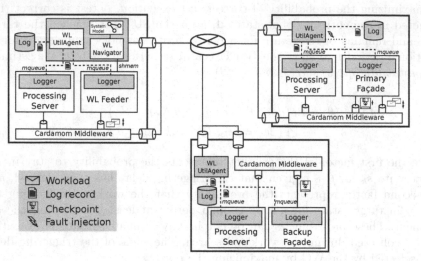

Fig. 4. The FDPS case study

Requests are associated to a specific flight track, which is identified by an *FDP-ID* number: for each FDP-ID, the Façade dispatches no more than one request at time towards the PSs, and enqueues other requests for the same FDP-ID until the request under processing has been processed. The state of requests for each FDP is stored in an FDP Table of the Façade. Because the PSs are managed with a mono-threaded policy, the middleware in turn enqueues the requests forwarded to a PS if that PS is busy. The FT Service performs a *warm replication* of the Façade process: the FDP Table is checkpointed at each update and transmitted to backup replicas, which are activated in the case of failure of the primary replica. In our experimental setup, the application is installed on a LAN of RHEL Linux PCs connected through a 100Mbps Ethernet network; the FDPS deploys 3 Processing Servers, one active Façade and one backup Façade replica. The hosts adopted for the experiments were configured by disabling services that could interfere with the FDPS and the WG. In particular, we had to disable the NTP synchronization service, which modifies the system clock and can affect our synchronization algorithm [14].

In a previous study [3], we adopted fault injection to assess the fault tolerance of the warm replication mechanism implemented in the FDPS based on the CARDAMOM FT Service. The warm replication mechanism should copy the state of the FDP Table to a backup replica, and its effectiveness can be affected by the amount of data that has to be copied to the backup replica (i.e., the number of requests enqueued by the Façade) and by requests sent to PSs (both under processing and enqueued by the middleware). In this case study, we perform fault injection experiments in different states of the FDPS, by taking into account the number of requests enqueued by the Façade and sent to PSs. We include in the system model of the FDPS (and thus in the definition of the global state) the number of requests enqueued at the Façade and at each PS. The system model, described in [28], was not included here due to space constraints.

In this experiment the workload configuration w_c has a pair of parameters for each FDP-ID i, namely T_{m_i} and D_i: the first one specifies the period between two consecutive requests sent by the client for the i-th FDP-ID; the second one represents a delay that is introduced in the PSs during the processing of requests for the i-th FDP-ID. The WL Feeder generates a *stream of requests* for each FDP-ID according to w_c. These parameters are communicated by the Workload Navigator to the Workload Feeder through a UNIX shared memory, whereas the Feeder transmits the delays D_i to the Processing Servers in the request messages. Fig. 4 also includes the implementation of our state-driven testing approach in the FDPS (the shaded boxes in the figure). The *Logger*s are small libraries linked to FDPS processes; the CORBA objects were instrumented to collect application events by invoking Loggers, which in turn send event logs to the *WL UtilAgent*s using UNIX message queues. We log events that represent transitions in the system model. In particular, we log the invocation of CORBA methods (by invoking the Logger at the beginning of the CORBA method), and the accesses to request queues (e.g., by invoking the Logger when a private method for enqueuing requests is called). Event logs are processed by the WG (both during the search and the testing phases), which are translated into a sequence of global states: for instance, when a client request for the FDP #1 is received, a new global state is added in the sequence of global states, with $A1 = 1$ in the marking of the PN (see [28]). As an alternative to instrumenting the SUT, the events required by the system model can be obtained by system logs, if available. The *WL UtilAgent*s are processes that perform all the tasks required by the Workload Generator, such as log collection and off-line log analysis, and by the Test Executor, such as triggering a test. For instance, in our experiments we used the WL UtilAgent to inject faults in the Façade. We adopted the *process crash* as fault model, which is often adopted to evaluate the fault tolerance of distributed systems [2,7]. In our setup, the Test Executor is a process that forces a process crash, by killing the Façade process using UNIX signals. It is important to note that our approach can be adopted for injecting arbitrary fault models, depending on the type of system and on evaluation goals.

In the search phase, we configured the fitness function ([14], eqq. 4, 5) with parameters $\alpha = 10.4$ and $\varepsilon = 24$. A workload configuration w_c represents an individual for the genetic algorithm, with $2 \cdot \#FDPs$ chromosomes (i.e., the parameters T_{m_i} and D_i). At each iteration, the GA generates a new population of individuals (where each population consists of 8 individuals) from an old population, by repeatedly applying the following two rules:

– two individuals are randomly selected, with a probability based on their fitness; with probability $c = 90\%$ (*crossover rate*), the two individuals are

split in two parts (at a random point of the vector) and mixed (*crossover*), thus obtaining a new pair of individuals;
- with probability $m = 35\%$ (*mutation rate*), each parameter of the newly generated individuals are replaced (*mutation*), by randomly selecting a new value according to a normal distribution centered around the old value.

6   Experimental Evaluation

We conducted a set of fault injection experiments on the FDPS, in order to evaluate the feasibility and effectiveness of the approach. In these experiments, we evaluate the ability and the speed of the WG to bring the system into a given target state, and the ability to correctly trigger fault injection while the system is in the target state. The target states are defined using a set of *constraints*, that is, conditions that the global state needs to satisfy: in the case of a system model based on Petri nets, the target states are represented by a set of conditions on the marking of the Petri net. If several global states satisfy the constraints, they are considered equally useful from the perspective of testing the DS. The Workload Generator is adopted for bringing the system in three different targets states, where each experiment introduces an additional constraint to the constraints of the previous experiment. Introducing additional constraints makes the search for a workload increasingly difficult, since each constraint reduces the set of target states. The experiments are defined as follows:

Experiment #1: The workload should bring the distributed system into a global state in which two out of three PSs are busy, and one out of three PS is idle. This condition is expressed by a constraint stating that the sum of tokens in the places WRK_i (where $WRK_i = 1$ if the i-th PS is busy, and 0 otherwise [28]) should be exactly 2:

$$\sum_{i=1}^{3} WRK_i = 2 \tag{3}$$

Experiment #2: In addition to the previous constraint (Equation 3), the workload should bring the system into a global state such that the Façade should have enqueued at least 6 and at most 30 requests in its FDP Table (Equation 4). Both constraints should hold at the same time in order to reach the target state. The second constraint states that the sum of tokens in the places A_j, representing the number of requests in the FDP queue j (with six FDP queues in total) [28], should be between 6 and 30 tokens:

$$6 \leq \sum_{j=1}^{6} A_j \leq 30 \tag{4}$$

Experiment #3: The set of target states is further restricted, by (i) including the constraint of experiment #1 (Equation 3), (ii) replacing the constraint of Equation 4 with the more restrictive condition of Equation 5, and (iii) adding the condition that there should be at least one request enqueued by the PSs (Equation 6). All three constraints should hold at the same time in order to reach the target state. Equation 5 states that each FDP queue j should have between at least 1 and at most 5 enqueued requests (instead, Equation 4 disregards how enqueued requests are distributed across FDP queues); in Equation 6, tokens in the place BF represent requests enqueued by the middleware for the PSs, which should be more than zero [28]:

$$1 \leq A_j \leq 5 , \qquad j = 1, 2, \ldots, 6; \tag{5}$$

$$BF > 0 \tag{6}$$

It is important to note that, even though the first and the third constraints (Equations 3 and 6) appear to be contradictory, it is in fact possible to satisfy them at the same time. These constraints state that there should be an idle Processing Server, while the other two PSs should be busy and have requests enqueued for them, i.e., the enqueued requests should not be forwarded to the idle PS. This condition is actually possible since the request scheduler selects the PS for an incoming request on a round-robin basis, regardless of whether the selected PS is busy and whether there are idle PSs. Therefore, this condition is hard-to-reach, but possible.

We imposed a minimum sojourn time in the target state of $\tau = 0.3s$, which is large enough to allow our Test Executor module to be triggered and to kill the Façade process. We fixed the number of FDP-IDs to six, and set the domains for the parameters of the workload configuration ranging from 500ms to 5s, with a step of 500ms. Finally, we set $|\mathcal{R}_W| = 3$.

In Fig. 5, we depict the sojourn time in the target state attained by generated workloads. At each iteration of the genetic algorithm, a generation (i.e., set of solutions) is obtained by mutating and combining solutions from the best solutions of the previous generation (on the basis of the fitness function). We evaluated, for each generation, the sojourn time attained by the best solution of each generation. In every experiment, the WG was able to find a workload able to bring the system into the target state for an uninterrupted time period of at least $1.5s$; the convergence to a "good" solution was very quick in the case of experiments #1 and #2, and in the case of experiment #3, which imposed more restrictive constraints to the target state, the algorithm converged after 14 iterations, which were executed in about 3 hours.

For each experiment, we selected the workload with the highest uninterrupted sojourn times across all generations of the search, and then we used that workload for fault injection experiments.

The table 1 shows the probabilities of correct test execution and compares them with the estimations obtained by $pts_{\mathcal{S}_G,\tau}(\theta^*)$ (eq. 1). The probabilities of correct test executions has been obtained by performing 100 fault injection experiments on the system for each target state, and by evaluating whether faults where injected in the correct global state: most of fault injection experiments

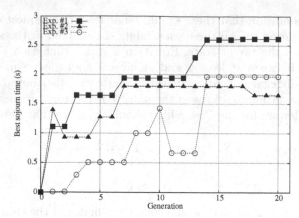

Fig. 5. Sojourn time of the best solution at each generation of the genetic algorithm

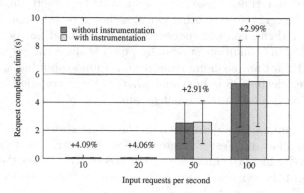

Fig. 6. Overhead of event log collection and processing

were correctly performed, with a probability of correctly reproducing the experiment of 60% in the worst case. In every case, the estimated probability of test success was close to the probability actually experienced during experiments, with a difference less than 10%. Since the probability of test success is high, it is likely that the test is performed in the correct state on the first try, or after a small number of repetitions.

Table 1. Probability of injecting a fault in the correct global state

	Exp. #1	Exp. #2	Exp. #3
Experimental test success probability	82.6%	82.9%	57.1%
Predicted test success probability	92.2%	75.0%	60.0%

We analyzed the overhead of our WG approach on fault injection experiments, by evaluating the performance loss due to our instrumentation. The only instrumentation we introduced was the logging of events in the FDPS, and the collection of these events in order to trigger the injection of faults. Fig. 6 shows the average response time of the FDPS over 20 executions, at different rates of input requests, when logging and collection are disabled and enabled, respectively. The increase of the request completion time is 4% in the worst case, and is less significant at higher rates of input requests. Therefore, the performance overhead incurred during execution with instrumentation can be considered negligible, meaning that the program behavior remains realistic during an experiment.

7 Conclusion

The global state is a major concern in the verification of a distributed system. State-driven testing of distributed systems proves to be challenging due to system complexity, the use of OTS components, the clock drift and the non-determinism of distributed systems. We proposed an approach for state-driven testing of complex distributed systems, that automates the search for a state-driven workload, and perform tests in a desired global state with probabilistic guarantees.

Acknowledgments. This work has been supported by the project "Embedded Systems in Critical Domains" (CUP B25B09000100007), by the TENACE PRIN Project (n. 20103P34XC) funded by the Italian Ministry of Education, University and Research, and by the Finmeccanica industrial group in the context of the project "Iniziativa Software CINI-Finmeccanica".

References

1. Lee, I., Iyer, R.: Faults, Symptoms, and Software Fault Tolerance in the Tandem GUARDIAN90 Operating System. In: Proc. Symp. on Fault-Tolerant Computing, pp. 20–29 (1993)
2. Chandra, R., Lefever, R., Joshi, K., Cukier, M., Sanders, W.: A Global-State-Triggered Fault Injector for Distributed System Evaluation. IEEE Trans. Parallel and Distributed Sys. 15(7), 593–605 (2004)
3. Natella, R., Cotroneo, D.: Emulation of transient software faults for dependability assessment: A case study. In: Proc. Eur. Dependable Comp. Conf., pp. 23–32 (2010)
4. Lefever, R., Cukier, M., Sanders, W.: An experimental evaluation of correlated network partitions in the Coda distributed file system. In: Proc. Intl. Symp. Reliable Distributed Systems, pp. 273–282 (2003)
5. Vieira, M., Madeira, H.: A dependability benchmark for OLTP application environments. In: Proc. 29th Intl. Conf. on Very Large Data Bases, pp. 742–753 (2003)
6. Arlat, J., Aguera, M., Amat, L., Crouzet, Y., Fabre, J., Laprie, J., Martins, E., Powell, D.: Fault injection for dependability validation: A methodology and some applications. IEEE Trans. Software Eng. 16(2), 166–182 (1990)
7. Meling, H., Montresor, A., Helvik, B., Babaoglu, O.: Jgroup/ARM: a distributed object group platform with autonomous replication management. Soft.: Pract. Exp. 38(9), 885–923 (2008)
8. Tsai, T., Hsueh, M., Zhao, H., Kalbarczyk, Z., Iyer, R.: Stress-Based and Path-Based Fault Injection. IEEE Trans. Computers 48(11), 1183–1201 (1999)
9. Kiskis, D., Shin, K.: SWSL: A synthetic workload specification language for real-time systems. IEEE Trans. Soft. Eng. 20(10), 798–811 (1994)

10. Duraes, J., Madeira, H.: Generic faultloads based on software faults for dependability benchmarking. In: Proc. Intl. Conf. Dependable Systems and Networks, pp. 285–294 (2004)
11. Kalakech, A., Kanoun, K., Crouzet, Y., Arlat, J.: Benchmarking the Dependability of Windows NT4, 2000 and XP. In: Proc. Intl. Conf. Dependable Systems and Networks, pp. 681–686 (2004)
12. Zhang, F., Cheung, T.Y.: Optimal transfer trees and distinguishing trees for testing observable nondeterministic finite-state machines. IEEE Trans. Soft. Eng. 29(1), 1–14 (2003)
13. Nachmanson, L., Veanes, M., Schulte, W., Tillmann, N., Grieskamp, W.: Optimal strategies for testing nondeterministic systems. ACM Soft. Eng. Notes 29(4), 55–64 (2004)
14. Natella, R., Scippacercola, F.: Issues and Ongoing Work on State-Driven Workload Generation for Distributed Systems. In: Vieira, M., Cunha, J.C. (eds.) EWDC 2013. LNCS, vol. 7869, pp. 96–110. Springer, Heidelberg (2013)
15. Utting, M., Pretschner, A., Legeard, B.: A taxonomy of model-based testing approaches. Software Testing, Verification and Reliability 22(5), 297–312 (2012)
16. Bourhfir, C., Dssouli, R., Aboulhamid, E., Rico, N.: Automatic executable test case generation for extended finite state machine protocols. In: Kim, M., Kang, S., Hong, K. (eds.) Testing of Communicating Systems. IFIP, pp. 75–90. Springer-Verlag US (1997)
17. Kerbrat, A., Jéron, T., Groz, R.: Automated test generation from SDL specifications. In: Proc. 9th SDL Forum, pp. 135–152 (1999)
18. Kiskis, D.L., Shin, K.G.: A synthetic workload for a distributed real-time system. Real-Time Systems 11(1), 5–18 (1996)
19. Weyuker, E.J., Vokolos, F.I.: Experience with performance testing of software systems: issues, an approach, and case study. IEEE Trans. Soft. Eng. 26(12), 1147–1156 (2000)
20. Arlat, J., Aguera, M., Crouzet, Y., Fabre, J., Martins, E., Powell, D.: Experimental evaluation of the fault tolerance of an atomic multicast system. IEEE Trans. Reliab. 39(4), 455–467 (1990)
21. Basile, C., Wang, L., Kalbarczyk, Z., Iyer, R.: Group communication protocols under errors. In: Proc. Intl. Symp. Reliable Distributed Systems, pp. 35–44 (2003)
22. Dawson, S., Jahanian, F., Mitton, T., Tung, T.: Testing of fault-tolerant and real-time distributed systems via protocol fault injection. In: Proc. Fault Tolerant Computing Symp., pp. 404–414 (1996)
23. Hoarau, W., Tixeuil, S.: A language-driven tool for fault injection in distributed systems. In: Wksp. Grid Comp., pp. 194–201 (2005)
24. Helvik, B.E., Meling, H., Montresor, A.: An approach to experimentally obtain service dependability characteristics of the Jgroup/ARM system. In: Dal Cin, M., Kaâniche, M., Pataricza, A. (eds.) EDCC 2005. LNCS, vol. 3463, pp. 179–198. Springer, Heidelberg (2005)
25. Joshi, K.R., Cukier, M., Sanders, W.H.: Experimental evaluation of the unavailability induced by a group membership protocol. In: Bondavalli, A., Thévenod-Fosse, P. (eds.) EDCC 2002. LNCS, vol. 2485, pp. 140–158. Springer, Heidelberg (2002)
26. Poirier, B., Roy, R., Dagenais, M.: Accurate offline synchronization of distributed traces using kernel-level events. ACM SIGOPS Operating Systems Review 44(3), 75–87 (2010)
27. Scippacercola, F.: State-Driven Workload Generation in Distributed Systems. Master's thesis, Università degli Studi di Napoli Federico II (2012)
28. Cotroneo, D., Natella, R., Russo, S., Scippacercola, F.: State-driven testing of distributed systems: Appendix. Technical report (2013),
    http://www.mobilab.unina.it/techreports.html

Self-stabilizing Resource Discovery Algorithm*

Seda Davtyan[1], Kishori M. Konwar[2], and Alexander A. Shvartsman[1]

[1] Department of Computer Science & Engineering,
University of Connecticut, Storrs CT 06269, USA
{seda,aas}@engr.uconn.edu
[2] University of British Columbia, Vancouver BC V6T 1Z3, Canada
kishori@interchange.ubc.ca

Abstract. Massive distributed cooperative computing in networks involves marshaling large collections of network nodes possessing the necessary computational resources. In order for the willing nodes to act in a concerted way they must first discover one another. This is the general setting of the Resource Discovery Problem (RDP). There are solutions for this problem that achieve impressive efficiency in the absence of failures, however, their correctness and performance cannot be guaranteed in the presence of failures. In practical environments it is important to have solutions that can cope with intermittent failures, and, in particular to design self-stabilizing algorithms for the problem. This paper presents a self-stabilizing algorithm that solves RDP in a deterministic synchronous setting. The approach is formulated in terms of evolving *knowledge graphs*, where vertices represent the participating network nodes, and edges represent one node's knowledge about another. Ideally, the diameter of such a graph is one, i.e., each node knows all others. The algorithm works in rounds as it evolves the knowledge graph by nodes sharing knowledge through gossip messages with the goal of reducing its diameter. We prove that the algorithm is *self-stabilizing*, that is, the algorithm is able to tolerate arbitrary perturbations in the nodes' local states and is guaranteed to solve the problem once such failures subside. We show that the algorithm has stabilization time of $O(\log D)$, and it takes at most $2\log D + 10$ complete round to stabilize, where D is the diameter of the initial knowledge graph. The corresponding message complexity is $O(|V|^2 \cdot \log D)$, where V is the set of participating nodes.

Keywords: Resource Discovery, Self-Stabilization, Distributed Algorithm.

1 Introduction

A large collection of networked computers may need to cooperate in implementing a distributed system, for example, to provide a shared data service, or to perform a set of tasks. The necessary first step in such settings is to discover the relevant resources in the network. This step can be formulated as the Resource Discovery Problem, where each willing resource must find all other available resources. This problem was introduced by Harchol-Balter, Leighton, and Lewin [7]

---

* This work is supported in part by the NSF award 1017232.

R. Baldoni, N. Nisse, and M. van Steen (Eds.): OPODIS 2013, LNCS 8304, pp. 129–144, 2013.
© Springer International Publishing Switzerland 2013

in the context of an application at Akamai Technologies with the motivation to build an Internet-wide content-distribution system that would speed up the access to web pages of major content providers. Before the computing nodes start cooperating in implementing the service, they need to find each other. Similar problems appear in peer-to-peer Internet systems where a large number of users share files without having to rely on centralized servers. Such systems are highly dynamic, with nodes constantly joining and leaving the network, making it desirable to efficiently discover the nodes that are willing to cooperate.

In studying message-passing algorithms for such problems, the commonly used efficiency metrics are the time and communication complexities. Kutten, Peleg, and Vishkin [12] provided a very efficient deterministic algorithm for the problem. However it does not provide strong fault-tolerance properties and does not deal with dynamic situations, and so its correctness and performance cannot be guaranteed in the presence of failures. The authors suggested that in order for their algorithm to handle dynamic networks their algorithm could be re-run from time to time. However, this is not easy because of the associated problem of *detecting termination without any a priori knowledge of the number of the participating nodes*; this is referred to as *Lipton's question*.

Our goal is to design algorithms that are able to deal with intermittent failures, and in particular we are interested in *self-stabilizing* solutions. Here the algorithm must automatically bring a system into a legitimate state in spite of transient failures. The self-stabilization requirement is that a legitimate state is reached from an arbitrary state in a finite time, cf. [4,5,15]. Our recent work [3] presented a self-stabilizing solution for the Resource Discovery Problem in synchronous settings. The stabilization time of the algorithm is linear in the diameter of the graph that represents the initial collective knowledge of the nodes about other nodes. Here we aim to substantially reduce the stabilization time.

Contributions. We present a self-stabilizing algorithm that solves the Resource Discovery Problem (RDP) in deterministic synchronous settings. The solution is formulated in terms of evolving *knowledge graphs*, where vertices represent the participating nodes, and edges represent one node's knowledge about another. Ideally, the diameter of such a graph is one, i.e., each node knows all others. We assume that initially each node is aware of only one neighbor other than itself, with whom it is able to communicate directly. Starting with this knowledge, nodes share their knowledge through gossip messages, allowing the nodes to discover one another and to reduce the graph diameter. We specify the algorithm and prove that it is self-stabilizing, that is, the algorithm is able to tolerate arbitrary perturbations to the nodes' local states so that it is guaranteed to solve the problem once such failures subside. We show that the algorithm has stabilization time $O(\log D)$, where D is the diameter of the initial knowledge graph. The corresponding message complexity is $O(|V|^2 \cdot \log D)$, where V is the set of participating nodes; this set is determined by the environment and it is unknown to the nodes. We now detail the setting and our contributions.

Let $G = (V, E)$ be the directed graph induced by the fixed initial knowledge of the nodes, i.e., E contains the edges (v, u) exactly when node v knows about node

u. We assume that the initial connectivity graph G is at least weakly connected and that initially each node has the knowledge of only one other node. As nodes communicate they learn about other nodes, and we model the global knowledge as the evolving connectivity graph. Following [12] we say that an algorithm solves RDP if it establishes and maintains the global state where there exists a node $v \in V$, called the root node, such that every node $u \in V$ recognizes v as the root node, and moreover v knows all nodes in V. Additionally we require that once the root node v knows V, every other node $u \neq v$ also learns V.

The nodes communicate using point-to-point messages. Sending (or multicasting) messages requires that the sending node has the knowledge of the destination nodes. The communication is synchronous in the sense that there is a known upper bound d on message delays; if a message is sent to a node in V, then it is delivered within d time units. Nodes have access to synchronous timers that can be used to implement message time-outs. Local computation takes negligible time relative to d. We do not assume that all nodes begin participating in the computation simultaneously; instead we allow the nodes to join the computation at arbitrary times. At a high level, the computation is structured in terms of synchronous rounds, however the activities within each round are not synchronized across the nodes.

The nodes are subject to arbitrary perturbations to their local (volatile) states; this includes arbitrary patterns of crash and restart events that occur in matched pairs, with the associated corruption of local states. The static code of each node, its constants, and the clock are incorruptible. All other variables are subject to corruption. Moreover, a corrupted variable may contain a value that is syntactically *indistinguishable* from a valid value. This is in contrast with some works in self-stabilization, where failures cause erasures of variable values, making such failures easily detectable, cf. [6]. Other works, e.g., [14], assume that any node identifier must represent an actual node in the system. Finally, we allow the adversary to corrupt messages in transit.

This work makes the following contributions.

1. We formulate models of computation and adversity, and we formally define the resource discovery problem (RDP) for our network setting. We formalize the properties of self-stabilizing solutions (in terms of the closure and the convergence conditions [2]). Our formalization enables one to reason rigorously about algorithms solving the problem.

2. We present an algorithm for the resource discovery problem, where we use the Timed Input/Output Automata formalism [8] to precisely specify its behavior. The algorithm works in rounds and it handles node joins, transient state and channel corruptions, and crash/restarts of nodes. (Note that if a node crashes permanently, thus possibly partitioning the connectivity graph, the algorithm solves the problem for every connected component.)

3. We formulate an invariant that implies that RDP is solved. We rigorously prove the self-stabilization properties of the algorithm: (*a*) in the absence of failures the invariant is maintained once it is established, and (*b*) when the failures subside, the invariant is eventually established. We reason about the performance of the algorithm and show that its stabilization time is $O(\log D)$, and that it takes at most $2 \log D + 10$ complete rounds to stabilize; this asymptotically meets the lower bound [7]. The associated message complexity is $O(|V|^2 \cdot \log D)$.

4. We consider the overall formalization, the formal treatment of the algorithm specification, and the rigorous reasoning about its properties to be an important contribution. We believe that our approach can provide valuable tools for methodical study of self-stabilizing algorithms.

Related Work. Harchol-Balter, Leighton, and Lewin [7] presented several algorithms for the resource discovery problem; their randomized algorithms have time complexity $O(\log^2 n)$ and message complexity $O(n \log^2 n)$, both with high probability, where n is the number of participating nodes. Law and Siu [13] gave a randomized algorithm for strongly connected initial graphs; one variant of the algorithm has time complexity $O(\log n)$ and message complexity $O(n^2)$, and another variant has time complexity $O(\log^2 n)$ and message complexity $O(n)$.

Kutten, Peleg, and Vishkin [12] gave a deterministic algorithm for RDP; its time complexity is $O(\log n)$ and message complexity is $O(n \log n)$. Kutten and Peleg [11] extended [12] to asynchronous networks and gave an algorithm with time $\Delta T + O(\log n)$, where ΔT is the difference between the wake-up times of the last and first vertices to be awakened; the message complexity is $O(n \log n)$.

Abraham and Dolev [1] provide upper and lower bounds for the asynchronous RDP and proved $\Omega(n \log n)$ message complexity lower bound when the size of the network is unknown. When each node knows the size of the connected component they provide an algorithm with message complexity $O(n\alpha(n, n))$, where $\alpha(n, n)$ is the inverse of the Ackermann's function.

Konwar et al. [9] considered RDP in a static synchronous setting and studied it under different assumptions about the ability of the nodes to communicate. They showed lower and upper bounds on the number of rounds needed to solve RDP. Konwar et al. [10] considered dynamic settings where the set of participants changes over time. They studied the number of communication rounds needed to solve the problem under a variety of assumptions about joins and failures.

Dolev and Herman [6] pursued a super-stabilizing approach to designing algorithms that maintain topological structures (e.g., a spanning tree) in the presence of perturbations. Nor, Nesterenko, and Scheideler [14] consider a self-stabilizing algorithm for skip-list construction in asynchronous networks. They prove that in their model one must constrain the states from which self-stabilizing solutions can be constructed: the state information has to form a weakly connected graph and it must only contain identifiers that are present in the system.

We note that, while it may be possible to adapt some existing self-stabilizing algorithms (e.g., [6] and [14]) to solve RDP, our algorithm, to the best of our knowledge, is the first to handle state perturbations that result in corruptions that can be indistinguishable from valid states.

Document Structure. Section 2 describes models of computation and failures, and self-stabilization properties. In Section 3 we present our algorithm.

In Section 4 we prove its correctness and self-stabilization. We conclude in Section 5. For paucity of space, additional technical details and proofs of some lemmas are given in the addendum available from the authors.

2   Models and Definitions

Model of Computation. We consider a universe of processes, with unique identifiers from a well-ordered set U. Let $V \subseteq U$ be the subset of processes that participate in the computation; this set is chosen by the environment. We let v_0 stand for $\min\{v : v \in V\}$. The set V, its cardinality, and v_0 are unknown to the processes, but each process in V is aware of one other process in V.

The processes communicate over a fully connected synchronous network. There is a known upper bound d on message delays. If a node expects a message from another node and the message is sent, then it is delivered within d time units. Nodes have access to local timers that can be used to implement message timeouts. Local computation takes negligible time relative to d.

We define a *round* to be some constant period of time sufficient for a process to send/multicast messages, to perform some local computation, and to accept any incoming messages. Let t be a time duration sufficient for implementing a round; t is established at compile time with the knowledge of the delay upper bound d. For our purposes it suffices to set t to $2d$.

The round structure provides only a coarse notion of synchrony. Distinct processes may execute *different* sequences of instructions during rounds, and the algorithm *cannot* assume that the individual instructions at different processes within concurrent rounds are synchronized. Lastly, we do not assume that all processes begin participating in the computation simultaneously; instead we allow the processes to join the computation at arbitrary times.

Failure Model. The processes are subject to *transient failures*. A transient failure is an event that corrupts the state of the system, but it does not change the algorithmic behavior of the system: the static code of each process and any constants are incorruptible. A failure may *arbitrarily* perturb state variables, including the program counter. A corrupted variable may contain a value that is syntactically *indistinguishable* from a valid value (this is in contrast with some works in self-stabilization, where failures cause erasures of variable values, thus making such failures detectable, cf. [6], or where it is assumed that a state contains only identifiers of the nodes that are present in the system, e.g., [14]). Thus, in our model it is possible for a state variable to contain incorrect information even though it appears to hold a valid value of a correct type. Messages in transit can also be *arbitrarily* corrupted.

Local States, Configurations, and Transitions. The *local state* of a process consists of the values of its variables and its program counter. We denote by s_v the state of node v. A *configuration* is a cross product of the local states.

Definition 1. *A system S is a triple (C, A, τ), where C is a set of configurations, A is a set of actions, and τ is a transition function $\tau : C \times A \to C$. An execution of S is a sequence $c_0, a_0, c_1, a_1, c_2, \ldots$ such that for all $i \geq 0$, $\tau(c_i, a_i) = c_{i+1}$.*

We denote some transition from configuration c_i to c_{i+1} by $c_i \xrightarrow{\tau} c_{i+1}$ and we let $c \xrightarrow{*}_{\tau} c'$ stand for the fact that c' can be reached from c by zero or more transitions. We denote the state variable X of node v in configuration c by $c.X_v$.

Self-stabilization. Self-stabilization is the ability of a system to recover from transient failures following their cessation. The impact of a failure is that the transition from configuration c to configuration c' may not obey the transition function τ, that is, a failure may cause $c' \neq \tau(c)$. In addition to local state corruption we assume that a system can start in any configuration. In designing solutions resilient to transient failures we use self-stabilization techniques, formalizing self-stabilization in terms of closure and convergence properties (cf. [2]).

Definition 2. *(Self-stabilization) Let problem P be to establish and maintain invariant $\psi()$, given as a predicate on configurations. System $S = (C, A, \tau)$ is a self-stabilizing solution for problem P, if the following two conditions hold:*
Closure: $\forall c \in C, \forall a \in A : \psi(c) \implies \psi(\tau(c, a))$, *i.e., τ maintains the invariant.*
Convergence: $\forall c \in C : \exists c' \in C : c \xrightarrow{*}_{\tau} c' \wedge \psi(c')$, *i.e., $\psi()$ can be established in the absence of failures.*

Resource Discovery Problem. We let each process v have a constant $nb_v \in V$, where $v \neq nb_v$, representing the knowledge of node v of some other node (a neighbor). This induces a directed graph.

Definition 3. *Given the set V and nb_v for all $v \in V$, we define the **connectivity graph** as the directed graph $G = (V, E)$, where $E = \{(u, v) : nb_u = v\}$.*

The connectivity graph is at least *weakly-connected*, representing the assumption that any process has the knowledge of at least one other process (as in the original formulation in [7]). Each process v has three local variables, $prt_v \in V$, $C_v \in 2^V$, and $world_v$, where $prt_v = u$ means that v considers u to be its *parent*, $u \in C_v$ means that v considers u to be its *child*, where C_v is the set of all children of v, and finally $u \in world_v$ means that v knows u. We now define our problem.

Definition 4. *Given the weakly-connected graph G, the **Resource Discovery Problem** (RDP) is to establish and maintain the following invariant on configurations: $(\exists v \in V : (C_v = V) \wedge (\forall u \in V : prt_u = v)) \wedge (\forall u \in V : world_u = V)$, that is, (1) there exists a node $v \in V$ such that $C_v = V$, and (2) for every node $u \in V$ we have $prt_u = v$, and (3) for every node $u \in V$ we have $world_u = V$.*

For convenience we let $G^u = (V, E^u)$ be the undirected graph induced by $G = (V, E)$, called the *initial knowledge graph*. Let D be the diameter of G^u and $dist(u, v)$ be the length of the shortest path from node u to v in G^u.

Measures of Efficiency. We assess the efficiency of the algorithm in terms of *stabilization time* and *stabilization message complexity*. The stabilization time is measured in terms of the worst case number of rounds following the cessation of perturbations needed to establish the resource discovery invariant.

Message complexity deals with the number of point-to-point messages, where in the case of multicast, each instance of multicast is assessed as the number of the resulting point-to-point messages. The stabilization message complexity is measured in terms of the worst case number of point-to-point messages sent among the *participants* to establish the resource discovery invariant following the cessation of perturbations. Note that local state corruptions may cause messages to be sent to an *arbitrary* subset of processes in U. In assessing stabilization message complexity we charge to the environment any messages sent by an algorithm prior to the cessation of perturbations, and any messages sent to bogus destinations as the result of state corruptions. This is because the adversary may cause an arbitrary number of messages sent in each round; in particular, in the case when $|V| = o|U|$, bogus messages may dominate message complexity, rendering any algorithm inefficient. In the analysis we show that, following the cessation of perturbations, after at most two complete iterations no messages are sent to bogus destinations.

Programming Notation. We use *Timed Input/Output Automata* (TIOA) [8] formalism to specify and reason about our algorithm. A timed automaton is a labeled state transition system. The state of the automaton is defined by its state variables. The discrete transitions of the automaton are defined in terms of *actions*, where each action is of the type *input, output,* or *internal.* The state of the timed automaton may change in two ways: by *discrete transitions* that change the state atomically, and by *trajectories* that describe the evolution of the state over intervals of time. The overall system is composed of the automata for all processes and its state is composed of the states of all automata.

The automata must be input-enabled and must not block time passage. A timed automaton *executes* by performing a sequence of alternating trajectories and discrete transitions, in which the states match up properly. We consider only executions where during any finite time period no infinite number of actions occur. We also consider only *fair* executions, where during each (algorithm-specific) round every locally-controlled enabled action (i.e., internal and output actions) occurs by the end of that round, and for every message sending action that is enabled at the beginning of the round the corresponding message receiving action occurs before the end of the round.

Communication Primitives. Nodes communicate via multicast, where each multicast results in a point-to-point message from the source to each destination. A multicast is invoked using the $\mathsf{msend}(m, I)_i$ action, where m is the message, I is the set of destinations, and i is the node invoking the multicast. Multicast messages are received through the $\mathsf{mrecv}(m, u)_i$ action, where m is the message, u is the source node, and i is the node receiving the message. Actions $\mathsf{msend}(m, I)_i$ and $\mathsf{mrecv}(m, u)_i$ are implemented in a straightforward way using point-to-point send/receive. We denote by $Channel_{i,j}$ the conventional synchronous channel from node i to node j (the implementation is not specified here). We assume that each $Channel_{i,j}$ contains state variable $S_{i,j}$ storing messages m in transit from node i to node j.

Modeling State and Message Corruption. The adversary can perturb the state of any node and corrupt any message in transit. We model state corruption at node i by means of the action $\mathsf{perturb}_i$ that is always enabled and whose effects contain the HAVOC command (borrowed from Lampson's SPEC language) that arbitrarily changes the state of node i: "$\mathsf{perturb}_i$: Effect: HAVOC".

Action $\mathsf{corrupt}(m, m')_{i,j}$ models the corruption of a message in transit: "$\mathsf{corrupt}(m, m')_{i,j}$: Precondition: $m \in S_{i,j}$; Effects: $S_{i,j} \leftarrow (S_{i,j} - \{m\}) \cup \{m'\}$," i.e., any message m in transit (in $S_{i,j}$) can be replaced by some message m'.

While such explicit modeling of message corruption is straightforward, doing so would complicate the reasoning about our algorithm. However, we can model message corruptions in terms of state perturbations. In fact, for any execution that corrupts messages there is an execution that does not corrupt messages and instead suitably perturbs local states, so that the two executions are indistinguishable for any node. (The explicit construction is given in the addendum.)

3   Description of Resource Discovery Algorithm *RDS*

The algorithm, which we call RDS, has an iterative structure consisting of two synchronous rounds. We refer to the first round as the gossip phase and to the second round as the confirm phase. In both phases nodes propagate information to other nodes, while in the confirm phase nodes additionally validate the identities of the nodes contacted in the gossip phase. The unique root is ultimately discovered as the node with the smallest identifier. Each node computes a local minimum based on its knowledge and it considers the node with the smallest identifier to be its parent. Conversely, any node includes among its children every node that considers it to be the parent. Each node also maintains knowledge about neighbors based on the initial graph G. The communication takes the form of a constrained "gossip," where in the gossip phase a node multicasts to its parent, children, a neighbor in G, and to all other nodes it discovers, and in the confirm phase each node responds to the messages received in the gossip phase and to its neighbors. Thus, a node receiving such responses validates the identities of the nodes it contacted in the gossip phase. Because failures can corrupt the state of any node, if a node does not hear from its parent during an iteration it decides that something is wrong and resets its state.

The behavior of each node $i \in V$ is specified as a timed I/O automaton, called RD_i. The specification in Figure 1 defines data types, constants, signature, and state variables, and Figure 2 contains the definition of the transitions and the trajectory. The full system, called RDS, is the composition of automata RD_i for $i \in V$, the multicast implementation, and the $Channel_{i,j}$ automata for $i, j \in V$.

Algorithmic Details. We now detail the state and operation of RD_i. When a state variable x appears outside of the scope of its definition we use notation x_i. The main variables are $active_i$, C_i, $world_i$, $Nbrs_i$, and prt_i. Boolean $active_i$ indicates whether node i is active or not, set C_i contains the children of node i, set $world_i$ contains the universe known to i, set $Nbrs_i$ contains the identifiers of the nodes that i considers to be neighbors in G^u, lastly prt_i is the identifier of the node that node i views as its parent. The remaining variables are $phase_i$,

Data-types:
  U, the set of node identifiers M, the set of messages

Constants:
  $nb : U$ outgoing neighbor of i $t : real > 0$

Derived Constants:
  $\widehat{N} = \{i\} \cup \{nb\}$

Signature:
  Input: Output: Internal:
    $\mathsf{mrecv}(m, u)_i,\ m \in M,\ u \in U$ $\mathsf{msend}(m, I)_i,\ m \in M,\ I \subset U$ $\mathsf{restart}_i$
    join_i reset_i
    $\mathsf{perturb}_i$ $\mathsf{end\text{-}round}_i$

State:
  $active : bool$ $prt : U$ $C : 2^U$ set of children of i
  $phase : \{\mathsf{gossip}, \mathsf{confirm}\}$ $R : 2^U$ $New\_C : 2^U$
  $clock : real$ $Dest : 2^U$ $Nbrs : 2^U$ set of all neighbors of i
  $do\_msend : bool$ $world : 2^U$

Fig. 1. Data types, signature, and state of RD_i at node i for $i \in V$

New_C_i, do_msend_i, R_i, and $Dest_i$. Here $phase_i$ controls the phase (gossip or confirm). Set New_C_i is used to keep C_i up to date. Boolean do_msend_i enables multicast exactly once in each round. Set R_i contains the identifiers of all nodes that contacted node i in the current iteration. And finally set $Dest_i$ contains the identifiers of the target nodes for multicast in the current phase.

We next describe the transitions. The environment may activate node i by using input action join_i, and it may disable and/or corrupt the state of node i by means of input action $\mathsf{perturb}_i$, where HAVOC assigns arbitrary values to the state variables. If HAVOC sets $active$ to $false$, the action models a crash of the node. Internal action $\mathsf{restart}_i$ is always enabled, modeling the assumption that each node $i \in V$ is eventually active. Nodes gossip by sending and receiving messages through actions msend_i and mrecv_i (detailed later).

Local operation of a node is structured in terms of rounds. Variable $clock$ represents the time of the synchronous system. Recall that failures cannot change the synchronous nature of the system, and thus this is the only variable that is not affected by transient failures. The variable records the passage of time consistently at all nodes: the trajectory evolves $clock$ at the same rate as real time $(d(clock) = 1)$. We establish the compile-time constant t to be sufficient for a node to multicast outgoing messages, to perform the needed local computation, and to accept any incoming messages (this constant is readily obtained from the structure of the algorithm and from the knowledge of the worst case message delivery delay d). The constant t is used to control the duration of rounds. The trajectory specification says that time "stops" when $clock \% t = 0$ for an active node. The value of $clock$ is used to determine whether an active node is in the gossip or confirm phase. When $clock \% 2t = 0$ the node enters the gossip phase, and when $clock \% 2t = t$ the node enters the confirm phase.

A round ends with either action $\mathsf{end\text{-}round}_i$ or action reset_i. Action $\mathsf{end\text{-}round}_i$ is enabled every t time units when $clock \% t = 0$ at the conclusion of each round if the node's state suggests that its parent is active (this does not mean that

Transitions:

Input join_i
  Effect:
    $active \leftarrow true$

Input perturb_i
  Effect:
    HAVOC

Output $\text{msend}(\langle N, p, ch, W \rangle, I)_i$
  Precondition:
    $active$
    do_msend
    $N = \widehat{N}$
    $p = prt$
    $ch = C$
    $W = world$
    $I = Dest$
  Effect:
    $do\_msend \leftarrow false$

Input $\text{mrecv}(\langle N, p, ch, W \rangle, s)_i$
  Effect:
    if $active$ then
      $R \leftarrow R \cup \{s\}$
      if $phase = \text{gossip}$ then
        if $i \in N$ then
          $Nbrs \leftarrow Nbrs \cup \{s\}$
        if $p = i$ then
          $New\_C \leftarrow New\_C \cup \{s\}$
      if $phase = \text{confirm}$ then
        $world \leftarrow world \cup W \cup \{s\}$

Trajectories
  stop when
    $active \wedge clock \% t = 0$
  evolve
    $d(clock) = 1$

Internal restart_i
  Effect:
    $active \leftarrow true$

Internal end-round_i
  Precondition:
    $active$
    $clock \% t = 0$
    $clock \% 2t = t \vee prt \in R$
  Effect:
    if $clock \% 2t = 0$ then /* gossip phase */
      $prt \leftarrow \min \{u : u \in R \cup \widehat{N}\}$
      $R \leftarrow \emptyset$
      $New\_C \leftarrow \emptyset$
      $Nbrs \leftarrow \widehat{N}$
      $Dest \leftarrow \{prt\} \cup Nbrs \cup C \cup world$
      $phase \leftarrow \text{gossip}$
    else /* confirm phase */
      $world \leftarrow R \cup Nbrs$
      $C \leftarrow New\_C$
      $Dest \leftarrow R \cup Nbrs$
      $phase \leftarrow \text{confirm}$
      $do\_msend \leftarrow true$
      $clock \leftarrow clock + \epsilon$

Internal reset_i
  Precondition:
    $active \wedge clock \% 2t = 0 \wedge prt \notin R$
  Effect:
    $prt \leftarrow \min\{u : u \in \widehat{N}\}$
    $world \leftarrow R \leftarrow \emptyset$
    $C \leftarrow New\_C \leftarrow \emptyset$
    $Nbrs \leftarrow \widehat{N}$
    $Dest \leftarrow \{prt\} \cup Nbrs \cup C$
    $phase \leftarrow \text{gossip}$
    $do\_msend \leftarrow true$
    $clock \leftarrow clock + \epsilon$

Fig. 2. Transitions of RD_i at node i for $i \in V$

perturbations did not occur). Action reset_i is enabled every $2t$ time units when $clock \% 2t = 0$ and the parent does not respond during the iteration. In this case the node gives up, resets its state and starts anew.

In more detail, when $clock \% 2t = 0$ action end-round_i concludes the current phase and starts a new iteration with the gossip phase. For this phase the variables are updated as follows: prt_i is set to the smallest identifier among the nodes in $\widehat{N}_i$ and those that sent a message to i in the previous iteration. Note that while updating prt_i we also consider $\widehat{N}_i$ just in case a failure causes R_i to be empty. Sets R_i and New_C_i are set to $\emptyset$, since those sets reflect the corresponding knowledge of node i in the current iteration. Set $Nbrs_i$ is set to the neighbors in graph G (i.e., self and its fixed neighbor), and finally $Dest_i$ is set to the destinations for the multicast in this new phase. Essentially, here the node establishes a parent and cleans up its state so as to not rely on variable values that may have been corrupted.

When $clock \% 2t = t$, action end-round$_i$ concludes the current phase and commences the confirm phase. In the confirm phase node i propagates its knowledge to all nodes from which it received a message in the previous gossip phase and to the neighbors in G^u that were discovered in that phase. Node i also sets C_i to the set of nodes that considered it the parent in the previous phase. Furthermore, node i sets $world_i$ to the set of nodes from whom messages were received in the gossip phase united with the neighbors of i in graph G^u.

Note that the preconditions of actions end-round$_i$ and reset$_i$ are mutually exclusive. Each of these actions also cause $clock$ to advance by ϵ ($\epsilon \ll d$), modeling the passage of time after it was "stopped" by the trajectory, and enable msend by setting do_msend to $true$.

We now detail msend$_i$ and mrecv$_i$. Action msend$_i$ is enabled at the beginning of every round and its invocation multicasts a message from node i to the nodes in $Dest_i$. The message contains the set of neighbors in graph G (i.e., self and its fixed neighbor), the parent, the set of children, and $world_i$. Recall that destinations are established at the end of the previous phase. The action sets do_msend_i to $false$ to prevent multiple invocations of msend$_i$ in a round.

Action mrecv$_i$ updates the state based on the messages received. First, the set R_i accumulates the identifiers of the nodes from whom messages are received. Additionally, if a message is received from node s and s considers node i to be a neighbor in the gossip phase, then s is added to $Nbrs_i$. If node s considers node i to be its parent, it is added to New_C_i. In the confirm phase node i also updates the set $world_i$ by including node s and $world_s$ received from s.

4   Algorithm Analysis

We now prove the self-stabilization properties of algorithm RDS and analyze its performance. We start by stating the algorithm's invariant $\psi()$ that directly implies the RDP invariant in Definition 4.

Resource Discovery Invariant. Configuration $c \in C$ is a legitimate configuration if invariant $\psi(c)$ holds, where the invariant is defined as follows.
  (1) For every node $v \in V$ we have $active_v = true$ and $prt_v = \min\{u : u \in V\}$.
  (2) There exists a node $v_0 \in V$ such that $v_0 = \min\{u : u \in V\}$ and $C_{v_0} = V$, while for every other node $w \neq v_0$, with $w \in V$, we have $C_w = \emptyset$.
  (3) For every node $v \in V$ we have $world_v = V$.

We use $\tilde{\tau}$ to denote the transition function of RDS, with $\tilde{A}$ denoting the corresponding set of actions. We use τ to denote the transition function that excludes transitions corresponding to actions join and perturb that are caused exclusively by the environment; we use A to denote the corresponding set of actions. In reasoning about the self-stabilization properties of executions we consider only those executions where join and perturb occur only in some finite execution prefix.

Knowledge Graph. Let c be some configuration of the system. The knowledge graph at c is a derived state variable $c.\mathcal{G} = (c.\mathcal{V}, c.\mathcal{E})$, where (1) $\mathcal{V} = \{v \in V : c.active_v = \text{true}\}$ is the set of nodes that joined the computation, and (2) $c.\mathcal{E} = \{(u, v) : u, v \in c.\mathcal{V} \wedge v \in c.world_u\}$ is the set of edges.

In this definition of $c.\mathcal{E}$ an edge (u, v) models the fact the u knows v, however, this does not imply that v knows u. Where the configuration c is implied by the context, we use the simplified notation $\mathcal{G} = (\mathcal{V}, \mathcal{E})$. We denote by $\mathcal{G}^u$ the undirected version of graph $\mathcal{G}$.

We start with Lemmas 1 and 2 showing that in two iterations of the algorithm the state variables of every node $v \in V$ that appear in the invariant ψ are free of bogus identifiers. Lemma 2 also shows that reset action is not invoked after two iterations, allowing the algorithm to converge.

Lemma 1. *Consider an execution prefix of RDS that ends with configuration c. Any fair extension of the execution of a sufficient length using only the actions from A reaches configuration c' in at most one complete iteration, where in c': (1a) for every node $v \in V$ we have $active_v = \text{true}$, and (1b) for any two distinct nodes $u, v \in V$ we have $phase_u = phase_v = \text{gossip}$, and (2) in any further execution extension following c', any invocation of action $msend_v$ in a state with $phase_v = \text{confirm}$, results in a message sent to all neighbors of v in G^u.*

Lemma 2. *Consider an execution prefix of RDS that ends with configuration c. Any fair extension of the execution of a sufficient length using only the actions from A reaches configuration c^* in at most two complete iterations, where for every node $v \in V$ the following hold: (a) $C_v \subseteq V$ and $world_v \subseteq V$, and (c) $prt_v \in V$, and (d) action $reset_v$ is not invoked following configuration c^*.*

Next we show that after a constant number of rounds following the cessation of failures no message is sent to bogus destinations.

Lemma 3. *Consider an execution prefix α of RDS that ends with configuration c. Any fair extension of α of a sufficient length that uses only the actions from A reaches a configuration c^* in at most two complete iterations, where no messages are sent to bogus destinations by any node $v \in V$ following c^*.*

Proof sketch. Let execution α_1 be an extension of α, reaching configuration c^* exactly as in Lemma 2. The proof follows from the algorithm and from Lemmas 1 and 2. This is because according to the algorithm messages are only sent to the nodes in $Dest$. Note that after c^* for every node $i \in V$ we have $prt_i \in V$. On the other hand, sets $Nbrs_i$, C_i, and $world_i$ are reset in every iteration and any node v added to these sets after c^* is an active node that belongs to V.   □

We now reason about the algorithm's self-stabilization properties.

Theorem 1. (Closure) *Consider any execution prefix of RDS consisting of complete iterations, where c is the final configuration. If c is legitimate, then any extension of the execution by up to one complete iteration using only the actions from A results in $c \xrightarrow{*}_{\tau} c'$, where c' is a legitimate configuration.*

The proof of Theorem 1 is by induction on the length of the execution extension. We next address convergence, starting with preparatory lemmas. Lemma 4 shows that in gossip phase if a node v knows u at the beginning of the phase, then by the end of the gossip phase node u knows v.

Lemma 4. *Consider an execution prefix of RDS that ends with configuration c. Any fair extension of the execution of a sufficient length using only the actions from A reaches a configuration c^* in at most two complete iterations, where following configuration c^* for any two distinct nodes $u, v \in V$ the following holds: if $phase_u = phase_v =$ gossip and $u \in world_v$ then $v \in R_u$ and $v \in world_u$ following the invocation of* end-round$_u$ *action.*

The following lemma shows that every node $v \in V$ retains its knowledge about the network from one iteration to the next.

Lemma 5. *Consider an execution prefix α of RDS that ends with configuration c. Let execution α_1 be an extension of α, reaching configuration c_1 exactly as in Lemma 2. Let execution α_2 be an extension of α_1 by 0 or more complete iterations reaching configuration c_2. Furthermore, let α_3 be an extension of α_2 by exactly one iteration reaching configuration c_3. Then for every node $v \in V$ we have $c_2.world_v \subseteq c_3.world_v$.*

The next lemma shows that in at most $\log D + 3$ complete iterations every node $v \in V$ knows v_0.

Lemma 6. *Consider an execution prefix α of RDS that ends with configuration c. Any fair extension of the execution of a sufficient length using only the actions from A reaches a configuration c^* in at most $\log D + 3$ complete iterations, such that in configuration c^* for every node $u \in V$, we have $v_0 \in world_u$.*

Proof. Let execution α_1 be an extension of α, reaching configuration c_1 exactly as in Lemma 2. Based on the specification of action end-round and according to Lemmas 1 and 2 it is clear that in configuration c_1 for every node $v \in V$ we have $Nbrs_v \subseteq world_v$, where $Nbrs_v$ is the set of neighbors of v, including v itself, in the initial knowledge graph $G^u = (V, E^u)$. Hence, $E^u \subseteq c_1.\mathcal{E}^u$, where $c_1.\mathcal{E}^u$ is the set of edges of the evolving knowledge graph $\mathcal{G}^u$ in configuration c_1.

Let us consider any path $u = u_0, u_1, ..., u_{k-1}, u_k = v_0$ in graph $c_1.\mathcal{G}^u$, where $0 \leq k \leq D$. Consider any three consecutive nodes u_{j-1}, u_j and u_{j+1}, for $0 < j < k$. From configuration c_1 it follows that (u_j, u_{j-1}) and (u_j, u_{j+1}) are in $c_1.\mathcal{E}^u$. Let us extend α_1 by one complete round reaching configuration c_2, after the invocation of action end-round$_v$ for every node $v \in V$. Let α_2 be the extended execution. From Lemma 4 it follows that for any node u_j, such that $0 < j < k$, we have $u_{j-1}, u_{j+1} \in c_2.world_{u_j}$, this is because in configuration c_1 we have $u_j \in world_{u_{j-1}}$ and $u_j \in world_{u_{j+1}}$ for all $0 < j < k$. Note also, that from the specification of end-round it follows that $world_v \subseteq Dest_v$ for every node $v \in V$.

Let us further extend α_2 by a complete round reaching configuration c_3. Let α_3 be the extended execution. From the specification of actions msend and mrecv it follows that $u_{j-2}, u_{j+2} \in world_{u_j}$. This is because, as we argued above, in configuration c_2 we have $u_{j-2}, u_j \in world_{u_{j-1}}$ and $u_j, u_{j+2} \in world_{u_{j+1}}$. Hence, in configuration c_3 there exists a path $u = u_0, u_2, ..., u_{2i}, ..., u_k = v_0$ for k even, or $u = u_0, u_2, ..., u_{2i}, ..., u_{k-1}, u_k = v_0$ for k odd between nodes u and v_0 in the undirected knowledge graph $c_3.\mathcal{G}^u$. Observe that the length of above path between the nodes u and v_0 is $\lceil \frac{k}{2} \rceil$.

Hence, based on Lemma 5 and from above, after at most $\log D$ subsequent finite extensions of α_3 by a complete iteration configuration c^* is reached in which for every node $u \in V$ we have $v_0 \in world_u$. This completes the proof. $\square$

Theorem 2. (Convergence) *Consider an execution prefix of RDS that ends with configuration c. Any fair extension of the execution of a sufficient length using only the actions from A reaches a configuration c_l in at most $\log D + 5$ complete iterations, such that c_l is a legitimate configuration.*

Finally we reason about the stabilization time and message complexity.

Theorem 3. *Any execution prefix of RDS ending in an arbitrary configuration can be infinitely extended to solve the resource discovery problem. The stabilization time of the algorithm is $O(\log D)$, taking at most $2\log D + 10$ complete rounds to stabilize. The stabilization message complexity is $O(|V|^2 \cdot \log D)$.*

Proof. From the proofs of Lemma 2, Lemma 6, and Theorem 2 it follows that after the cessation of transient failures, and given that no new nodes join the computation, any fair execution extension of a sufficient length takes at most $\log D + 5$ complete iterations to reach a legitimate configuration from any configuration c, and hence $2\log D + 10$ complete rounds. Theorem 1 (closure) guarantees that all subsequent configurations are legitimate. Thus the algorithm establishes the resource discovery invariant (Definition 4), taking at most $2\log D + 10$ complete rounds to stabilize, and then maintains the resource discovery invariant in perpetuity.

We now assess the stabilization message complexity. Consider some execution that includes an arbitrary configuration after which no actions from $\tilde{A} - A$ occur. The proof of Lemma 3 reasons that in a constant number of rounds no node identifiers from $U - V$ occur in any local state. Prior to this the algorithm may send messages to arbitrary sets of nodes due to corruptions of local states. We charge such messages to the environment (per model assumptions) and do not include them in the message complexity. Here we only consider messages sent to the nodes in V. From the specification of action end-round it follows that in both gossip and confirm phases the number of accountable messages sent by each node $v \in V$ is at most $|V|$. Therefore, since the stabilization time is $O(\log D)$, the message complexity of RDS is $O(|V|^2 \cdot \log D)$. $\square$

5   Discussion

We considered the distributed resource discovery problem in the self-stabilization context, we formalized the setting, presented a solution, and rigorously reasoned about its properties. In our formulation, the set of nodes to be discovered is established by the environment, and its size is unknown. Nodes join the computation at arbitrary times, and the states of the participants and messages in transit can be arbitrarily perturbed. The algorithm solves the problem once the perturbations subside; the participants establish and maintain the invariant, where there is a unique node that knows the set of participants, who in turn know the identity of that unique node. Our synchronous algorithm solves the problem starting from any arbitrary state. Its stabilization time is $O(\log D)$, and the message complexity of our solution is $O(|V|^2 \cdot \log D)$, where D is the diameter of the weakly-connected initial knowledge graph induced by the local knowledge of each participant. If the graph is not connected, or if permanent crashes disconnect the graph then our algorithm solves the problem for each connected component. Future work will deal with more virulent adversarial behaviors.

References

1. Abraham, I., Dolev, D.: Asynchronous resource discovery. In: Proceedings of the 22nd ACM Symposium on Principles of Distributed Computing, pp. 143–150 (2003)
2. Arora, A., Gouda, M.G.: Closure and convergence: A foundation of fault-tolerant computing. IEEE Trans. Software Eng. 19(11), 1015–1027 (1993)
3. Davtyan, S., Konwar, K., Shvartsman, A.A.: Brief announcement: Self-stabilizing resource discovery algorithm. In: Proceedings of the ACM Symposium on Principles of Distributed Computing (2013)
4. Dijkstra, E.W.: Self-stabilizing systems in spite of distributed control. Communications of the ACM 17(11) (1974)
5. Dijkstra, E.W.: A belated proof of self-stabilization. Distributed Computing 1(1) (1986)
6. Dolev, S., Herman, T.: Superstabilizing protocols for dynamic distributed systems. Chicago Journal of Theoretical Computer Science (1997)
7. Harchol-Balter, M., Leighton, F.T., Lewin, D.: Resource discovery in distributed networks. In: Proceedings of the 18th Symposium on Principles of Distributed Computing, pp. 229–237 (1999)
8. Kaynar, D., Lynch, N., Segala, R., Vaandrager, F.: The theory of timed i/o automata, 2nd edn. Morgan & Claypool (2011)
9. Konwar, K.M., Kowalski, D.R., Shvartsman, A.A.: Node discovery in networks. In: Anderson, J.H., Prencipe, G., Wattenhofer, R. (eds.) OPODIS 2005. LNCS, vol. 3974, pp. 206–220. Springer, Heidelberg (2006)
10. Konwar, K., Kowalski, D.R., Shvartsman, A.A.: The join problem in dynamic network algorithms. In: Proceedings of the International Conference on Dependable Systems and Networks, pp. 315–324 (2004)
11. Kutten, S., Peleg, D.: Asynchronous resource discovery in peer to peer networks. In: Proceedings of the 21st IEEE Symposium on Reliable Distributed Systems (SRDS 2002), pp. 224–231 (2002)
12. Kutten, S., Peleg, D., Vishkin, U.: Deterministic resource discovery in distributed networks. In: Proceedings of the 13th ACM Symposium on Parallel Algorithms and Architectures, pp. 77–83 (2001)

13. Law, C., Siu, K.-Y.: An $o(\log n)$ randomized resource discovery algorithm. In: Brief Announcements of the 14th International Symposium on Distributed Computing, Technical Report FIM/110.1/DLSIIS/2000, Technical University of Madrid, pp. 5–8 (2000)
14. Nor, R.M., Nesterenko, M., Scheideler, C.: Corona: A stabilizing deterministic message-passing skip list. In: Défago, X., Petit, F., Villain, V. (eds.) SSS 2011. LNCS, vol. 6976, pp. 356–370. Springer, Heidelberg (2011)
15. Tel, G.: Distributed algorithms. Cambridge University Press (2000)

Hybrid Distributed Consensus*

Roy Friedman[1], Gabriel Kliot[2], and Alex Kogan[3],**

[1] Department of Computer Science, Technion, Haifa, Israel
[2] Microsoft Research, Redmond, WA
[3] Oracle Labs, Burlington, MA

Abstract. Inspired by the proliferation of cloud-based services, this paper studies consensus, one of the most fundamental distributed computing problems, in a hybrid model of computation. In this model, processes (or nodes) exchange information by passing messages or by accessing a reliable and highly-available register hosted in the cloud. The paper presents a formal definition of the model and problem, and studies performance tradeoffs related to using such a register. Specifically, it proves a lower bound on the number of register accesses in deterministic protocols, and gives a simple deterministic protocol that meets this bound when the register is *compare-and-swap* (CAS). In addition, two efficient protocols are presented; the first one is probabilistic and solves consensus with a single CAS register access in expectation, while the second one is deterministic and requires a single CAS register access when some favorable network conditions occur. A benefit of those protocols is that they can ensure both liveness and safety, and only their efficiency is affected by the probabilistic and timing assumptions.

Keywords: Consensus, cloud computing, message passing, lower bounds.

1 Introduction

Distributed consensus [19] is one of the most fundamental distributed computing problems. Over the last few decades, it was intensively explored in different computation models, including message passing and shared memory models. In this paper, we consider a novel hybrid model, where computing parties, or nodes, may exchange information by passing messages and/or by accessing a shared reliable and highly-available register. Our work is inspired by the proliferation of *cloud computing*, which may provide services that implement such a register.

Cloud computing is an emerging paradigm in which various services can be placed in a data center equipped with a management middleware that ensures the service's availability, fault-tolerance, and scalability in an almost transparent manner. Consequently, when designing contemporary distributed systems, it is tempting to resort to centralized architectures, in which the crux of the system is executed as a cloud service, which is

---

* This work is partially supported by the Israeli Science Foundation grant 1247/09 and by the Technion Hasso Plattner Research School.
** The work of this author on the paper was done while he was with the Department of Computer Science, Technion.

R. Baldoni, N. Nisse, and M. van Steen (Eds.): OPODIS 2013, LNCS 8304, pp. 145–159, 2013.

accessed in a client/server fashion by the participating nodes. As an example, the distributed consensus problem can be solved with a single cloud based *compare-and-swap* (CAS) register[1] as follows. Assume the range of decision values is D; let $\perp$ be a value not in D and initialize the CAS register to $\perp$. The consensus protocol is simply to have every node p_i invoke CAS with $\perp$ and its initial value v_i and then decide based on the value read from the CAS register [25]. It is easy to verify that this simple protocol solves consensus despite any number of benign failures among the (non-cloud) participants in a completely asynchronous environment (outside the cloud), while as we discuss later, such a CAS register can be easily implemented by existing cloud services.

The problem with this well known solution is that it requires each process (or node) participating in the consensus protocol to access the cloud hosted register at least once on each instantiation of the consensus protocol. This imposes a high load on the cloud servers, limiting their scalability, and may incur high monetary cost to the nodes themselves [30]. Hence, we are motivated to develop protocols that exploit a cloud hosted register to obtain simplicity and guaranteed fast termination while minimizing the number of accesses to the register.

Specifically, we make the following contributions: First, we present the formal model of hybrid distributed computing and a formal definition of efficient distributed consensus within this model. Second, we prove a lower bound on the number of cloud hosted register accesses required to solve consensus in the benign crash failure model. We show that whenever the number of potential failures is f (for $f < n$), any deterministic protocol that solves consensus in a hybrid asynchronous system requires invoking at least $f + 1$ cloud-based register operations. The proof itself relies on the proof of the famous FLP result [19]. However, unlike the FLP model, here we cannot immediately deduce the case of $f > k$ directly from $f = k$ (for any k). This is because whenever we increase f, we allow the protocol to use more CAS operations, giving it more power than it has with smaller values of f. Thus, we develop a novel inductive argument to show our result.

Third, we develop three efficient protocols for solving consensus despite benign failures. The first protocol always invokes exactly $f + 1$ CAS operations, thereby meeting the lower bound, as described above, for deterministic protocols in a hybrid asynchronous system. The second protocol utilizes an Ω-like oracle [2][2]. However, the reliance on the Ω-like oracle is only in order to ensure efficiency, so that only a single CAS operation will be invoked when certain network conditions are met. Termination and safety are always ensured, regardless of whether the oracle/black-box really provides its semantics or not. Our third protocol is probabilistic, and it ensures that the expected number of CAS invocations will be 1. Here again, termination and safety are always ensured, and randomization only affects the expected efficiency of the protocol.

Finally, in the Appendix, we show how to apply the hybrid approach to the non-blocking atomic commit problem [9]. In the full version of this paper, we also discuss how the approach can be extended to cover Byzantine failures [29].

---

[1] We assume the common semantics for the CAS register; see formal definition in Section 2.2.

[2] Later in the paper, we use the term *black-box* rather than oracle, to emphasize the fact that both the liveness and safety properties of the protocol are guaranteed even if the black-box fails to provide its semantics.

The rest of this paper is organized as follows: The model and basic definitions appear in Section 2. The main results appear in Section 3. We compare our work with related work in Section 4 and conclude with a discussion in Section 5.

2 Preliminaries

2.1 Basic Model and Assumptions

We assume an asynchronous distributed system as modeled in [7], but enhance it with a cloud hosted register. That is, a hybrid asynchronous system consists of a set of n *nodes* as well as a register R implemented by means of a cloud service. The register may support any set of operations that have a sequential specification [26], exposed through a well defined interface. The register is assumed to be highly available and fault-tolerant, meaning that any invocation of one of the operation on the register by any of the nodes is guaranteed to terminate with a response within a finite time[3].

The nodes themselves can be modeled as deterministic automata similar to what is done in [7], and their state therefore advances by taking *steps*. The communication between nodes is performed by sending and receiving messages over a *network*. Each step of a node includes receiving 0 or more messages, performing some computation, and then generating 0 or more messages to be sent to other nodes and/or to the cloud hosted register. The node that generates a message is called the *sender* and the node that receives the message is called a *receiver*. The receiver is always known to the sender and vice versa. The network is further assumed to be reliable, meaning that messages transmitted are eventually delivered once and only once, they are delivered without being altered and messages that are delivered were indeed sent by their sender. Note that these properties can be easily provided on top of weaker networks, e.g., by adding summaries to protect against data corruption and by an ACK or NACK based retransmission mechanism in *fair-lossy* networks [8].

Yet, the nodes and the network are assumed to be asynchronous in the sense that there is no bound on the time for performing a step of a node and the time between the sending of a message until its delivery at the receiver, also known as the *latency* of the message. Nodes can have access to a local clock, but the local clocks of different nodes are not synchronized.

External observers of the system may have access to a global time. Hence, this global time can only be useful for external analysis of events in the system. For convenience, we further assume that the range of the global time T is the set of natural numbers $\mathcal{N}$.

For our lower bounds, we further borrow the well known definitions of a *protocol execution*, a *protocol configuration*, an *execution prefix*, an *execution extension*, and *indistinguishable executions* from the textbook of [7]. Intuitively, we can assume a sequential *scheduler* that may schedule one node at a time in any order it wishes to. A protocol execution is the sequence of steps taken by each process whenever it is scheduled by the scheduler. Such a step depends on the node's code (the protocol) and its state

---

[3] We note that in reality even a highly reliable cloud service can become temporarily unavailable due to, e.g., network congestion. The liveness of hybrid consensus protocols discussed in this paper depends on the liveness of the cloud service implementing R.

at the beginning of the step. An execution prefix is a prefix of the sequence of operations composing an execution. A configuration for a given execution prefix is the collection of states of each node at the end of this prefix as well as the messages already sent but not received during this prefix. An initial configuration is the collection of initial states of each process. An execution extension is a possible sequence of steps that can be obtained from a given configuration based on the protocol and the scheduler. Finally, two executions (or execution prefixes) are indistinguishable if the processes participating in them receive exactly the same messages in each of their steps.

2.2 Benign Failures

In the benign crash failure model, up to f nodes may *fail* by crashing anytime during the execution of their protocol. A crashed node stops executing its steps and, in particular, stops sending messages. A node that fails is called *faulty*. Nodes that do not fail are called *non-faulty*, or *alive*.

2.3 Additional Services

CAS register: Our consensus protocols make use of a cloud hosted *compare-and-swap* (CAS) register object providing its usual semantics. That is, its interface includes a single method, whose signature is

```
object oldValue = compareAndSwap(object expectedValue,
                        object newValue).
```

When invoked, this method atomically sets the value of the object to `newValue` if and only if its value at the time of invocation is equal to `expectedValue`. The method always returns the value of the object as it was just before its execution. The register is initialized with a special $\perp$ value, which cannot be a valid input of any node participating in the consensus protocol.

We focus on CAS since it is supported by existing cloud APIs, such as Windows Azure's REST API for accessing Azure Table and Blob Storage services (using the IF-MATCH header)[4] and Amazon Web Services Conditional Put for SimpleDB[5]. It can also be implemented by Yahoo's PNUTS `Test-and-Set-Write` operation [16]. Yet, the results for the benign failures model can be applied to any other object whose consensus number is ∞. Also, note that our lower bound does not assume any specific interface supported by the cloud hosted register. In particular, the lower bound holds for a register that supports any set of operations that have a sequential specification.

The $\tilde{\Omega}$ black-box: In one of our deterministic consensus protocols, we make use of a $\tilde{\Omega}$ black-box, which provides the following service. When invoked, it always returns the id of a single process that is presumed to be alive. Yet, whenever the system starts behaving in a synchronous way, then eventually all invocations of this black-box by

---

[4] http://msdn.microsoft.com/en-us/library/dd179427.aspx
[5] http://docs.aws.amazon.com/AmazonSimpleDB/latest/
DeveloperGuide/ConditionalPut.html

all nodes return the id of the same process, which is also indeed alive. Clearly, any known implementation of an Ω failure detector in a system that eventually becomes synchronous, or in other words has a *Global Synchronization Time* (GST) [12, 18], can be used to implement a $\tilde{\Omega}$ black-box regardless of the timing assumptions. Examples of such implementations include, e.g., [2]. Thus, implementing $\tilde{\Omega}$ is out of scope.

Given an execution σ of a protocol using a specific implementation of a black-box BX of the class $\tilde{\Omega}$, if there exists a time t_1 such that from t_1 onward, every call by any node to BX returns the same node id and this returned node is non-faulty in σ, then we say that the *execution stabilization time*, denoted $EST(\sigma,BX)$, is t_1. Otherwise, we define $EST(\sigma,BX)$ to be ∞. Denote t_0 the global starting time of the protocol. Whenever $EST(\sigma,BX) \leq t_0$ we say that BX is *well behaved* in σ.

Random generator: In our randomized protocols, we assume that each node has access to a random numbers generator, which in the analysis is assumed to return truly uniformly distributed results. In practice, as the correctness of the protocols does not depend on this assumption, the assumption can be safely relaxed to any modern pseudo-random number generator that is common in modern computers with negligible impact on the actual performance of the protocols.

3   Hybrid Consensus with Benign Failures

3.1   Problem Statement

In the consensus problem, each node p_i starts with an initial value v_i from some range V. Each node p_i is required to compute a decision value. A protocol solving the consensus problem must satisfy the following properties:

Validity: The value decided on by each non-faulty node is one of the initial values.
Agreement: The decision values of all non-faulty nodes that decide are the same.
Termination: Eventually, every non-faulty node decides on some value.

As mentioned in the introduction, given that we assume the presence of a highly-available (CAS) register object, it is possible to solve consensus despite any number of nodes' crash failures by having each node access the register once with its initial value. Yet, this imposes scalability and economical problems as each instantiation of such a protocol involves a total of n CAS invocations. Consequently, we introduce the following definitions of *hybrid efficient consensus protocols*.

Definition 1 (Hybrid k-efficient execution). *Given an execution σ of a protocol for solving distributed consensus in which f nodes are faulty, σ is called* hybrid k-efficient *if the total number of register accesses in σ is at most k.*

Definition 2 (Hybrid k-efficient protocol). *A protocol for solving distributed consensus is called* hybrid k-efficient *if all its executions are hybrid k-efficient.*

Definition 3 (Hybrid efficient probabilistic protocol). *A randomized protocol for solving distributed consensus is called* hybrid efficient probabilistic *if the expected number of register accesses in an arbitrary execution of the protocol is 1.*

3.2   Lower Bounds

For simplicity, we show the proof for the binary case, i.e., when the only allowed values are 0 and 1. Extending it to a larger domain is trivial. Also, before stating and proving the lower bound, we repeat the known definition of *valency* [7,19], which plays a crucial role in the proof. Specifically, a configuration of a protocol solving consensus in called *bi-valent* if it has at least one execution extension in which the decision value is 0 and at least one execution extension in which the decision value if 1. A configuration is *univalent* if in all of its execution extensions nodes decide on the same value; if the value is 0, the configuration is called 0-valent and it is said to be 1-valent otherwise.

Theorem 1. *In a hybrid asynchronous system prone to f benign failures, there does not exist a hybrid f-efficient protocol.*

Before going into the proof's details, let us remark that, as mentioned in the introduction, unlike the standard asynchronous system model, here the fact that consensus cannot be solved in a hybrid f-efficient manner for $f = 1$ does not immediately imply that it holds for $f > 1$. This is because by the definition of a hybrid efficient protocol, increasing f also increases the power of the system by allowing the protocol to invoke more operations on the shared cloud hosted register. The proof below builds upon the FLP proof as it appears in the textbook of [7][6].

Proof. We prove the theorem by induction on f, the number of allowed failures and register accesses. As for the base of the induction, when $f = 1$, only a single register access is permitted. Clearly, when the register can only be accessed once, only the process that accessed the register knows the result of that access. (Note that in a trivial case when the register always returns the same value so that processes know the result of the access without actually accessing the register, the existence of a hybrid 1-efficient protocol would immediately contradict the FLP result [19]). Hence, this process can intuitively simulate the register access without anyone noticing.

More formally, if there exists a hybrid f-efficient protocol $\mathcal{P}$ with $f = 1$, then define a corresponding protocol $\mathcal{P}'$ in which whenever a process accesses the register in $\mathcal{P}$, then this same process would execute locally in $\mathcal{P}'$ the same computation as in the function supported by the register according to its sequential specification for a single invocation. By the assumption about $\mathcal{P}$, all its executions terminate and ensure the validity and agreement properties of consensus. Moreover, by construction, each execution σ of $\mathcal{P}$ has a corresponding execution σ' of $\mathcal{P}'$ that is indistinguishable from it. Hence, each such execution σ' also terminates and ensures validity and agreement. In other words, $\mathcal{P}'$ solves consensus in an asynchronous environment prone to failures with $f = 1$ (without accessing the cloud hosted register). This contradicts the famous FLP result [19].

Induction Step: Assume that the theorem holds for $f = k$, we will show that it holds for $f = k + 1$ as well. To that end, assume by contradiction that there exists a hybrid

---

[6] To be precise, in [7] there is only a proof for the read/write shared memory model. However, an earlier version [5] includes a complete proof for the message passing model that follows the same steps and terminology.

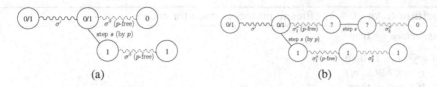

Fig. 1. Illustrations for the lower bound proof. Valences of configurations are indicated by values in boxes. Question marks specify configurations that might be either bi-valent or univalent.

f-efficient protocol $\mathcal{P}$ with $f = k + 1$. Hence, by the induction hypothesis, some of the executions of $\mathcal{P}$ must involve $k + 1$ register accesses (else, $\mathcal{P}$ solves consensus with only k operations on the register despite $f = k + 1$ failures).

We claim that $\mathcal{P}$ has at least one bi-valent initial configuration. The proof of this claim is exactly the same as the proof of the corresponding claim in the FLP result. Now, consider all executions of $\mathcal{P}$ that start from bi-valent initial configurations. Clearly, there is at least one such execution σ of $\mathcal{P}$ (that starts from a bi-valent configuration) in which at most k nodes have failed prior to the invocation of the $k + 1$ operation on the register. The existence of this execution can be easily proved by contradiction using a simple indistinguishability argument on the last failure prior to invoking the $k + 1$ operation on the register – this proof is eliminated for lack of space.

We further claim that the configuration immediately after invoking the $k+1$ operation on the register has to be uni-valent. Otherwise, we remain with a bi-valent configuration in an asynchronous system and no additional operations on the register can be invoked. Here we can apply the same arguments as in the FLP result showing that there has to be at least one such execution that is either infinite or violates agreement.

Thus, there has to be some step s taken by some process p in σ such that the configuration prior to s is bi-valent and the configuration after s is uni-valent. Also, s is either the $k + 1$ invocation of the register mentioned above, or a prior step in σ. Consider the prefix σ' of σ that ends just before s. Note that due to the determinism of $\mathcal{P}$ and the asynchrony of the system, for every (possibly empty) valid extension $\tilde{\sigma}$ of σ' that does not include any step of p, $\tilde{\sigma}s$ is also a valid extension of σ'.

Assume, w.l.o.g., that the configuration immediately after s is 1-valent. Since the configuration at the end of σ' is bi-valent, σ' has a valid extension σ'' such that $\sigma'\sigma''$ ends in a 0-valent configuration; denote by σ'' the shortest such extension. If σ'' does not include any step by p, then $\sigma'\sigma''s$ is a valid extension of σ'. Moreover, it is by definition 0-valent. Similarly, due to the fact that σ'' does not include any operation by p and the asynchrony of the system, $\sigma'so''$ is also a valid extension of σ', yet is 1-valent. However, processes cannot distinguish between executions that extend $\sigma'\sigma''s$ and $\sigma'so''$ (see illustration in Figure 1a). A contradiction.

On the other hand, if σ'' does include a step by p, then σ'' can be written as $\sigma_1''s\sigma_2''$, where σ_1'' must include at least one operation and none of the operations in σ_1'' are by p while σ_2'' may include zero or more operations by any processes. Thus, we have that $\sigma'\sigma_1''s\sigma_2''$ is 0-valent. However, since σ_1'' does not include any operation by p and due to the asynchrony of the system, $\sigma'so_1''\sigma_2''$ is also a valid extension of σ', which is 1-valent. Yet, processes cannot distinguish between executions that extend $\sigma'\sigma_1''s\sigma_2''$ and $\sigma'so_1''\sigma_2''$ (see illustration in Figure 1b). A contradiction.

Algorithm 1. An $(f + 1)$-efficient protocol - code for node i

---

 1: *undecided* := **true**

 2: **if** $i \leq (f + 1)$ **then**
 3:     $r :=$ CAS($\bot, v_i$)
 4:     **if** $r == \bot$ **then**
 5:         decide(v_i)
 6:     **else**
 7:         decide(r)
 8:     **end if**
 9: **end if**

10: decide(v)
11: **if** *undecided* **then**
12:     *undecided* := **false**
13:     broadcast(DEC, v)
14:     **return** v
15: **end if**

16: **upon receiving** broadcast(DEC, v)
17:     decide(v)

3.3 Upper Bounds

A Hybrid $(f + 1)$-efficient Protocol. Our first protocol utilizes the CAS register implemented by cloud-based services as following. Each node having id smaller than or equal to $f+1$ invokes CAS. The first node to succeed, i.e., the first node to receive the special $\bot$ value as a response from CAS, decides on its value. Other nodes that invoke CAS and get some non-$\bot$ value decide on that value. The rest of the nodes simply wait until a *decide* message broadcasted by one of the deciders reaches them.

The pseudo-code of this simple protocol is given in Algorithm 1. In the following theorem we prove that our first protocol is hybrid $(f + 1)$-efficient.

Theorem 2. *Protocol 1 is hybrid $(f + 1)$-efficient.*

Proof. The safety properties of the protocol follow trivially from the properties of the CAS register and the fact that it is always invoked with the initial value of the process that calls it. As for termination, since at most f nodes can be faulty, in every execution there is at least one live process that will invoke CAS and therefore terminate. The failure efficiency of the protocol follows trivially from the code, where only the first $f + 1$ nodes invoke CAS.

Notice that the code for handling DEC messages, and in particular Line 13 involve n (unreliable) broadcasts. These can be replaced by a more efficient reliable broadcast protocol, e.g., [27]. As this is a known trick, details are omitted for clarity and brevity.

An $\tilde{\Omega}$-Based Protocol. The problem with Protocol 1 is that in each execution of the protocol, the CAS register object is invoked $f + 1$ times regardless of the actual number

---

Algorithm 2. $\tilde{\Omega}$-based protocol - code for node i

---

```
 1: for {j = 1; j <= limit && undecided; j++} do
 2:    if p_i == Ω̃.get() or j == limit then
 3:       r := CAS(⊥, v_i)
 4:       if r == ⊥ then
 5:          decide(v_i)
 6:       else
 7:          decide(r)
 8:       end if
 9:    end if
10:    wait Δ time
11: end for
       {The code for decide and for handling DEC messages is the same as in Algorithm 1}
```

---

of faulty nodes. It also requires knowing f up front. This brings the question of whether we can devise a protocol that will terminate with fewer CAS invocations (preferably, just one), at least when the environment behaves "favorably". We answer this affirmatively by presenting Protocol 2, which relies on an $\tilde{\Omega}$ black-box to limit the number of CAS invocations whenever it is well behaved.

The pseudo-code is given in Algorithm 2. There, a node repeatedly queries the $\tilde{\Omega}$ black-box and accesses the CAS register only when the black-box returns its id, or a threshold of iterations controlled by the configuration parameter *limit* has passed. Notice also that due to the properties of the CAS register object, safety and termination are always ensured with this protocol, regardless of the behavior of $\tilde{\Omega}$. The latter only impacts the performance of the protocol in terms of running time and the number of CAS invocations (or cloud accesses). In particular, if the failure detector is guaranteed to meet its specification (even at an arbitrary, unknown but finite, eventual time), then *limit* can be eliminated (or set to ∞).

Between iterations, each node waits for Δ time, which is the node's estimate for the time required for a computation step of the protocol including the expected latency of decision messages to propagate through the network. Clearly, a bad estimation of Δ does not hurt the correctness of the protocol, only its performance. Specifically, overestimating Δ may slightly increase the running time of the protocol as it delays polling the failure detector while underestimating Δ may result in redundant polling of the failure detector and potentially reaching *limit* iterations and invoking CAS redundantly simply due to a node that did not wait long enough for a decision message (DEC) to arrive.

The following lemma shows that in any execution in which the black-box is well behaved, only a single node needs to invoke the CAS register at the cloud.

Lemma 1. *Every execution of the protocol in which $\tilde{\Omega}$ is well behaved is hybrid 1-efficient.*

Algorithm 3. Randomized protocol - code for node i

```
1:  for {j = 1; j <= limit && undecided; j++} do
2:      if Random(n) == 0 or j == limit then
3:          r := CAS(⊥, v_i)
4:          if r == ⊥ then
5:              decide(v_i)
6:          else
7:              decide(r)
8:          end if
9:      end if
10:     wait Δ time
11: end for
    {The code for decide and for handling DEC messages is the same as in Algorithm 1}
```

Proof. Consider an execution σ of the protocol. From the assumption that $\tilde{\Omega}$ is well behaved in σ, a single non-faulty process will evaluate the condition in Line 2 to true and therefore a single node will invoke CAS, decide, and transmit its decision to all others. Hence, all other alive nodes will decide and terminate without invoking CAS.

A Randomized Protocol. The randomized protocol is presented in Algorithm 3. It utilizes a random generator that accepts a parameter k and returns a uniformly selected integer value in the range $[0, \ldots, (k-1)]$. This protocol also relies on the configuration parameter called *limit*, which limits the maximal number of iterations that a node is willing to wait before invoking CAS deterministically. In each iteration, every node chooses to invoke CAS with probability $1/n$, as listed in Line 2. As before, between iterations, each node waits for Δ time. Here, underestimating Δ may result in redundant invocations of CAS simply due to a node that did not wait long enough for a decision message (DEC) to arrive.

Clearly, the worst case running time of the protocol is $\Delta \cdot limit$ and in the worst case, there will be n invocations of CAS. For the following analysis, assume that Δ is correctly estimated. That is, once some node invokes CAS in a given iteration, then all nodes will decide in this iteration.

Denote by X the number of iterations required by the algorithm. When *limit* is set to ∞, the probability p that none of the nodes will invoke CAS in an arbitrary iteration is given by $(1 - 1/n)^n$. Obviously, the probability that at least one node will invoke CAS is $1 - p$. Thus, the expected number of iterations until at least one node invokes CAS (and decides) is given by

$$E(X) = \sum_{j=1}^{\infty} [(1-p) \cdot j \cdot p^{j-1}] = \frac{1}{1-p} = \frac{1}{1 - (1 - \frac{1}{n})^n}.$$

Notice that when n is very large, $(1 - \frac{1}{n})^n$ approaches e^{-1}, in which case the above expression becomes roughly 1.588. Moreover, the lower the value of *limit* is, the closer the expected number of iterations becomes to 1.

It is worth noting that when *limit* is set to ∞, the probability that the algorithm will require more than $a > 0$ rounds is given by $Pr(X > a) = \left[1 - \frac{1}{n}\right]^{n \cdot a}$. It follows then

$$Pr(X > \ln n) = \left[1 - \frac{1}{n}\right]^{n \cdot \ln n} < e^{-\ln n} = \frac{1}{n}.$$

In other words, with high probability, the algorithm requires $\ln n$ rounds or less.

We now calculate the expected number of CAS invocations. From the assumption on Δ, it is enough to consider the first iteration in which CAS was invoked (i.e., the last iteration of the protocol). Notice that the invocations of CAS in each individual iteration can be viewed as a set of n Bernoulli trials, each with probability $1/n$ of success. Hence, the expected number of invocations in each iteration is 1. As this is true for each iteration, it is also true for the iteration in which CAS was indeed invoked.

Note that safety and termination are always deterministically ensured for the probabilistic protocol we have presented. The randomization aspect only controls the expected running time and the number of CAS invocations. In other words, probability only affects performance, but not correctness. It is always possible to play with the parameter *limit* and the distribution of the random number generator in order to tradeoff faster termination vs. fewer CAS invocations.

4 Related Work

Since the famous FLP impossibility result for solving consensus was introduced [19], a plethora of papers on how to circumvent it have been published. Some follow the line of Chandra and Toueg as well as Lamport by enriching the environment with failure detector oracles [12, 13, 28], while others weaken the termination guarantees to being probabilistic, e.g., [6, 10, 11] to list a few.

In Disk-Paxos [20] and Byzantine Disk-Paxos [1], consensus is solved by relying of shared disks, such that each process can write to a certain block and read a certain fraction of other processes' blocks. These works also rely on a shared storage service. However, they only require read-write semantics from the shared storage, and hence must also rely on a leader oracle to ensure termination. The inspiration for these works are *storage area networks* (SAN).

Motivated by SAN and *active storage technology*, in Active Disk-Paxos [14], multiple fail prone *read-modify-write* registers are used to implement a *ranked-register* abstraction, which is then used to implement a Paxos style consensus protocol. The main benefit of this is that it enables solving consensus with an unbounded number of clients. In our work, we merely use the number of clients in the random protocol to ensure that with high probability only a single client will access the CAS object. As this number is not required for correctness, we can replace it with a rough estimate on the actual number of clients rather than an exact figure. Also, similarly to the original Disk-Paxos works, Active Disk-Paxos does not try to minimize the number of storage accesses.

Past work has investigated the minimal synchrony, and in particular the minimal number of synchronous links required to solve consensus [3, 4, 24]. The idea in this line of work is that synchrony is hard to ensure (whereas total lack of synchrony prevents deterministic solutions for consensus [19]). Thus, synchronous links are likely to be expensive or have lower bandwidth than asynchronous ones, which motivates investigating how to use them in the most parsimonious manner.

Wormholes is another approach for solving consensus, both Byzantine and benign, by relying on special secure, synchronous, and temper-proof channels [17]. As was shown in [17], wormholes can greatly reduce the complexity of solving consensus and improve the ratio of faulty processes required to solve the problem. There, too, wormholes are assumed to be expensive and offer lower bandwidth than the "standard" network. Consequently, wormholes should be used judiciously.

Both lines of research (minimal synchrony and wormholes) share the vision of adding some expensive service to enable and simplify solving consensus. Both also investigate how to use such a service wisely. In that sense, it is related to our work. However, in their case, the service is a special type of a communication link, whereas in our case it is a cloud hosted service.

Golab et al. [21] use the remote memory references (RMRs) metric to measure the performance of algorithms that solve consensus and other related problems in two asynchronous shared memory models. They consider blocking algorithms, and distinguish between local and remote memory accesses, where the latter traverse the processor-to-memory interconnect. They show that in this setting the consensus problem can be solved using only a constant number of RMRs, while the progress is guaranteed only when all active process are alive.

Guerraoui and Schiper discuss a related idea of a consensus service, which might be used by client processes to solve an agreement problem [23]. The consensus service is implemented by a set of server processes, which might be the same as or distinct from client processes. The paper concentrates on the generality of the suggested service, showing how it can be used to solve a series of agreement problems. Even though Guerraoui and Schiper consider communication costs of the described protocols, they do not distinguish between messages sent by clients and by servers, and thus do not strive to optimize the communication between clients and servers.

5 Discussion

In this paper we have focused on solving consensus in hybrid systems, where processes communicate by message passing and by accessing a shared highly-available register. In particular, inspired by the proliferation of cloud-based services, we have studied the problem of minimizing the number of register accesses for better scalability and cost reduction in cloud assisted implementations of the suggested hybrid model. The hybrid approach brings several benefits. First, in the case of benign failures, it enables solving consensus in an otherwise asynchronous environment with $f < n$ failures. The protocols are very simple and terminate quickly. Also, our randomized and $\tilde{\Omega}$-based deterministic protocol enable solving consensus with a single register access in expectation or in the "typical" case, respectively.[7] We have also shown a lower bound on the number of register accesses for deterministic protocols as well as a protocol that

---

[7] Let us reiterate that in these protocols, termination and safety are always ensured regardless of any synchrony assumptions and only the efficiency of the protocol depends on the behavior of the black-box or timeout setup.

satisfies this bound. A shortcoming of the $\tilde{\Omega}$-based and the randomized protocol is that when the black-box in not well behaved in the former or the timeout is not accurate in the latter, the protocol may require up to n accesses to the cloud hosted register. Limiting this number to $f + 1$ in such cases is left for future work.

For the Byzantine case, we conjecture that the lower bound for deterministic protocols is $3f + 1$ register accesses and $f < 3n$. In the full version of this paper, we show an algorithm that meets this presumed bound. We also show there a probabilistic protocol that terminates with just one cloud access by correct nodes in expectation while always ensuring termination and safety. In addition to proving the conjectured lower bound, an open problem related to Byzantine failures is the impact of the semantics of the register on the minimal number of register accesses in the deterministic case. That is, can stronger objects reduce the required number of register accesses?

Acknowledgments. We would like to thank Sergey Bykov and Alan Geller from Microsoft Research for their help and advice.

References

1. Abraham, I., Chockler, G.V., Keidar, I., Malkhi, D.: Byzantine disk paxos: optimal resilience with byzantine shared memory. In: Proc. of the 23rd Annual ACM Symposium on Principles of Distributed Computing (PODC), pp. 226–235 (2004)
2. Aguilera, M.K., Delporte-Gallet, C., Fauconnier, H., Toueg, S.: On implementing omega with weak reliability and synchrony assumptions. In: Proc. of the 22nd Annual ACM Symposium on Principles of Distributed Computing (PODC), pp. 306–314 (2003)
3. Aguilera, M.K., Delporte-Gallet, C., Fauconnier, H., Toueg, S.: Communication-efficient leader election and consensus with limited link synchrony. In: Proc. of the 23rd Annual ACM Symposium on Principles of Distributed Computing (PODC), pp. 328–337 (2004)
4. Aguilera, M.K., Delporte-Gallet, C., Fauconnier, H., Toueg, S.: Consensus with byzantine failures and little system synchrony. In: Proc. of the International IEEE Conference on Dependable Systems and Networks (DSN), pp. 147–155 (2006)
5. Attiya, H.: Lecture notes for course #236357: Distributed algorithms (spring 1993); Technical report, Department of Computer Science, The Technion (January 1994)
6. Attiya, H., Censor-Hillel, K.: Lower bounds for randomized consensus under a weak adversary. SIAM J. Comput. 39(8), 3885–3904 (2010)
7. Attiya, H., Welch, J.: Distributed Computing: Fundamentals, Simulations, and Advanced Topics, 2nd edn. John Wiley and Sons, Inc. (2004)
8. Basu, A., Charron-Bost, B., Toueg, S.: Simulating reliable links with unreliable links in the presence of process crashes. In: Babaoğlu, Ö., Marzullo, K. (eds.) WDAG 1996. LNCS, vol. 1151, pp. 105–122. Springer, Heidelberg (1996)
9. Bernstein, P., Hadzilacos, V., Goodman, H.: Concurrency Control and Recovery in Database Systems. Addison-Wesley, Reading (1987)
10. Bracha, G.: An $o(\lg n)$ expected rounds randomized byzantine generals protocol. In: Proc. 17th Annual ACM Symposium on Theory of Computing (STOC), pp. 316–326 (1985)
11. Canetti, R., Rabin, T.: Fast asynchronous byzantine agreement with optimal resilience. In: Proc. 25th Annual ACM Symposium on Theory of Computing (STOC), pp. 42–51 (1993)

12. Chandra, T., Toueg, S.: Unreliable failure detectors for asynchronous systems. J. ACM 43(4), 685–722 (1996)
13. Chandra, T.D., Hadzilacos, V., Toueg, S.: The weakest failure detector for solving consensus. J. ACM 43, 685–722 (1996)
14. Chockler, G., Malkhi, D.: Active disk paxos with infinitely many processes. In: Proc. of the 21st Annual ACM Symposium on Principles of Distributed Computing (PODC), pp. 78–87 (2002)
15. Chu, F.: Reducing ω to $\diamond s$. Information Processing Letters 67(6), 298–293 (1998)
16. Cooper, B.F., Ramakrishnan, R., Srivastava, U., Silberstein, A., Bohannon, P., Jacobsen, H.-A., Puz, N., Weaver, D., Yerneni, R.: Pnuts: Yahoo!'s hosted data serving platform. Proc. of VLDB Endowment 1, 1277–1288 (2008)
17. Correia, M., Neves, N.F., Lung, L.C., Veríssimo, P.: Low complexity byzantine-resilient consensus. Distributed Computing 17, 237–249 (2005)
18. Dwork, C., Lynch, N., Stockmeyer, L.: Consensus in the Presence of Partial Synchrony. Journal of the ACM 35(2), 288–323 (1988)
19. Fischer, M.J., Lynch, N.A., Paterson, M.S.: Impossibility of distributed consensus with one faulty process. Journal of the ACM 32, 374–382 (1985)
20. Gafni, E., Lamport, L.: Disk paxos. Distributed Computing 16, 1–20 (2003)
21. Golab, W., Hadzilacos, V., Hendler, D., Woelfel, P.: Constant-RMR implementations of CAS and other synchronization primitives using read and write operations. In: Proc. ACM Symposium on Principles of Distributed Computing, PODC (2007)
22. Guerraoui, R.: Non-Blocking Atomic Commit in Asynchronous Distributed Systems with Failure Detectors. Distributed Computing 15, 15–17 (2002)
23. Guerraoui, R., Schiper, A.: The generic consensus service. IEEE Transactions on Software Engineering 27(1), 29–41 (2001)
24. Hamouma, M., Mostefaoui, A., Trédan, G.: Byzantine consensus with few synchronous links. In: Tovar, E., Tsigas, P., Fouchal, H. (eds.) OPODIS 2007. LNCS, vol. 4878, pp. 76–89. Springer, Heidelberg (2007)
25. Herlihy, M.: Wait-free synchronization. ACM Trans. Prog. Lang. Syst. 13(1), 124–149 (1991)
26. Herlihy, M., Wing, J.: Linearizability: A correctness condition for concurrent objects. ACM Trans. on Programming Languages and Systems 12(3), 463–492 (1990)
27. Kaashoek, M.F., Tanenbaum, A.S., Hummel, S.F.: An efficient reliable broadcast protocol. ACM SIGOPS Operating Systems Review 23, 5–19 (1989)
28. Lamport, L.: The part-time parliament. IEEE Transactions on Computer Systems 16(2), 133–169 (1998)
29. Lamport, L., Shostak, R., Pease, M.: The byzantine generals problem. ACM Transactions on Programming Languages and Systems 3(4), 382–401 (1982)
30. Wang, H., Jing, Q., Jiao, S., Chen, R., He, B., Qian, Z., Zhou, L.: Distributed systems meet economics: Pricing in the cloud. In: Proc. USENIX HotCloud (2010)

A   Non-Blocking Atomic Commit

Non-blocking atomic commit (NBAC), originating in the area of databases [9], is a related problem to consensus, but with a twist. That is, in NBAC, each process starts with a 'yes' or 'no' value and the processes also need to decide on the same output value, which is either 'commit' or 'abort'. However, if at least one of the initial values is 'no', then the only allowed decision value is 'abort'. Further, if all initial values are 'yes', then the only allowed decision value is 'commit' unless at least one process has

---

Algorithm 4. A hybrid NBAC protocol - code for node i

1: send(v_i) to everyone
2: **wait** until received v_j from every node or $?P ==$true or $(\exists j, v_j ==$no$)$
3: **if** received v_j from every node and $(\forall j, v_j ==$yes$)$ **then**
4:     decide(*hybrid-efficient-consensus*(commit))
5: **else**
6:     decide(*hybrid-efficient-consensus*(abort))
7: **end if**

---

failed, in which case it is also permissible to decide 'abort'. As was shown in [22], when $n > 2f$, any solution to the NBAC problem requires a failure detector of the class $\Diamond S + ?P$, where $?P$ is a failure detector that eventually returns 'true' if and only if at least one process has failed. We also remind the reader that the failure detector class $\Diamond S$ is equivalent to Ω [15]. Clearly, it is possible to define a hybrid k-efficient (or efficient probabilistic) protocol for solving NBAC in a similar manner to what has been done above for consensus (cf. Section 3.1).

As was shown in [22], NBAC can be easily solved using a reduction to consensus. For self-containment, this reduction is repeated in Figure 4, in which we replaced the invocation of consensus with an invocation to *hybrid-efficient-consensus*, which stands for any of the hybrid k-efficient or efficient probabilistic protocols mentioned above. In the listing in Figure 4, the initial vote of each node is denoted v_i. It is easy to verify that when invoked with a $?P$ failure detector in Line 2 and the consensus protocol of Lines 4 and 6 is hybrid k-efficient (or efficient probabilistic), then the overall NBAC protocol becomes hybrid k-efficient (or efficient probabilistic), solves NBAC for any $f < n$, and only relies on $?P$ for ensuring the safety and liveness properties of NBAC.

Notice that the use of a register does not eliminate the need for a $?P$ failure detector. This is because in NBAC, it may not be safe to decide 'commit' before hearing from every process or knowing with certainty that at least one of the processes has failed. Hence, the benefits of the hybrid approach for the NBAC problem is in increasing the resilience to $n-1$ failures while eliminating the need for the $\Diamond S$ (or Ω) failure detector.

Speculative Concurrent Processing
with Transactional Memory in the Actor Model

Yaroslav Hayduk, Anita Sobe, Derin Harmanci,
Patrick Marlier, and Pascal Felber

University of Neuchatel, Switzerland
first.last@unine.ch

Abstract. The *actor* model has been successfully used for scalable computing in distributed systems. Actors are objects with a local state, which can only be modified by the exchange of messages. One of the fundamental principles of actor models is to guarantee sequential message processing, which avoids typical concurrency hazards, but limits the achievable message throughput. Preserving the sequential semantics of the actor model is, however, necessary for program correctness.

In this paper, we propose to add support for speculative concurrent execution in actors using *transactional memory* (TM). Our approach is designed to operate with message passing and shared memory, and can thus take advantage of parallelism available on distributed and multi-core systems. The processing of each message is wrapped in a transaction executed atomically and in isolation, but concurrently with other messages. This allows us (1) to scale while keeping the dependability guarantees ensured by sequential message processing, and (2) to further increase robustness of the actor model against threats due to the rollback ability that comes for free with transactional processing of messages. We validate our design within the Scala programming language and the Akka framework. We show that the overhead of using transactions is hidden by the improved message processing throughput, thus leading to an overall performance gain.

Keywords: Concurrency, actors, transactional memory, speculative processing.

1 Introduction

The actor model, initially proposed by Hewitt [1], is a successful message-passing approach that has been integrated into popular frameworks [2]. The actor model introduces desirable properties such as encapsulation, fair scheduling, location transparency, and data consistency to the programmer. It also perfectly unifies concurrent and object-oriented programming. While the data consistency property of the actor model is important for preserving application safety, it is arguably too conservative in concurrent settings as it enforces sequential processing of messages, which limits throughput and hence scalability.

R. Baldoni, N. Nisse, and M. van Steen (Eds.): OPODIS 2013, LNCS 8304, pp. 160–175, 2013.

In this paper, we address this limitation by proposing a mechanism to boost the performance of the actor model while being faithful to its semantics [2]. The key idea is to apply speculation, as provided by transactional memory (TM), to handle messages concurrently as if they were processed sequentially. In cases where these semantics might be violated, we rely on the rollback capabilities of TM to undo the operations potentially leading to inconsistencies.

We see a high potential for improvement in scenarios where actors maintain state that is read or manipulated by other actors via message passing. With sequential processing, access to the state will be suboptimal when operations do not conflict (e.g., modifications to disjoint parts of the state, multiple read operations). TM can guarantee safe concurrent access in most of these cases and can handle conflicting situations by aborting and restarting transactions.

Speculation can also significantly improve performance when the processing of a message causes further communication. Any coordination between actors requires a distributed transaction, which we call *coordinated transaction*. We combine coordinated transactions and TM to concurrently process messages instead of blocking the actors while waiting for other transactions to commit.

We have implemented our approach in the Scala programming language and the integrated Akka framework [3]. Since this implementation cover changes to the Akka framework only, the developer is not affected at all. We evaluate our approach using a distributed linked list benchmark already used with other concurrent message processing solutions [4]. We show that concurrent message processing and non-blocking coordinated processing can considerably reduce the execution time for both read-dominated and write-dominated workloads.

The rest of the paper is organized as follows. We first give an overview of the actor models, transactional memory, and related work in Section 2. We then discuss the limits of the sequential message processing in Section 3 and propose improvements to sequential message processing in Section 4. We describe our implementation in Section 5 and present evaluation results in Section 6. We finally conclude in Section 7.

2   Background and Related Work

Actor models are inherently concurrent. They are widely used for implementing parallel, distributed, and mobile systems. An actor is an independent, asynchronous object with an encapsulated state that can only be modified locally based on the exchange of messages. It comprises a mailbox in which messages can be queued, as well as a set of dedicated methods for message processing [5].

The actor model provides *macro-step semantics* [6] by processing messages sequentially. As a consequence, it also guarantees the following properties:

Atomicity. The state of an actor can only be observed before or after operations took place, therefore changes on the state are perceived either all at once or not at all.

Isolation. The actor model forbids any concurrent access to the local state of an actor. This means that any operation on the state of the actor is done as if it were running alone in the system.

These characteristics make actor models particularly attractive and contribute to their popularity. Numerous implementations of actor models exist in many languages like Java, C, C++, and Python. We decided to use Scala, which is a general-purpose language that runs on top of the JVM and combines functional and object-oriented programming patterns. The recent versions of Scala integrate the Akka Framework [3] for realizing actors. They also supports transactional memory (TM) [7], a programming model that provides atomicity, isolation, and rollback capabilities within transactional code regions [8]. The programmer simply has to demarcate the blocks of instructions that must execute atomically and the TM performs all the necessary bookkeeping to ensure that the target code is executed in a transaction, i.e., the consistency of data accessed within the block are not affected by concurrent accesses. TM provides built-in support for check-pointing and rollback, which we exploit for controlling concurrent message processing.

Existing actor frameworks such as surveyed by Karmani et al. in [2] do not include TM and differ regarding the way they handle parallelism. As an example, implementations of Habanero-Scala and Habanero-Java [4] introduce parallelism by mixing the actor model with a fork-join design (*async-finish* model). Actors can start concurrent sub-tasks (*async* blocks) for the handling of a single message.

Since the processing of a message terminates only when all sub-tasks are finished, this approach enforces sequential handling of messages. To alleviate this restriction and improve scalability, Habanero also allows messages to be processed in parallel. To ensure that the actor's state is not accessed concurrently, a *pause and resume* model that works similar to *wait and notify* is used. While processing a message, the actor can spawn external sub-tasks it must then pause to avoid intermediate modification to its state. When the sub-tasks finish with changing the state, the actor resumes its operation and can process further messages. While this approach avoids concurrent access to an actor's state, it must be used carefully as it provides no protection against synchronization hazards such as data races and deadlocks.

Parallel actor monitors (PAM) [9] support concurrent processing by scheduling multiple messages in actor queues. Using PAM, the programmer must understand the concurrency patterns within the application and define application-specific schedulers. This may prove particularly challenging for applications where concurrency patterns vary during execution. In contrast, our approach (see Section 4) removes any programmer intervention and automatically allows concurrent executions when possible. Further, we do not break the original actor semantics at any time, while using an inappropriate scheduler with PAM can cause inconsistencies.

3   Problem Statement

Despite its inherent concurrency, the actor model requires sequential message processing. While this requirement is a deliberate choice to avoid synchronization hazards, it unnecessarily limits the performance (i.e., throughput) of individual actors, in particular when they execute on multi- or many-core computers.

We elaborate the problem of sequential processing with the help of the examples depicted in the left part of Figure 1. They involve three actors (A, B and C) performing operations as illustrated on their respective time lines (horizontal dashed line). The transfer of a message between actors is indicated by an arrow and its processing is indicated by thick solid lines, where we explicitly mark the beginning and end of processing with brackets. In our examples, the actors are responsible for maintaining a distributed linked list of ordered integers. Actor B stores the first part of the list and actor C the second part, while actor A acts as a client and performs operations on the list (e.g., search, insert, remove).

Sequential processing in the actor model limits performance in two ways:

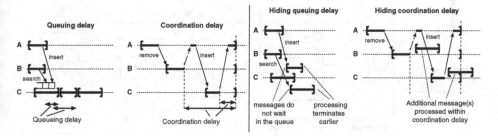

Fig. 1. Sequential (left), concurrent and non-blocking coordinated processing (right) and their effect on execution time

Queuing Delay: In the first block of Figure 1 we depict the delay that is introduced upon arrival of multiple messages on a single actor. If actor A and B send messages to actor C, which is busy, both messages are stored in a queue. The queuing delay is the time a message has to wait at a given actor from arrival to the start of its processing.

Coordination Delay: In the second block of Figure 1 we depict a common communication pattern.

Consider that actor A wants to move the value x from the list of actor B to the list of actor C. For doing so, it sends messages remove(x) to actor B and insert(x) to actor C. To fulfill the macro-step semantics in actor A, the list operations have to be part of a coordinated transaction, which commits when all three actors successfully finish their task. The coordinated commit protocol defines a barrier on which actors B and C block until they can resume processing other messages. Hence, the coordination delay describes the time actors have to wait after finishing their own tasks until the distributed transaction commits.

4   Message Processing Model

Queuing delays are inherent to the structure of the actor model and its sequential processing operation; their reduction may become particularly important for actors that receive many messages. Further, upon coordination delay, actors block and thus cannot perform any useful work. We claim that these delays are unnecessarily long and can be significantly reduced by thoughtful changes to the message processing of actors.

Our main idea is based on the observation that we can guarantee atomicity and isolation if we encapsulate the handling of messages inside transactions. Thanks to the rollback and restart capability of transactions, several messages can be processed concurrently, even if they access the same state. We call this approach *concurrent message processing*. Additionally, we exploit the characteristics of transactions to avoid blocking actors while waiting for a coordinated commit (*non-blocking coordinated processing*).

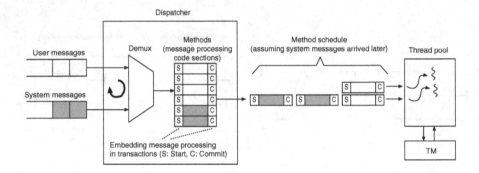

Fig. 2. Implementation of concurrent message processing

To explain the principle of our two optimizations, consider the same example as in Figure 1 with actor A performing operations on a list stored on actors B and C. The right part of Figure 1 illustrates how the delays caused by sequential processing can be reduced.

Queuing Delay: By processing several messages concurrently on a single actor, we can reduce the queuing delay as shown in the first block of Figure 1. Therefore, if A and B send a message to C, which is currently busy, the messages do not have to wait. If the transactions do not conflict and can immediately commit, the queuing delay is avoided.

Coordination Delay: Actor A wants to move a value from the list of actor B to the list of actor C, which requires both an insertion and a removal action. This is typically achieved using a coordinated transaction. To ensure consistency, however, participating actors cannot process new messages until the coordinated transaction commits. By treating messages speculatively, one can avoid blocking the actors and allow concurrent execution of non-conflicting transactions (e.g., as in actors B and C in the second block of Figure 1), therefore hiding the coordination delay.

5 Implementation

To hide the delays as explained before, we extend Scala version 2.10.2. Specifically, we concentrated on two parts of Scala: the Akka framework version 2.10.0 and the Scala-STM [10] library version 0.7. Akka provides a clean and efficient implementation of the actor model for the JVM. Scala-STM supports transactional memory in Scala and, while it adds some overhead for checkpointing and concurrency control, it is particularly non-invasive and well integrated in the language. In the following we describe the specific changes we made to Akka and Scala-STM to realize the proposed optimizations.

Concurrent Processing of Messages. The concurrent message processing only involves changes of the message handling provided by the Akka framework. Specifically, we changed the behavior of the actor's mailbox. In the original Akka implementation a dispatcher is responsible for ensuring that the same mailbox is not scheduled for processing messages more than once at a given time. Another particularity of Akka is that every actor has one mailbox with two queues: the first one stores user messages, i.e., messages received from other actors, while the second one is used for maintaining system messages specific to Akka, which control lifecycle operations (i.e., start, stop, resume). Once a user message is scheduled, the dispatcher checks first if there are any system messages. Then, all existing system messages are treated before the user message. The same is done after the processing of the user message. As system messages are rare, actors spend most of their time processing user messages.

To facilitate concurrent message processing, we reimplemented the mailbox and message treatment as shown in Figure 2. System messages are handled as in the sequential case, before and after user messages, but instead of processing user messages one at a time, we process them concurrently in batches. Each user message from a batch is submitted to a thread-pool for execution. The actual message processing is performed concurrently inside a transaction, as indicated by the start (S) and commit (C) events in the figure. For the transactional handling of messages we use the default Scala-STM. If the concurrent operations do not conflict, we can hide the queuing delay as illustrated in Figure 1.

Non-blocking Coordinated Transactions. The non-blocking coordinated transaction alters the commit behavior of the Scala-STM. By default, a coordinated transaction is blocking (see Figure 1). All actors participating to a coordinated transaction must reach a commit barrier before any other message can be processed.

Consider the case of a transaction that executes a block of code corresponding to the processing of a message. After the transactional code is executed, the STM makes an attempt to commit the changes, possibly rolling back and trying again upon failure. In the process of a commit, several steps are performed:

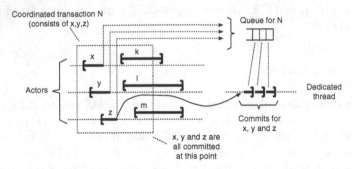

Fig. 3. Sketch of the implementation of non-blocking coordinated processing

(1) locks for the variables accessed in the transaction are obtained; and (2) if the transaction belongs to a coordinated transaction, an external *decider* is consulted. The coordinated transaction's commit barrier blocks as long as some of its transactions are still executing. Once they have all successfully completed, the commit barrier is unlocked and control is returned to the caller. After the external decider returns, the transaction does final sanity checks and flushes outstanding writes to main memory. In our implementation, illustrated in Figure 3, we perform the following operations instead of blocking the thread when waiting for other parties to arrive at the barrier. We first save the current transaction descriptor in a queue (*queue for N*). Then, we return from the atomic block immediately, bypassing any additional logic associated with the commit operation. By doing so, we do not fully commit the transaction; we instead suspend its commit at the point where it would normally block.

While an actor is busy with a coordinated transaction, it can handle other messages concurrently, hiding the coordination delay as illustrated on transactions k, l, and m in Figure 3. If concurrent messages are independent, they can commit immediately. If there is a write-write conflict, the processing of one of the messages rolls back. Read-write conflicts represent a special case: if the coordinated transaction reads a value and a concurrent message wants to write the same value, we delay the commit of the write until the coordinated transaction completes.

In a system comprising multiple actors, it is likely that several coordinated transactions execute concurrently. Each coordinated transaction uses its own queue to store its suspended *pre-committed* transactions (N corresponds to the identifier of the coordinated transaction in the figure). Hence, we do not mix pre-committed transactions belonging to different coordinated transactions. To resume the commit, a dedicated thread is notified when all the parties belonging to the same coordinated transaction have completed their work.

6   Evaluation

Our optimizations are expected to be most useful in applications where state is shared among distributed actors. Hence, to evaluate our approach, we use a benchmark application provided by Imam and Sarkar [4] that implements a stateful distributed sorted integer linked-list. The architecture considers two types of actors: *request* and *list* actors. Request actors only send requests such as `lookup`, `insert`, `remove`, and `sum`. List actors are responsible for handling a range of values (buckets) of a distributed linked list. In a list with l actors, where each actor can store at most n elements representing consecutive integer values, the ith list actor is responsible for elements in the $[(i-1) \cdot n, (i \cdot n) - 1]$ range, e.g., in a list with 4 actors and 8 entries, each actor is responsible for two values. A request forwarder matches the responsible list actors with the incoming requests. We extend this benchmark to evaluate different facets of our proposed optimizations. We evaluate different workloads, different numbers of actors holding elements of the list, etc. For the `sum` operation, each actor holds a variable that represents the current sum of all its list elements, called `partial_sum`, which is updated upon insertion and removal. When computing the sum of the whole list, we only accumulate the partial sum of each list actor without the need of traversing all list elements. While the lookup, insert, and remove operations execute on a single list actor, the sum operation needs to traverse all the list actors in order to return the partial sums. Hence, the sum operation involves multiple list actors. The original benchmark did not initially ensure atomicity of the `sum` operation; we therefore changed the implementation so that the computation of the sum is performed within a coordinated transaction.

We execute the benchmark on a 48-core machine equipped with four 12-core AMD Opteron 6172 CPUs running at 2.1GHz. Each core has private L1 and L2 caches and a shared L3 cache. The sizes of both instruction and data caches are 64KB at L1, 512KB at L2, and 5MB at L3.

We apply each of the extensions—concurrent message processing and non-blocking coordinated processing—to a read-dominated workload and then to a write-dominated workload. Each sample corresponds to the geometric mean of 7 runs. We first evaluate both extensions separately to better assess their benefits and drawbacks, i.e., for the first results, non-blocking coordinated processing does not include concurrent message processing. Then, we conduct experiments with both approaches combined. Their performance is compared against sequential message processing, i.e., using default Akka/Scala constructs without transactions for read and write operations. The sum operation is put into a coordinated transaction as provided by Akka/Scala. We finally complete our evaluation with a comparison against the Habanero-Scala implementation.

Our experiments are either read-dominated (lookup) or write-dominated (insert, remove). These workloads additionally contain a number of sum operations, which are treated as coordinated messages. More precisely, each actor performs $R = x\%$ reads, $W = y\%$ writes, and $S = z\%$ sum operations, where $R + W + S = 100\%$. Since sum requests are likely to be rare in comparison with other operations, we keep this parameter constant at $S = 1\%$. For read-dominated workloads, we choose $R = 97\%$ and $W = 2\%$. The write-dominated workload is configured with $R = 1\%$ and $W = 98\%$.

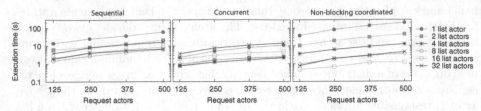

Fig. 4. Execution time for sequential, concurrent, and non-blocking message processing on a read-dominated workload

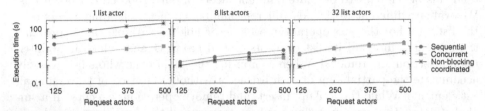

Fig. 5. Execution time for sequential, concurrent, and non-blocking message processing using 1, 8, and 32 list actors on a read-dominated workload

Insert and remove operations are handled independently and are chosen at random, but each evaluation run gets the same input. We vary the number of list and request actors, each of the latter sending 1,000 requests. The request actors wait for a response before sending the next message. The list can contain a maximum of 41,216 integers split evenly between actors. For instance, if there are 32 list actors, each will be responsible for 1,288 buckets. The list is pre-filled to 20 % of its capacity.

6.1   Read-Dominated Workload

Figure 4 presents the results for the three separate scenarios: sequential message processing, concurrent message processing, and non-blocking coordinated processing. On the x-axis we show the effects of increasing the number of request actors (125–500), while the y-axis displays the execution time in seconds (log scale), i.e., the time needed to finish processing all requests. The lower the execution time, the better. One can see in all cases that adding more request actors leads to higher execution times, which is not surprising because the workload becomes higher.

The concurrent execution time (Figure 4, center) is lower than the sequential scenario up to 16 list actors, which indicates that allowing multiple messages to be processed concurrently introduces an immediate performance gain. In our linked-list example, messages sent from request actors R1 and R2 to a list actor L1 will be treated within a transaction. If there are no conflicts—which is likely in a read-dominated workload—concurrent transactions commit successfully. Therefore, the time required to process a batch of messages will be equal to the execution time of the longest associated transaction $(\max{(T_1, T_2, \ldots)})$ instead of the sum of all execution times $(\sum T_i)$. The performance with 16 and 32 actors starts to degrade because the workload provides less exploitable concurrency. With 32 list actors the performance is even worse than with a single list actor.

When considering non-blocking coordinated message processing (Figure 4, right), the explanation for the increase of the execution time for concurrent processing with 16 and 32 list actors is clear: when the number of list and request actors is high, coordinated transactions are likely to fail because of increased contention. Non-blocking coordinated message processing allows us to reduce this execution time considerably. Actors can process other messages while the sum operation is in progress. The reduction of execution time is especially high for read-dominated workloads, because a lookup operation and the read of the partial sum are non-conflicting operations. With large numbers of list and request actors, however, the likelihood of insert and remove operations increases significantly.

Figure 5 shows a more detailed comparison for an increasing number of list actors. The left graph in presents the execution time for a single list actor. One can see that concurrent message processing improves the execution time considerably, while non-blocking coordinated processing exhibits worse performance than sequential message processing. Indeed, since there is only one list actor, no coordinated transactions are executed, i.e., the sum operation only returns the partial sum of the current list actor. Hence, the execution time of the non-blocking coordinated processing shows the overhead of executing all operations inside transactions. When increasing the number of list actors, this overhead is compensated by the benefits of non-blocking coordinated processing. When increasing the number of list actors to at least 8 (Figure 5, center), the contention of coordinated transactions increases and non-blocking processing performs even better than concurrent message processing. When the number of coordinated transactions and write-write conflicts becomes too high, concurrent message processing yields performance similar to sequential processing, as can be observed in the right graph of Figure 5 for 32 list actors. In contrast, non-blocking coordinated transactions lead to significantly lower execution times than both sequential and concurrent message processing.

To summarize our findings so far, concurrent message processing has the highest impact if the number of list actors is low because each will have more messages to process, i.e., the penalty from serialization is more important and the workload provides more exploitable concurrency. The opposite trend can be observed with non-blocking coordinated transactions: they benefit most when the number of list actors is high because coordinated transactions become longer, i.e., the penalty of the blocking operation is higher and contention is relatively low. Therefore, the combination of both techniques is expected to provide good overall performance for all considered scenarios.

6.2 Write-Dominated Workload

We expect to observe more conflicts with a write-dominated workload because each insert and remove operation also modifies the value of the partial sum. As a consequence the execution time generally increases in comparison with the read-dominated load, as shown by the graphs in Figure 6.

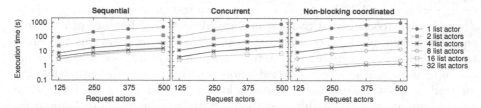

Fig. 6. Execution time for sequential, concurrent, and non-blocking message processing on a write-dominated workload

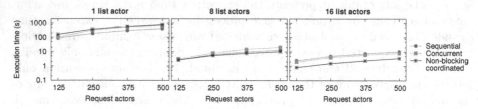

Fig. 7. Execution time for sequential, concurrent, and non-blocking message processing using 1, 8, and 16 list actors on a write-dominated workload

Sequential processing (Figure 6, left) performs similarly to the read-dominated workload case. The execution time first improves when adding more list actors. Then, we observe similar execution times for 8 and 16 list actors, and the degradation starts for 32 actors as the impact of coordinated transactions becomes more significant. Concurrent processing (Figure 6, center) provides better overall performance, but the best improvement is obtained with 16 list actors when there is sufficient exploitable concurrency. Finally, with non-blocking coordinated processing, performance improves with the number of list actors. The execution time is better than concurrent processing starting from 4 list actors. The reason is that sum operations are read operations. Thus, for coordinated transactions a write-write conflict is not possible. We expect that write operations in coordinated transactions conflicting with other writes lead to execution times close to concurrent processing.

Figure 7 shows the execution times of sequential, concurrent, and non-blocking coordinated processing for various sizes of list actors. With a single list actor (Figure 7, left), we observe that concurrent and non-blocking coordinated processing perform poorly due to the many write-write conflicts and resulting aborts.

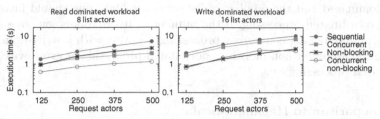

Fig. 8. Execution time for sequential, concurrent, non-blocking, and combined message processing

With 8 list actors (Figure 7, center) the performance of non-blocking coordinated processing becomes close to sequential processing, whereas concurrent processing still has a higher execution time. Finally, with 16 list actors the advantage of non-blocking coordinated processing is obvious, while concurrent processing now performs similarly to sequential processing because the coordination delays dominate.

Summing up the write-dominated workload results, we conclude that concurrent processing becomes less beneficial when the likelihood of write conflicts is high. In some cases, the high number of roll backs becomes high enough that sequential processing should be preferred. Coordinated transactions have a high influence on the execution time and significantly improve performance with many list actors. Therefore, a write-dominated workload can benefit more from non-blocking coordinated transactions than concurrent ones, and it is debatable whether the latter extension should be used at all when the number of write-write conflicts becomes very high.

6.3   Non-blocking Concurrent Processing

We conducted the same experiment as before, but combined concurrent and non-blocking coordinated message processing, which we call *non-blocking concurrent processing*, for the read and write-dominated workload. In the read workload, concurrent processing leads to lower execution times when the number of list actors is below 16. Indeed, one can observe that the performance of non-blocking concurrent processing is even better than non-blocking coordinated processing (Figure 8, left). When we increase the number of list actors, we see again the same behavior as for the write-dominated workload. The combination is thus useful when both concurrent and non-blocking coordinated processing lead to a lower execution time than sequential processing.

The results for the write-dominated workload show that concurrent processing does not have much influence on the execution time (Figure 8, right). In the 16 buckets scenario, pure concurrent processing has execution times similar to sequential processing, while non-blocking concurrent processing results in performance close to non-blocking coordinated processing.

To fully exploit the capabilities of the proposed mechanisms, it is therefore necessary to properly understand the nature of the workload. If it is read-dominated and the number of list actors is high, one should favor non-blocking coordinated processing. If the number of list actors is low, one should rather use non-blocking concurrent processing. Finally, with a write-dominated load, one should prefer non-blocking coordinated processing or even switch back to sequential processing.

6.4   Comparison to Habanero-Scala

In the original Habanero-Scala benchmark shown in Figure 20 in [4], the authors ran their experiments with 64 list actors, each responsible for 400 buckets. Additionally, the authors used a workload with a balanced mix of reads and writes (50:50). 50 read actors are used to access the list elements, with 32,000 accesses by actor. For these settings, Habanero-Scala (*LightActor* implementation) performed slightly better than default sequential processing (Akka). Note that the LightActor implementation of the list benchmark does not spawn any sub-tasks. It use a *finish* construct instead of the default countdown latch for coordinating list and request actors of the list. Thus, the difference in performance is due to their lightweight implementation of actors, a custom task scheduler, and a different thread-pool implementation.

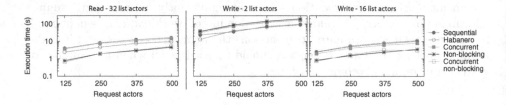

Fig. 9. Execution time comparison with Habanero-Scala

To show the full capabilities of our approach, we executed our experiments with Habanero-Scala. For this, we use the default Habanero-Scala constructs for read and write operations (no transactions). The behavior of the sum operation is provided by a barrier implemented in the list actors using the *DataDriveFuture* construct (including sub-tasks) of Habanero-Scala.

For the read-dominated workload, Habanero-Scala performs similarly to sequential message processing, except for 32 buckets where Habanero-Scala performs better by approximately 40%. There, the difference is higher, because of the penalty of the blocking coordinated sums used in the sequential implementation. As seen in the left graph of Figure 9, our approach outperform Habanero-Scala by 50 to 70%.

In the write-dominated workload, Habanero-Scala performs again similarly to sequential message processing. With our speculative extensions, the contention is too high for less than 8 list actors. With 16 list actors the improvement of our approach is significant (Figure 9, right).

6.5   Discussion

Our approach, which combines concurrent and non-blocking coordinated processing, guarantees the same correctness as the original actor model. We perform the processing of messages within transactions, which means that concurrent operations on the actor's state will execute atomically and in isolation. Therefore, conflicting operations will be serialized, but non-conflicting messages should be processed concurrently.

It is important to note that the actor model does not impose any order on the processing of messages that are in its incoming queues. Therefore the non-deterministic order in which transactions will commit does not break the semantics of the actor model. If ordering were required, we could extend the STM as proposed in [11] to enable parallel processing but commit transactions in order.

As we rely on transactional memory to process messages equivalently to a serial execution and we preserve the original actor model, concurrent processing also provides the same correctness guarantees. The same applies for non-blocking coordinated processing. All messages are handled within transactions, which means that the conflicts are handled as for concurrent message processing. However, for read-write conflicts (e.g., concurrent sum and insert operations) the order will be preserved by our delayed commit mechanism. An issue that currently limits our approach is that the code of "transactified" message processing should not contain any action that is not under the control of the TM, such as I/O or OS library code (irrevocable code). Strategies for supporting irrevocable actions are left to the responsibility of the underlying TM and are not specific to our extensions.

Our approach has the important benefit of being adaptable. The transactional processing of a message can be aborted at any time without side effects on the current state of the actors. This implies that each of our extensions can be enabled or disabled at any time during execution, and one can switch from one extension to the other within the same execution. Such flexibility can be exploited to play with trade-offs between performance and resource utilization. On the performance side, messages need to be processed anyways, either sequentially or concurrently. Hence, if we have idle resources and we can process messages concurrently, the overall task can be completed in a shorter time. On the resource utilization side, the adaptability of our approach allows us, for example, to apply simple energy-efficiency strategies that enable or disable transactional execution, possibly even temporarily switching off some cores, in order to fit the consumption of the application to the desired energy requirements.

7   Conclusion

The actor model implements synchronization by the means of message passing. This decoupled communication paradigm is particularly scalable since it allows multiple actors to perform independent computations concurrently as they do not share state. However, each actor processes arriving messages sequentially.

To address this limitation, we proposed an approach that combines transactional memory (TM) and actors as implemented by the Akka Framework. Incoming messages are dequeued in batches and processed speculatively inside transactions. The atomicity, consistency, and isolation properties of TM guarantee that messages do not interfere when being processed concurrently. In addition, as the actor model does not impose any order on the handling of user messages in the incoming queues, our approach preserves its semantics.

To further improve concurrency, we also extended the coordinated transaction mechanism of Scala-STM to support non-blocking operations. The traditional design prevents actors involved in a coordinated transaction to process any additional message until the transaction commits. The resulting delays can be especially high when actors are distributed on several nodes and communication has non-negligible latency. We solve this issue by speculative concurrent message processing.

Together, these two mechanisms can significantly lower the queuing and coordination delays, and hence increase concurrency. We implemented both mechanisms in the Scala language using the integrated Akka framework. Experiments on a 48-core server show that our extensions provide important performance benefits over sequential processing on both read-dominated and write-dominated workloads.

Future work regards dynamic switching between sequential and concurrent processing and the investigation of a real-world application.

Acknowledgements. This research has been funded in part by the European Community's Seventh Framework Programme under the ParaDIME Project (www.paradime-project.eu), grant agreement no. 318693.

References

1. Hewitt, C., Bishop, P., Steiger, R.: A universal modular actor formalism for artificial intelligence. In: IJCAI 1973: Proceedings of the 3rd International Joint Conference on Artificial Intelligence, pp. 235–245 (1973)
2. Karmani, R.K., Shali, A., Agha, G.: Actor frameworks for the jvm platform: a comparative analysis. In: PPPJ 2009: Proceedings of the 7th International Conference on Principles and Practice of Programming in Java, pp. 11–20 (2009)
3. Haller, P.: On the integration of the actor model in mainstream technologies: the scala perspective. In: Proceedings of the 2nd Edition on Programming Systems, Languages and Applications Based on Actors, Agents, and Decentralized Control Abstractions, AGERE! 2012, pp. 1–6. ACM, Tucson (2012)
4. Imam, S.M., Sarkar, V.: Integrating task parallelism with actors. In: Proceedings of the ACM International Conference on Object Oriented Programming Systems Languages and Applications, OOPSLA 2012, Tucson, Arizona, USA, pp. 753–772 (2012)
5. Agha, G.A., Mason, I.A., Smith, S.F., Talcott, C.L.: A foundation for actor computation. Journal of Functional Programming 7(1), 1–72 (1997)
6. Karmani, R.K., Agha, G.: Actors. In: Padua, D. (ed.) Encyclopedia of Parallel Computing, pp. 1–11. Springer (2011)

7. Goodman, D., Khan, B., Khan, S., Luján, M., Watson, I.: Software transactional memories for scala. Journal of Parallel and Distributed Computing (2012)
8. Harris, T., Larus, J., Rajwar, R.: Transactional Memory, 2nd edn. Morgan and Claypool Publishers (2010)
9. Scholliers, C., Tanter, E., Meuter, W.D.: Parallel actor monitors. In: SBLP 2010: 14th Brazilian Symposium on Programming Languages, Salvador, Brazil (2010)
10. ScalaSTM, http://nbronson.github.com/scala-stm/
11. Brito, A., Fetzer, C., Sturzrehm, H., Felber, P.: Speculative out-of-order event processing with software transaction memory. In: DEBS 2008: Proceedings of the International Conference on Distributed Event-Based Systems, pp. 265–275 (2008)

An Optimal Broadcast Algorithm
for Content-Addressable Networks

Ludovic Henrio, Fabrice Huet, and Justine Rochas

Univ. Nice Sophia Antipolis, CNRS, I3S, UMR 7271, 06900 Sophia Antipolis, France
ludovic.henrio@cnrs.fr, fabrice.huet@unice.fr, justine.rochas@inria.fr

Abstract. Structured peer-to-peer networks are powerful underlying structures for communication and storage systems in large-scale setting. In the context of the Content-Addressable Network (CAN), this paper addresses the following challenge: how to perform an efficient broadcast while the local view of the network is restricted to a set of neighbours? In existing approaches, either the broadcast is inefficient (there are duplicated messages) or it requires to maintain a particular structure among neighbours, e.g. a spanning tree. We define a new broadcast primitive for CAN that sends a minimum number of messages while covering the whole network, without any global knowledge. Currently, no other algorithm achieves those two goals in the context of CAN. In this sense, the contribution we propose in this paper is threefold. First, we provide an algorithm that sends exactly one message per recipient without building a global view of the network. Second, we prove the absence of duplicated messages and the coverage of the whole network when using this algorithm. Finally, we show the practical benefits of the algorithm throughout experiments.

Keywords: Broadcast, Peer-to-Peer, Content-Addressable Network.

1   Introduction

In this work, we are interested in *Structured Overlay Networks* (SONs) where peers are organised in a well-defined topology and resources are stored at a deterministic location. The underlying geometric topology is used by communication primitives and ensures their efficiency. We are interested in CAN (Content-Addressable Network) [1] P2P networks, where peers are organised according to a multi-dimensionary cartesian space. This space is organised in a geometrical way; the geometrical organisation dictates the dependencies between peers, as we will see in Section 2.

This paper presents a broadcast algorithm for the CAN overlay network that prevents a peer from receiving the same message more than once. We call such a broadcast algorithm *efficient*, in the sense that it minimises the number of exchanged messages between peers. Of course, a broadcast algorithm also has to be *correct*, and reach every peer of the network.

R. Baldoni, N. Nisse, and M. van Steen (Eds.): OPODIS 2013, LNCS 8304, pp. 176–190, 2013.

In previous works, Bongiovanni and Henrio proved, using the Isabelle/HOL theorem prover, that an efficient broadcast protocol for CAN *existed* [2]. However, the algorithm that was exhibited to prove the *existence* of an optimal solution was naive and had a very high latency, making it unusable in practice. In this work, we are interested in the design and implementation of an effective broadcast protocol that, in practice, also has an acceptable latency.

The contribution of this paper is as follows:

- Firstly, we propose a new broadcast algorithm that greatly improves the state of the art.
- Secondly, we prove that this algorithm is both correct (it covers the whole network) and optimal in terms of exchanged messages.
- Thirdly, we set an experimental comparison of the algorithm with others in a realistic distributed environment and show its efficiency in practice.

This paper is organised as follows. First, Section 2 will show that several broadcast algorithms exist for CAN but none of them was able to completely remove duplicated messages purposes. Section 3 will present our broadcast algorithm, together with its proof of efficiency and correctness. Section 4 will present the evaluation of our algorithm over a distributed peer-to-peer network. Finally Section 5 concludes this paper.

2   Related Works and Objectives

2.1   Context and Motivation

A *CAN* [3] is a structured P2P network based on a d-dimensional Cartesian coordinate space labeled $\mathcal{D}$. This space is dynamically partitioned among all peers in the system such that each node is responsible for storing data, in the form of *(key, value)* pairs, in a sub-zone of $\mathcal{D}$. To store a (k, v) pair, the key k is deterministically mapped onto a point in $\mathcal{D}$ and the value v is stored by the node responsible for the zone comprising this point. The search for the value corresponding to a key k is achieved by applying the same deterministic function on k to find the node responsible for storing the corresponding value. These two mechanisms are performed by an iterative routing process starting at the query initiator and which traverses its adjacent neighbours (a peer only knows its neighbours), and so on and so forth until it reaches the zone responsible for the key to store/retrieve. One can find several definitions for a valid CAN, i.e. which shape can the zone of each peer have and how peers can be organised (see [2]). Here we rely on a very generic and simple definition: each zone is an hyperrectangle, and the only structure is the neighbouring relation: each peer only knows the peers whose zones are adjacent to its own zone. Additionally, a CAN is a torus, and the peers on the left border know the ones on the right border, but we will not use this feature in this paper. Figure 1 shows a 2-dimension CAN and some exchanged messages between neighbours.

Filali et al. [4] used a CAN to store large set of RDF data, and to perform queries taken from the BSBM benchmark [5]. They realised that the multicast

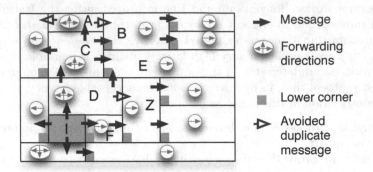

Fig. 1. M-CAN - Message forwarding

queries over several dimensions of the CAN did not scale properly because even the best performing broadcast algorithm generates a lot of duplicate messages (Section 4). These messages take valuable network resources, decreasing the overall performance. Our objective is to design an efficient broadcast algorithm that minimises the number of communications and that is only based on local information in a CAN.

2.2 Positioning

Problem Statement. The basic problem of optimal broadcast in a CAN is that, as a CAN is a P2P network, each peer only has information about the zone it manages, and the zones managed by its neighbours. Consequently, it is impossible to split the entire network into sub-spaces such that each zone exactly belongs to one sub-space: in Figure 1, the initiator has no knowledge about **Z** and cannot know that it must give the whole responsibility for zone **Z** to either **D** or **F**. Indeed, the initiator could decide that **F** is responsible for the lower half of **Z**, and that **D** is responsible for the upper half. In that case, **Z** would receive the message twice. It is possible to design an optimal algorithm based on sub-spaces, but this algorithm is inefficient because it almost never splits the space to be covered, and only one message is communicated at a time[1] [2]. Consequently, contrarily to the case of Chord [6], a broadcast algorithm for CAN that is both efficient and optimal cannot simply rely on the partitioning of the space to be covered.

Robustness. One can argue that having duplicated messages should increase the robustness of the algorithm in case of failure, but there are much more efficient ways to replicate the messages than an inefficient algorithm. A much better way to ensure robustness would be simply to perform two efficient broadcasts carrying the same message from two different initiators and along different directions. In M-CAN [7], for example, some nodes receive the message once, while

---

[1] More precisely the space to be covered is only split if it is not path-connected.

others can receive it an arbitrarily high number of times, in an unpredictable manner. This is clearly not the best way to ensure robustness.

Churns. Peers joining and leaving during a communication might require additional mechanisms to ensure that each peer correctly receives the message. Dealing with this issue generally relies on low-level synchronisations that depends on the implementation of CAN and is out of scope here. However, in order to tolerate churns between two broadcasts, our algorithm must rely only on the structure provided by CAN. For example, a classical additional structure for efficient broadcast is a *spanning tree* [8] but we do not use such a structure here because it is difficult and costly to maintain on an evolving CAN.

Multicast. A crucial question is whether the primitive we aim for is a broadcast or a multicast, i.e., whether it can be targeted at only some of the nodes. In M-CAN [7], the authors suggest to reduce the problem of multicast to the one of broadcast on another (CAN) network. While this approach is valid here, we are interested in multicast over a range of values, i.e. along hyperrectangles included in the CAN. Indeed, considering our definition of a CAN (each node is responsible for a hyperrectangle zone), the intersection between an hyperrectangle to be covered and a CAN remains a CAN, thus our algorithm is still valid to multicast on a range of coordinates, or to cover only a certain number of dimensions.

An alternative definition of CAN [3] keeps track of the history of joining nodes, which forms a tree. Using this tree as a spanning tree has two disadvantages: first, this would limit the contribution to a subset of all possible CAN. Second, this tree would not allow to perform range multicast because the restriction of a CAN to an hyperrectangle leads to disconnected branches.

Our approach is the only one that allows efficient multicast over any particular zone of a CAN, without relying on additional structures. Our algorithm additionally features the following characteristics (1) It can perform either *broadcast or range multicast.* (2) It avoids duplicates, while replication is generally needed in a peer-to-peer network; but for reliability reasons it should be added above an efficient algorithm *in a controlled way.* (3) It tolerates *churns in between two executions* of the algorithm as it only relies on the CAN structure; dealing with churns during communications could only be done specifically for a particular implementation of the CAN.

2.3   Related Works

A lot of work has been dedicated to broadcast and multicast on overlay networks. The availability of efficient algorithms depends mostly on the ability to build a spanning tree on the overlay. A tree-based system such as P-Grid [9] offers a natural support for broadcast. Others such as Chord [10], Tapestry [11] or Kademlia [12], can be seen as *k-ary trees.* Based on this observations, authors in [6] propose an efficient broadcast algorithm. Although this work is close to our own, it cannot be applied to CAN overlays, as building and maintaining a spanning tree is difficult and costly.

M-CAN [7] is an application-level multicast primitive which is almost efficient, but does not eliminate all duplicates if the space is not perfectly partitioned (i.e. if the zones managed by the peers have not an equal size). The authors measured 3% of duplicates on a realistic example. In a publish/subscribe context, Meghdoot [13], built atop CAN, proposes a mechanism that totally avoids duplicates but requires the dissemination to originate from one corner of the zone to be covered. In general, finding the corner of the area to be covered would introduce a significant overhead (in terms of messages), resulting in an inefficient broadcast.

Compared to those approaches, our algorithm can originate from any node of the CAN and still avoid duplicates. In this sense, we position our algorithm as an improvement of M-CAN that completely eliminates duplicates. Below, we describe more precisely the dissemination algorithm proposed by M-CAN, which is the closest work to our approach.

2.4   M-CAN

In the following, the broadcast starts from one particular node, that we will call the *initiator*. A message is sent along a given dimension (from 1 to D, where D is the dimension of the CAN), and according to a given *direction* (which is either *ascending* if the coordinates along the considered dimension are increasing, or *descending* in the other case). It is only possible to forward the message to a node that is a neighbour along the considered dimension and direction.

The basic steps of the M-CAN algorithm are as follows:

1. The initiator sends the message to all of its neighbours.
2. A node receiving the message from a neighbour along dimension i in direction *dir* will forward the message to neighbours:
   - along dimensions $1 \ldots (i\text{-}1)$
   - along dimension i in direction *dir*.

Figure 1 shows a 2-dimensional CAN where initiator **Init**, starts a broadcast. In this figure, since node **B** has received a message from **C** along dimension 1, in the *ascending* direction, node **B** will forward it only on the *ascending* direction in dimension 1. Node **C**, on the other hand, has received the message along dimension 2, in the *ascending* direction. Thus it will forward the message in both directions along dimension 1, and only in the *ascending* direction along dimension 2. In Figure 1, the set of directions that each node is responsible for is pictured with red circled arrows.

This algorithm can lead to duplicated messages. For example, node **B** receives the same message from **C** and **A**. A deterministic condition is used to remove some of the duplicates: a node only forwards the message if it abuts the lowest corner of the neighbour it wants to forward to. This deterministic condition is called the *corner criteria*. The lowest corner is defined here as the corner which touches the propagation dimension and minimises the coordinates in all other dimensions. According to this *corner criteria*, node **A** will not forward the message to node **B** since **A** does not touch **B**'s lowest corner.

However, this only removes duplicates arising from the first dimension and cannot be applied in higher dimensions, otherwise the correctness of the broadcast could not be ensured. This is why some duplicates are still left with this algorithm. For example in Figure 1, node **E** receives the message twice.

3   Efficient Broadcast Algorithm

Our algorithm extends M-CAN, and remove duplicated messages that arise in dimensions higher than one. For this, we introduce a *spatial constraint* that allows us to always apply the *corner criteria*: we always propagate on the first dimension of a constrained sub-CAN.

3.1   Principles

The algorithm reasons on a set of nodes, where each node manages a rectangular zone. Considering a dimension i, the lower bound and the upper bound of the zone managed by node N are denoted $N.LB[i]$ and $N.UB[i]$. We denote by D the dimension of the CAN. Each message is sent according to a dimension (between 1 and D), and according to a direction (either *ascending* or *descending*).

Remember that the *corner criteria* prevents duplicates along the first dimension on which all the nodes forward. To prevent duplicates in the second dimension, we constraint the algorithm to only send the message to nodes belonging to a particular hyperplane in the CAN space. Each of the nodes belonging to the hyperplane will be responsible for propagating the message along the first dimension. We define the hyperplane as a set of fixed values in each dimension but the last one. These values are arbitrarily chosen in the zone of the initiator and, together, form what we call the *spatial constraint*. This *spatial constraint* is then an hyperplane of dimension $d - 1$. The nodes belonging to this hyperplane form a sub-CAN of dimension $d-1$. So we can recursively apply our algorithm on this sub-CAN, with an hyperplane of dimension $d - 2$ as *spatial constraint*; and so on. When the hyperplane becomes a line, no duplicate can arise when following the propagation direction if we send the message to the only one neighbour that contains the line in this direction.

Here is how the algorithm works. When a message is received along dimension k, it is forwarded to neighbours along dimensions $1..k - 1$ in both directions, and along dimension k in only one direction (*ascending* or *descending*, identically to the reception). We then apply our additional condition: among the neighbours that are left, we send the message only to the ones that intersect the *spatial constraint* on dimensions $1..k - 1$, and that satisfy the *corner criteria* on dimensions $k + 1..d$. All dimensions but k are thus constrained either by the *spatial constraint* or by the *corner criteria*. We show that this ensures efficiency and correctness of the algorithm in Section 3.3.

Figure 2 illustrates the algorithm on the same configuration as Figure 1. In Figure 2, there is only one *spatial constraint* (on dimension 1) because the CAN only has two dimensions. In this case, it is set to the upper bound of the initiator

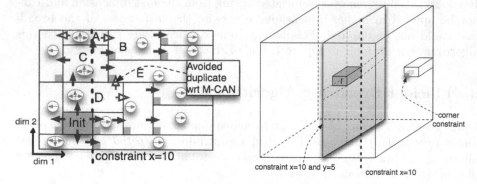

Fig. 2. Efficient broadcast in 2D   **Fig. 3.** Efficient broadcast in 3D

(node **Init**), where $constraint x = 10$. When **D** receives the message from **Init** along dimension 2, **D** only forwards the message to neighbours which intersect the line defined with $constraint\ x = 10$. Here, **D** only sends the message to **C**. **E** is also a neighbour of **D** along dimension 2 in the *ascending* direction, but **E** does not intersect the line. **E** will receive the message along dimension 1 afterwards. More formally, with a CAN of 2 dimensions, a node forwards the message to a neighbour if the following conditions are valid:

– when propagating along dimension 1:

$$Sender.LB[2] \leq Neighbor.LB[2] < Sender.UB[2]$$

– when propagating along dimension 2:

$$Neighbor.LB[1] \leq constraint[1] < Neighbor.UB[1]$$

As illustrated in Figure 3, this principle can be generalised to dimensions greater than 2. Thanks to our additional condition, we still have no duplicate. In dimension 3, the initiator first sends the message to the nodes intersecting a plane. In this plane, the problem is reduced to the example shown in Figure 2. In particular, one *spatial constraint* is used and a 2 dimensional *corner criteria* is applied. Then, when propagating along dimension 1, a three dimensional *corner criteria* is applied as depicted in Figure 3.

3.2   Broadcast Algorithm

We describe below the general algorithm in a more formal way. The data structures used in our algorithm are the following. A message embeds the *spatial constraint* that is transmitted without modification. The *spatial constraint* is a set of D coordinates that should represent a point belonging to the initiator node; for example it can be its lowest corner. constraint[i] denotes the ith coordinate of this constraint. As the *spatial constraint* is transmitted without modification together with the message, we denote it as a global value. Each message is sent and received along a given dimension (*dimension* $\in [1..D]$) and in a given direction (*direction* $\in \{ascending, descending\}$). Neighbours can be formally defined as follows:

Definition 1. *The neighbours of node N on dimension k and direction ascending are the set of nodes N' such that:*

$$N'.LB[k] = N.UB[k] \ \wedge \ \forall i \neq k. \ [N.LB[i], N.UB[i][\cap [N'.LB[i], N'.UB[i][\neq \emptyset$$

Symmetrically, neighbours of node N on dimension k and direction descending are the set of nodes N' such that:

$$N'.UB[k] = N.LB[k] \ \wedge \ \forall i \neq k. \ [N.LB[i], N.UB[i][\cap [N'.LB[i], N'.UB[i][\neq \emptyset$$

Algorithm 3.1. Efficient broadcast algorithm

```
1:  upon event reception of message M on dimension d0 and direction dir0 on node
2:      for each k≤d0 do
3:          if k=D+1 then
4:              direction← ∅
5:          else
6:              if k < d0 then
7:                  direction← {descending,ascending}
8:              else
9:                  direction←dir0
10:         for each dir in direction do
11:             for each neighbour on dimension k and direction dir do
12:                 for each i in 1 .. k − 1 do                    ▷ Spatial Constraint
13:                     if not ( neighbour.LB[i] ≤ constraint[i] < neighbour.UB[i]) then
14:                         skip neighbour
15:                 for each i in k + 1 .. D do                    ▷ Corner Criteria
16:                     if not( node.LB[i] ≤ neighbour.LB[i] < node.UB[i]) then
17:                         skip neighbour
18:                 send message on dimension k and direction dir to neighbour
19: end event
```

The detailed algorithm is given in Algorithm 3.1. Upon message reception along dimension $d0$, a node must forward it along lower dimensions (line 2) in both directions (line 7), and along dimension $d0$ in the same direction (line 9). For each neighbour in the considered dimensions and directions, their coordinates in dimensions lower than the propagating dimension are checked against the *spatial constraints* (line 12-14), and their coordinates in dimensions higher than the propagating dimensions are checked against the *corner criteria* (line 15-17). The *spatial constraint* condition on a dimension i checks that the neighbour's zone contains the ith value of the *spatial constraint* in the dimension i:

$$neighbour.LB[i] \leq constraint[i] < neighbour.UB[i]$$

The *corner criteria* on dimension i checks that, along dimension i, the lower bound of the neighbour in the dimension i is in the zone of the sender:

$$node.LB[i] \leq neighbour.LB[i] < node.UB[i]$$

If a neighbour verifies both conditions, the message is sent to it. This algorithm is initiated by sending a broadcast message to the initiator from an artificial dimension $D + 1$ (line 3).

3.3   Properties of the Algorithm

In the following, we prove the main properties of the algorithm. Those properties ensure that each node of the CAN receives the message exactly once. We first introduce two lemmas that are crucial to prove the properties of the algorithm.

Lemma 1. *If node N sends a message to node N' along dimension d and in direction dir then:*

$$\forall i < d.\, N'.LB[i] \leq constraint[i] < N'.UB[i]$$

and if N' is not the initiator (i.e., $d \leq D$) then:

- *either dir = ascending and $N'.LB[d] > constraint[d]$,*
- *or dir = descending and $N'.UB[d] \leq constraint[d]$.*

Proof. By recurrence on the length of the path needed to reach node N', i.e., on the number of messages needed to reach node N'.

The initiator artificially receives a message from outside the CAN on dimension $D + 1$; Here it is sufficient to verify:
$\forall i < D + 1.\, N'.LB[i] \leq constraint[i] < N'.UB[i]$
As the constraint must belong to the initiator node, this is trivial.

Now suppose that N' is not the initiator; node N sends a message to node N' on dimension d and from direction dir. First, as the message was sent from node N (possibly the initiator), by executing Algorithm 3.1, the algorithm ensures that $\forall i < d.\, N'.LB[i] \leq constraint[i] < N'.UB[i]$, else N' would have been skipped at line 16. Second, suppose $dir = ascending$ (the message is sent towards increasing coordinates). Then two cases are possible:

- N is the initiator and $d < D + 1$, then $N.LB[d] \leq constraint[d] < N.UB[d]$ (because the constraint belongs to the initiator's zone).
- N is not the initiator, thus there was a message sent from N_0 to N on dimension d' and direction dir'. By definition of the algorithm, we have two possibilities:

- $d = d'$ and $dir' = ascending$; by recurrence hypothesis $N.LB[d] > constraint[d]$; additionally, we always have $N.UB[d] > N.LB[d]$.
- $d < d'$; in that case, by recurrence hypothesis $N.LB[d] \leq constraint[d] < N.UB[d]$.

In all cases, we have $N.UB[d] > constraint[d]$. As N' is a neighbour of N on dimension d and direction $ascending$, by Definition 1, $N'.LB[d] = N.UB[d]$, consequently, $N'.LB[d] > constraint[d]$.

The case where $dir = descending$ is similar: we have by recurrence $N.LB[d] \leq constraint[d]$, and by the neighbouring definition $N'.UB[d] \leq constraint[d]$.

The following corollary is a direct consequence of the preceding lemma.

Lemma 2 (Corollary). *If node N sends a message to node N' on dimension d and direction dir then:*

$$(\forall i < d'. N'.LB[i] \leq constraint[i] < N'.UB[i]) \Rightarrow d' \leq d$$

$$(N'.LB[i] > constraint[i] \vee N'.UB[i] \leq constraint[i]) \Rightarrow i \geq d$$

From the two lemmas above, we can prove the efficiency and correctness of the algorithm. First, our broadcast algorithm is efficient in the sense that the same message is never received twice by the same node:

Theorem 1 (Efficiency). *Two nodes cannot send the message to the same third one.*

Proof. We prove the theorem by contradiction: we suppose node N_1 sends the broadcast message on dimension d_1 and direction dir_1 to node N and that N_2 sends the broadcast message on dimension d_2 and direction dir_2 to node N, with $N_1 \neq N_2$.

Let us first prove that $d_1 = d_2$ by contradiction too. Suppose without loss of generality that $d_1 < d_2$, then by Lemma 1 applied on the message from N_2 to N on dimension d_2, as $d_1 < d_2$ we have $N.LB[d_1] \leq constraint[d_1] < N.UB[d_1]$. Additionally, by Lemma 1 applied on the message from N_1 to N we have either $dir = ascending$ and $N.LB[d_1] > constraint[d_1]$ or $dir = descending$ and $N.UB[d_1] \leq constraint[d_1]$. In both cases there is a contradiction; thus $d_1 = d_2$. Also $dir_1 = dir_2$, else the application of Lemma 1 would also lead to a contradiction.

Secondly, suppose again that $dir = ascending$ (the case $descending$ is similar). By definition of Algorithm 3.1, the message was not skipped at line 21, neither by N_1 nor N_2, and so:
$$\forall i \in d_1 + 1..D. N_1.LB[i] \leq N.LB[i] < N_1.UB[i]$$
$$\wedge\ N_2.LB[i] \leq N.LB[i] < N_2.UB[i].$$

Additionally, as N is neighbour of N_1 and N_2 along dimension d_1 and direction $ascending$,

$N.LB[d_1] = N_1.UB[d_1] = N_2.UB[d_1]$ (Definition 1). Finally, we also have:
$\forall i \in 1..d_1 - 1. N_1.LB[i] \leq$ constraint$[i] < N_1.UB[i]$
$\qquad \wedge \ N_2.LB[i] \leq$ constraint$[i] < N_2.UB[i]$,
because N_1 and N_2 themselves received the message on a dimension greater or equal to d_1 and by Lemma 1. Now consider the point P of coordinates:
(constraint$[1], .., $constraint$[d_1 - 1], N.LB[d_1] - \varepsilon, N.LB[d_1 + 1], .., N.LB[D])$
where ε is a small value (e.g., half the smallest dimension of the smallest zone of the CAN). The arguments above allow us to prove that P is both in the zone of N_1 and in the zone of N_2, which is contradictory with the definition of a CAN: each point of the Cartesian space is managed by one and only one node. Hence N_1 and N_2 are necessarily the same node.

We proved that Algorithm 3.1 is efficient. Note that showing that the initiator does not receive the message twice needs a separate but similar proof. Finally, we can prove that this broadcast algorithm covers the whole network. Overall, we show that each node of the CAN receives the message exactly once.

Theorem 2 (Coverage). *Each node of the network receives the message.*

The proof of this theorem is provided in the research report associated to this paper [14]. It heavily relies on Lemma 1 and 2. The principle is simple: according to Lemma 1 and 2 we can deduce the node that is "responsible" for sending the message to each node. Consider the highest dimension on which a node does not intersect the constraint. If, on this dimension, this node is above the constraint, then the responsible node is a neighbour located along this dimension in the *descending* direction that should send it the message. Additionally, we say that a node N_1 is "closer" than another node N_2 to the constraint if either the highest dimension on which the node does not intersect the constraint is bigger for N_2 than for N_1 (i.e. N_1 meets more constraints), or if this dimension is the same, and the lower bound of N_1 is closer to the constraint than N_2 on this dimension. The proof works by contradiction: we consider N_0, the uncovered node the "closest" to the constraint. We then prove that the node N' "responsible" for sending the message to N_0 effectively meets all the constraints for sending the message and that it is "closer" to the constraint. If N' received the message it should have sent it to N_0, and if it did not, the N_0 is not the closest uncovered node as N' is closer. This is contradictory.

It is worth noticing that it is easy to make our algorithm robust to communication failures. Indeed, it is sufficient to perform two independent broadcasts from two different initiators, and reversing the role of each dimension, this way each node receives the message exactly twice and from different senders.

4   Evaluation

In this section we present experiments highlighting the performance of our algorithm. We show that, in realistic situations, it significantly reduces the volume of data exchanged. We have based our implementation on the EventCloud [15] platform. Entirely written in Java, EventCloud is a system that uses CANs as the underlying structure for event processing. It currently runs a flooding-based (naive) broadcast algorithm. We have added a version of our algorithm and an implementation of M-CAN to this framework, and studied the performance of these three algorithms.

4.1   Variation of the Number of Peers

Experimental Setup. We have experimented on a grid of four geographically distant clusters, using up to 200 physical machines. All the machines involved in the experiment have two 4-core CPUs and at least 16GB of memory. In each site, the machines are linked with a 1Gb/s Ethernet network. Inter-site communications rely on a 10Gb/s dark fiber.

The software setup was as follows. In all experiments we built CAN overlays with a variable number of peers (from 50 to 1500) and 5 dimensions. Applications that use CAN usually vary from two to an infinite number of dimensions, as in works [16,17]. The improvement due to our algorithm is greater as the CAN has more dimensions, as detailed in [14]. We considered that 5 dimensions would be a good compromise to show that, even with a small number of dimensions, our algorithm can already achieve a meaningful speedup. Each peer runs in its own Java Virtual Machine and we ensure that no machine executes more than 8 peers. The construction of the overlay was performed using the canonical algorithm described in [1]: when a new peer wants to join the overlay, it randomly chooses a point in the whole space. It then finds the peer responsible for the zone where this point lies, and takes half of it.

Since we wanted our experiments to represent realistic scenarios and to compare the different algorithms in similar conditions, we have used the following experimental protocol:

1. A CAN is randomly built with a given number of peers.
2. For each algorithm, ten broadcasts are started simultaneously from different peers chosen at random.
3. Step 1 and step 2 are repeated ten times.

Experimental Results. Figure 4 shows the average number of exchanged messages per broadcast algorithm. The horizontal lines highlight the optimal (minimum) number of messages required to cover the entire network. The naive broadcast algorithm produces a high number of duplicate messages. By contrast, the M-CAN algorithm improves a lot the naive algorithm but a non-negligible number of duplicate messages is still left, especially in large networks. With 1500 peers, 395 duplicate messages are recorded on average. Moreover, from the error bars, we can see that the M-CAN algorithm is unpredictable. The number of messages is very dependent on the CAN configuration and on the location of the broadcast initiator. This is why a particular execution can generate up to twice the optimal number of messages. On the other hand, our algorithm always requires the minimum number of messages in order to reach every peer in the network.

We have measured the total size of exchanged data for each algorithm. Note that the messages did not contain any useful payload, thus we only measure the

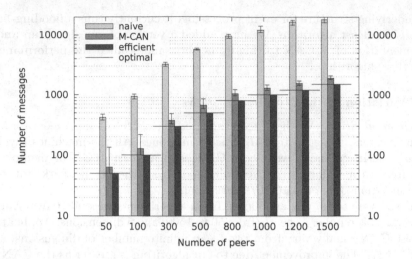

Fig. 4. Average number of messages and optimal number of messages with 5 dimensions

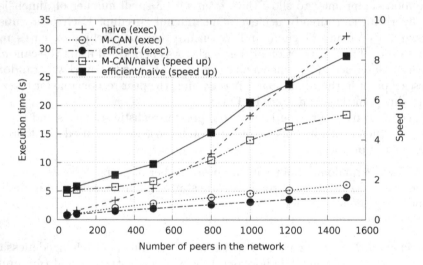

Fig. 5. Average execution time and speed up from the naive broadcast

cost of the broadcast operation. With 1500 peers in the network, the M-CAN algorithm generated 25.6 MB of data on average. Our algorithm generated only 20.3 MB of data on average, i.e. a 20% reduction. Aside, we have also experimented with various number of dimensions: from 2 to 15 [14]. The percentage of duplicate messages with 15 dimensions using M-CAN was 112%. But even with a small number of dimensions, from 3 to 5 dimensions, we measured in average 25% of duplicate messages with M-CAN. As a consequence, our algorithm is always significantly more efficient in terms of messages than M-CAN.

We have also measured the execution time, i.e. the time needed for each peer to receive at least one of the broadcast message. Figure 5 shows the average execution time of the three algorithms, and the speed up compared to the naive broadcast algorithm. The naive broadcast algorithm is significantly slowed down as the network grows. This is due to the quantity of duplicate messages that overload the network. On the other hand, both M-CAN and our algorithm maintain good performance as the network size increases. However, M-CAN exhibits a lower scalability because of the remaining duplicate messages. Compared to the naive broadcast on 1500 peers, M-CAN has only a speed-up of 5 whereas our algorithm reaches 8.

The previous experiments show that, although the number of duplicates with M-CAN is low, it still has a clear impact on realistic systems. Our algorithm, by totally avoiding duplicate messages, offers a significant improvement in terms of bandwidth and execution time, even when the CAN has a small number of dimensions.

5 Conclusion

In this article we have provided an algorithm for efficient broadcast over CAN peer-to-peer networks. We have proven that this algorithm covers the whole network, while preventing any node from receiving the same message twice. Moreover, it solely relies on the structure of the overlay and does not require to maintain a spanning tree, which would be too costly. To show the practical usefulness of our algorithm, we have implemented it in a large scale platform and performed extensive experiments using up to 1500 peers on 200 physical machines. Our experiments show that the algorithm scales and completely prevents duplicated messages. Compared to the previously best broadcast algorithm, we reduce the amount of data on the network by up to 20%. As a consequence, when performing a high number of parallel broadcast queries, we were able to show a significant speedup compared to existing solutions.

Overall, this article shows that CAN overlays can be used effectively as information dissemination architectures. One of the main advantages of our approach is that we rely on a very broad definition of CAN overlays: a CAN is a N-dimensional space partitioned into hyperrectangles. As a consequence, our algorithm can be adapted to many variants of CAN, as long as zones are hyperrectangles and neighbours correspond to adjacent zones.

Acknowledgments. This work was funded by the EU FP7 STREP PLAY (www.play-project.eu) and French ANR SocEDA (www.soceda.org). Experiments presented in this paper were carried out using the Grid'5000 experimental testbed, being developed under the INRIA ALADDIN development action with support from CNRS, RENATER and several Universities as well as other funding bodies (see www.grid5000.fr).

References

1. Ratnasamy, S., Francis, P., Handley, M., Karp, R., Shenker, S.: A Scalable Content-Addressable Network. In: Proceedings of the 2001 Conference on Applications, Technologies, Architectures, and Protocols for Computer Communications (SIGCOMM), pp. 161–172. ACM (2001)

2. Bongiovanni, F., Henrio, L.: A mechanized model for can protocols. In: Cortellessa, V., Varró, D. (eds.) FASE 2013 (ETAPS 2013). LNCS, vol. 7793, pp. 266–281. Springer, Heidelberg (2013)
3. Ratnasamy, S., Handley, M., Karp, R., Shenker, S.: A Scalable Content Addressable Network. In: Proceedings of the Third International COST264 Workshop on Networked Group Communications (August 2001)
4. Filali, I., Pellegrino, L., Bongiovanni, F., Huet, F., Baude, F., et al.: Modular p2p-based approach for rdf data storage and retrieval. In: Advances in P2P Systems (2011)
5. Bizer, C., Schultz, A.: The berlin sparql benchmark. International Journal on Semantic Web and Information Systems (2009)
6. El-Ansary, S., Alima, L., Brand, P., Haridi, S.: Efficient broadcast in structured P2P networks. In: Kaashoek, M.F., Stoica, I. (eds.) IPTPS 2003. LNCS, vol. 2735, pp. 304–314. Springer, Heidelberg (2003)
7. Ratnasamy, S., Handley, M., Karp, R.M., Shenker, S.: Application-Level multicast using Content-Addressable Networks. In: Crowcroft, J., Hofmann, M. (eds.) NGC 2001. LNCS, vol. 2233, pp. 14–29. Springer, Heidelberg (2001)
8. Perlman, R.: An algorithm for distributed computation of a spanningtree in an extended lan. In: Proceedings of the Ninth Symposium on Data Communications, SIGCOMM 1985. ACM (1985)
9. Aberer, K., Cudré-Mauroux, P., Datta, A., Despotovic, Z., Hauswirth, M., Punceva, M., Schmidt, R.: P-Grid: a self-organizing structured P2P system. ACM SIGMOD Record 32(3), 33 (2003)
10. Stoica, I., Morris, R., Karger, D., Kaashoek, M.F., Balakrishnan, H.: Chord: A Scalable Peer-to-Peer Lookup Service for Internet Applications. In: Proceedings of the 2001 Conference on Applications, Technologies, Architectures, and Protocols for Computer Communications (SIGCOMM), pp. 149–160. ACM, New York (2001)
11. Zhao, B.Y., Huang, L., Stribling, J., Rhea, S.C., Joseph, A.D., Kubiatowicz, J.D.: Tapestry: a resilient global-scale overlay for service deployment. IEEE Journal on Selected Areas in Communications 22(1), 41–53 (2004)
12. Maymounkov, P., Mazières, D.: Kademlia: A peer-to-peer information system based on the xor metric. In: Druschel, P., Kaashoek, M.F., Rowstron, A. (eds.) IPTPS 2002. LNCS, vol. 2429, pp. 53–65. Springer, Heidelberg (2002)
13. Gupta, A., Sahin, O.D., Agrawal, D.P., El Abbadi, A.: Meghdoot: content-based publish/subscribe over P2P networks. In: Jacobsen, H.-A. (ed.) Middleware 2004. LNCS, vol. 3231, pp. 254–273. Springer, Heidelberg (2004)
14. Henrio, L., Huet, F., Rochas, J.: An Optimal Broadcast Algorithm for Content-Addressable Networks – Extended Version. Research report (September 2013), http://hal.inria.fr/hal-00866228
15. INRIA: The EventCloud middleware (2012), http://www.play-project.eu/solutions/event-cloud
16. Li, M., Ye, F., Kim, M., Chen, H., Lei, H.: Bluedove: A scalable and elastic publish/subscribe service. In: IPDPS (2011)
17. Anceaume, E., Le Merrer, E., Ludinard, R., Sericola, B., Straub, G.: Fixme: A self-organizing isolated anomaly detection architecture for large scale distributed systems. In: Baldoni, R., Flocchini, P., Binoy, R. (eds.) OPODIS 2012. LNCS, vol. 7702, pp. 1–15. Springer, Heidelberg (2012)

On Local Fixing

Michael König and Roger Wattenhofer

Computer Engineering and Networks Laboratory,
ETH Zurich, 8092 Zurich, Switzerland
{mikoenig,wattenhofer}@ethz.ch

Abstract. In this paper we look at the difficulty of fixing solutions
of classic network problems. We study local changes in graphs (edge
resp. node insertion resp. deletion), and network problems (e.g. maximal
independent set, minimum vertex cover, spanning trees, shortest paths).
A change/problem combination is *locally fixable* if an existing solution
of a problem can be fixed in constant time in case of a local change in
the graph. We analyze a variety of well-studied classic network problems
with different characteristics.

Keywords: Local Fixing, Fault Tolerance, Graph Problems, Complexity Classes and Maximal Independent Set.

1 Introduction

Every driver knows about the buying vs. fixing dilemma: Is it worth it to repair
the old car, or should one instead rather buy a new model? This dilemma also
exists in the context of distributed computing: If a solution to a problem breaks
because of a small topology or input change, is it cheaper to fix the solution, or
should one rather compute a new solution from scratch? Clearly the answer to
this general question depends on many parameters, such as the studied problem,
or how broken a solution is, or the measure of cost for fixing and computing.

For the weighted matching problem, Lotker, Patt-Shamir, and Rosen proved
that fixing [21] is indeed strictly cheaper than computing [14]. Even more surprisingly, there are also examples where computing is cheaper than fixing. Kutten
and Peleg show that fixing a maximal independent set (MIS) is NP-complete
in a footnote in [17], whereas computing is known to take at most polylogarithmic time [22]. These two examples motivated our quest towards a better
understanding of the distributed complexity of fixing vs. computing.

In this paper we freeze two of the many parameters of the problem space.
First, we are only interested in whether graph changes can be fixed locally (in
constant time). Second, we assume that a solution is pretty much intact, i.e.,
the broken pieces are small, and well-separated in space or time. The topology
changes we are looking at in particular are deletions and insertions of single
nodes and edges, as they would happen in a moderately dynamic network. For
node changes we further differentiate between nodes with one or more edges.
We believe that this array of changes is a suitable model for typical failures in

R. Baldoni, N. Nisse, and M. van Steen (Eds.): OPODIS 2013, LNCS 8304, pp. 191–205, 2013.
© Springer International Publishing Switzerland 2013

real networks, where a single node might crash or a single edge could become disconnected. In our analysis we only cover one such change in the entire network, however, it is possible that several such changes happen, as long these changes are either well-separated in space (such that they do not influence each other) or time (such that there is enough time to fix one change before the next happens). We examine a diversity of well-studied classic network problems with different characteristics.

Our main findings are as follows: (i) Many problems that feature a constant or polylogarithmic distributed computing complexity can be fixed locally. However, there is no general rule, as there are exceptions. (ii) Global problems are generally not locally fixable. However, adding or removing leaves (nodes with a single edge) often seems to pose no difficulty. Again, there is no general rule, as there are exceptions. (iii) In addition, we show relations between different types of changes.

In summary, even though fixing is often cheaper than computing, in a mathematical sense the two are orthogonal. An overview of our concrete findings is given in Table 1.

Table 1. Overview of our results. On the left side we present the known lower and upper bounds to compute a solution for a given problem; these bounds are in the local model, where message size is not bounded. The problems are subdivided by their distributed complexity classes (local, polylogarithmic and global). On the right hand side we list the cost of fixing each problem/change combination (the shorthands for the changes are explained in Section 2.3). A "✔" entry means that the combination can be fixed locally (in constant time), a "✗" entry means that it is not possible to fix the combination locally, and "—" entries only appear in rows where the problem instance does not have edge weights, i.e., where weight changes are not defined. Note that there is a "✔" *and* a "✗" in every column, in all distributed complexity classes.

	Computation		Local Fixing						
	lower bound	upper bound	$+e$	$-e$	$w \to w'$	$+v_1$	$-v_1$	$+v_*$	$-v_*$
Γ_1-Count	$\Omega(1)$	$O(1)$	✔	✔	—	✔	✔	✔	✔
$o(n)$-MDS	$\Omega(1)$	$O(1)$ [16]	✗	✗	—	✗	✗	✗	✗
MIS	$\Omega(\sqrt{\log n})$ [14]	$O(\log n)$ [22]	✔[17]	✔[17]	—	✔[17]	✔[17]	✔	✔
$O(1)$-MWM	$\Omega(\sqrt{\log n})$ [14]	$O(\log n)$ [21]	✔	✔	✔	✔	✔	✔[21]	✔[21]
MM	$\Omega(\sqrt{\log n})$ [14]	$O(\log n)$ [12]	✔	✔	—	✔	✔	✔	✔
2-MVC	$\Omega(\sqrt{\log n})$ [14]	$O(\log n)$ [12]	✔	✔	—	✔	✔	✔	✔
$\Gamma_{\log n}$-Count	$\Omega(\log n)$	$O(\log n)$	✗	✗	—	✗	✗	✗	✗
ST	$\Omega(D)$	$O(D)$	✔	✗	—	✔	✔	✔	✗
MST	$\Omega(D)$	$O(D)$	✗	✗	✗	✔	✔	✗	✗
SPT	$\Omega(D)$	$O(D)$	✗	✗	✗	✔	✔	✗	✗
Flow	$\Omega(D)$	$O(D)$	✗	✗	✗	✔	✔	✗	✗
Leader	$\Omega(D)$	$O(D)$	✔	✔	—	✔	✔	✔	✔
Count	$\Omega(D)$	$O(D)$	✔	✔	—	✗	✗	✗	✗

2   Model

2.1   Distributed Computing

We are given a network modeled as a graph $G = (V, E)$, in which the nodes must base their computations and decisions on the knowledge about their local neighborhoods. More precisely, a distributed algorithm needs time t if each node $v \in V$ can decide based on its t-hop neighborhood $\Gamma_t(v)$. Nodes decide individually on their outputs without communication. Hence, the output of each node v is a function of $\Gamma_t(v)$.

This *neighborhood model*, first introduced by Linial [19], is related to the classic *message passing* model of distributed computing. In the message passing model, the distributed system is modeled as a communication network, again described by an undirected graph $G = (V, E)$. Each vertex $v \in V$ represents a node (host, device, processor, ...) of the network, and an edge $(u, v) \in E$ is a bidirectional communication channel that connects two nodes.

Initially, nodes have no knowledge about the network graph; they only know their own identifier and potential additional inputs. All nodes wake up simultaneously and computation proceeds in synchronous *rounds*. In each round, every node can send one message to each of its neighbors. A node may send different messages to different neighbors in the same round. Additionally, every node is allowed to perform local computations based on information obtained in messages of previous rounds. Communication is reliable, i.e., every message that is sent during a communication round is correctly received by the end of the round. A message passing algorithm has *time complexity* t if all nodes compute their output in t communication rounds.

If messages may be large, it is well known that the message passing model is equivalent to the neighborhood model, i.e., nodes can compute their output based on their t-hop neighborhood if and only if they can compute their output in t rounds of synchronous communication in the message passing model. This common t is known as the distributed time complexity.

Similarly, we can define the time t to fix a change to be either the size of the neighborhood $\Gamma_t(v)$ of a node v that is involved in the fix, or as the number of communication rounds t in a message passing algorithm to fix the change.

Various distributed complexity classes are known for t. The most important classes are

- *local* algorithms, where the time t is a constant independent of any parameter of the network, i.e., $t \in \Theta(1)$,
- *polylog* algorithms where the time t is polylogarithmic in the number of nodes n, i.e., $t \in \Theta(\text{polylog } n)$, and
- *global* algorithms which need $\Theta(D)$ time, where D is the diameter of the network.

Depending on the application, the boundary between local and polylog [23,27] or the boundary between polylog and global [19] are considered more important. In this paper we deal with all three classes. Regarding the fixing time, we are only interested in strictly local algorithms, i.e., a change must be fixed in constant time, in the $O(1)$-neighborhood. Regarding the computing time, we look at both the polylog and the global class in order to get a broader sense of the fixing vs. computing issue.

2.2 Network Problems

The different network problems we discuss are, grouped by complexity class:

- *local*
  - $o(n)$-Minimum Dominating Set
  - Counting the 1-neighborhood
- *polylog*
  - Maximal Independent Set
  - Maximal Matching
  - $O(1)$-Maximum Weighted Matching
  - 2-Minimum Vertex Cover
  - Counting the $\log n$-neighborhood
- *global*
  - Spanning Trees
  - Minimum Spanning Trees
  - Shortest Paths Tree
  - Maximum Flow
  - Leader Election
  - Counting the whole graph

For space reasons we omit the full problem definitions here and ask the interested reader to consult the full version.

2.3   Examined Graph Changes

We considered the following graph changes when examining the possibility of local fixing:

- *Edge insertion (+e):* adding a previously absent edge to the graph without changing the nodes of the graph.
- *Edge deletion (−e):* removing a previously present edge from the graph without changing the nodes of the graph.
- *Edge weight change (w → w′):* changing the weight of an already present edge in the graph without changing the nodes of the graph.
- *1-edge vertex insertion (+v₁):* adding a vertex to the graph plus a single edge connecting the new vertex to an existing one.
- *1-edge vertex deletion (−v₁):* removing a vertex which is only adjacent to one edge together with its edge from the graph.

– *Vertex insertion (+v*):* adding a vertex to the graph plus any amount of edges connecting the new vertex to existing ones.
– *Vertex deletion (−v*):* removing any vertex and all edges adjacent to it from the graph at once.

For weighted graphs inserted edges may have any positive weights assigned to them. The insertion and deletion of nodes without any edges is trivial for all the problems in question. We assume that after a change occurs all nodes directly adjacent to the change are notified of the exact kind of change that occurred.

Further, we allow treating a node "crash" (i.e., a sudden removal from the communication graph) as if the node gracefully "signed off" (organizing any necessary restructuring of the system prior to the node's departure). For this we let every node whose sudden removal would be critical create a "last will" and deploy it at its immediate neighbors. The last will contains the results a proper sign-off procedure would have had. To compute the last will, the sign-off procedure is simulated beforehand, which we require to be local (i.e., conclude within $O(1)$ rounds). Note that every time a state change in the graph could cause the results of a sign-off to change the respective last will must be computed and distributed anew. However, also note that this procedure does not affect the time complexity of computing or fixing a problem, as we require the computation of the last will to only take $O(1)$ rounds. We require last wills for some local fixability results in Sections 4.4 and 4.10.

Definition 1 (P^C Notation). *We write P^C to denote the problem of fixing a solution of the graph problem P after a graph change C. For instance, MIS^{+e} denotes the problem of fixing a maximal independent set after an edge insertion.*

3   Related Work

Distributed network algorithms have been studied ardently for almost 30 years. One of the most basic problems is the maximal independent set (MIS) problem. It was shown that the distributed computation of an MIS can be done in $O(\log n)$ time [2,22]. Closely related to the MIS problem is the maximal matching problem, as a maximal matching can essentially be computed by computing an MIS on the edges, and as such both algorithms are similar [12]. Since the vertices adjacent to a maximal matching are a 2-approximation for vertex cover, also 2-MVC can be solved in $O(\log n)$ time.

The study of distributed weighted matching is more recent, the first constant approximation in polylogarithmic time was shown less than a decade ago [28]. Later, [21] discovered that some of the steps of the algorithm of [28] can be executed in parallel, improving the distributed time complexity to $O(\log n)$. It was shown by [20] that one can even achieve a $(1 + \varepsilon)$-approximation in the same time, using a different method.

Kuhn et al. showed that a polylogarithmic approximation for MVC cannot be solved in less than polylogarithmic time [13,14]. Using reductions, one can immediately prove an $\Omega(\sqrt{\log n})$ lower bound for our problems with polylogarithmic distributed complexity. This lower bound was strengthening the earlier log-star lower bound by Linial [19], showing that all these problems (and some more) are indeed in the polylogarithmic distributed complexity class.

Our tree-based problems are in the global distributed complexity class, as one must send information across the whole network, and as such $\Omega(D)$ is a time lower bound. If message size is not bounded, just gathering all the information at all the nodes, and then computing the solution locally solves all problems in asymptotically optimal $O(D)$ time. Using a simple flooding process, one can compute a spanning tree in $O(D)$ time using small messages only. In the synchronous model, this spanning tree will be a shortest path tree. For the MST problem, it is not possible to get a solution in $O(D)$ time using short messages only [25,9,26]. For flow and other global problems, there are results which also suggest a distributed complexity polynomial in n [26,10]. Our overview table contains the results in the unbounded message size model, also known as the local model.

The subject of our paper is not so much the complexity of distributed *computing*, but rather the complexity of distributed *fixing*. Clearly, faults have played a major role in distributed computing since an early time. In fact, one may argue that distributed fixing was in fact studied even earlier, as early as in the 1970s when Dijkstra introduced the concept of *self-stabilization* [6,7]. In contrast to our work, a self-stabilizing algorithm must survive many failures, not just one, and as such it seems to be a difficult challenge. However, as shown 20 years ago [4,1,5], efficient self-stabilization often boils down to distributed *computation*. As such, surprisingly, computation and self-stabilization are more closely related than computation and fixing. See [8,18] for an overview. More recently, "self-healing" algorithms have gained attention [24,11].

Dynamic networks are another area related to our work, in which the graph topology is permanently changing, either because of changing environmental conditions (edge changes in wireless networks), mobility (edge changes because of moving nodes in mobile networks), algorithmic dynamics (edge changes due to algorithmic decisions in overlay networks), or churn (nodes constantly joining or leaving as in peer-to-peer systems). In dynamic networks no node is capable of maintaining up-to-date global information on the network. Instead, nodes have to perform their intended (global or polylogarithmic) task based on locally available information only, i.e., all computation in these systems is inherently local. In the last decade there was a tremendous rise in interest in dynamic networks, see [15] for an overview. This line of work is also more ambitious than ours in the sense that large fractions of the network can change concurrently. On the other hand, we restrict ourselves to constant time solutions.

Regarding fixing vs. computing, a most inspiring prior work is by Kutten and Peleg [17]. For the MIS problem, if $P \neq NP$, they show that fixing can be much harder than computing. For this, they consider a model, in which each node is in one of three states: ('1') in the MIS, ('0') not in the MIS, or ('?') forgot whether or not in the MIS. They then study how long it takes to compute the missing node states. 3SAT can be reduced to this problem in a straightforward way. We briefly describe the construction here, because Kutten and Peleg only mention it in a footnote, and did not bother to describe it in detail. Every clause of a 3SAT instance is represented by a node in state '0'. Every variable is represented by two connected nodes (one for true, one for false), both in state '?'. For each clause, there are 3 edges between the clause node and the variable nodes of the variables in the clause. We conclude that fixing an MIS in their model is NP-complete.

In a more relaxed model they consider fixing an MIS where every node knows whether it is in the MIS but may be in a conflicting state, i.e., be in the MIS while having a neighbor in the MIS or not being in the MIS while having no neighbors in the MIS. They present a transformation for MIS algorithms yielding a $O(\log x)$ randomized and a $2^{O(\sqrt{\log x})}$ deterministic fixing algorithm, where x is the number of nodes in conflicting states. Our model has a certain overlap with this model: In case a topology change in our model only puts a constant number of nodes into a conflicting state, their method also offers a local fix.

Another milestone is Chapter 4 of the previously mentioned paper by Lotker, Patt-Shamir, and Rosen [21], where they prove that their technique can be adapted to dynamic graphs. In fact, not only do they introduce our notion of topology changes, but they also show that a single node insertion or deletion with any amount of adjacent edges in a maximum weighted matching solution can indeed be fixed in constant time, keeping a constant approximation ratio. Since this beats the lower bound regarding the computational complexity for this problem, it is a nice example that fixing can be strictly easier than computing.

4   Results

An overview of our results can be found in Table 1. For space reasons we will omit some of the lemmas and proofs here and ask the interested reader to consult the full version, which contains proofs for all of the listed results.

In the following we will make use of the two graph classes defined below.

Definition 2 (Paths, Rings). *A path graph with n vertices is given by $G = (V, E)$:*

$$V = \{0, \ldots, n-1\},$$

$$E = \bigcup_{i=1}^{n-1} (i-1, i) .$$

A ring graph additionally has the edge $(0, n-1)$.

4.1   Graph Change Relationships

The different graph changes we are studying are related. The following lemmas summarize some implications that can be made.

Lemma 1. *For any graph problem P:*

- *If we can fix P^{+v_*} locally, we can also fix P^{+v_1} locally.*
- *If we can fix P^{-v_*} locally, we can also fix P^{-v_1} locally.*

Lemma 2. *For any weighted graph problem P, if we can fix both P^{+e} and P^{-e} locally, we can also fix $P^{w \to w'}$ locally.*

Lemma 3. *For any graph problem P, if we can fix both P^{+v_*} and P^{-v_*} locally, we can also fix P^{+e}, P^{-e} and $P^{w \to w'}$ locally.*

4.2   Vertex Counting

In this section we will discuss the problem of each node knowing the number of nodes in its r-neighborhood for different values of r.

While very straightforward, Γ_1-Count is a typical example of a problem which can be computed and also fixed in constant time:

Lemma 4. *Γ_1-Count^{+e}, Γ_1-Count^{-e}, Γ_1-Count$^{+v_1}$, Γ_1-Count$^{+v_*}$, Γ_1-Count$^{-v_1}$ and Γ_1-Count$^{-v_*}$ are local.*

Proof. After any change all directly adjacent nodes can simply recompute their count values. This requires $O(1)$ rounds. Any node not adjacent to a change will still have a valid count. The lemma follows.

For $r \in \omega(1)$, i.e., non-constant r, counts can generally not be fixed in constant time anymore:

Lemma 5. *For any $r \in \omega(1)$: Γ_r-Count$^{+v_1}$, Γ_r-Count$^{+v_*}$, Γ_r-Count$^{-v_1}$ and Γ_r-Count$^{-v_*}$ are not local.*

Proof. Adding or removing a node with any (positive) amount of edges anywhere requires updating the node counts in all nodes up to r hops away. This requires $r \notin O(1)$ rounds. The lemma follows.

Lemma 6. *For any $r \in \omega(1)$: Γ_r-Count^{+e} and Γ_r-Count^{-e} are not local if $r < D$; Γ_r-Count^{+e} and Γ_r-Count^{-e} are local for $r \geq D$.*

Proof. Consider a path graph. Removing edge $(0,1)$ or adding edge $(0, n-1)$ requires updating the node counts in all nodes with indices 0 through $r-1$. This requires $r - 1 \notin O(1)$ rounds. The first part of the lemma follows.

If $r \geq D$ every node is counting all nodes in the graph, since we are not considering graph changes which disconnect the graph. Adding and removing edges would not change any node counts in that case. The second part of the lemma follows.

4.3 Minimum Dominating Set

In this section we will discuss the problem of approximating minimum dominating sets. This problem does not allow for any local fixing and was chosen to give an example for this particular phenomenon. We are considering only non-trivial approximations, i.e., within $o(n)$ of the minimum dominating set.

Note that although we can compute an $o(n)$-MDS from scratch in constant time [16], fixing one within $O(1)$ hops of a graph change is an entirely different problem!

Lemma 7. $o(n)$-MDS^{+v_1} *is not local.*

Proof. First, we will show, that no algorithm can solve k-MDS^{+v_1} in any constant number of steps c ("locally"), for any k with $1 \leq k \leq \frac{n+1}{c} - 2$. Let us define:

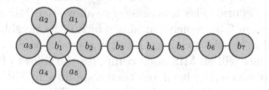

Fig. 1. Example graph with x=5 and y=7

$$x = \lfloor (k-1)(c+1) \rfloor,$$
$$y = 3c + 1,$$
$$G = (V, E),$$
$$V = (a_1, a_2, \ldots, a_x, b_1, b_2, \ldots, b_y),$$
$$E = \{(a_i, b_1) \mid 1 \leq i \leq x\} \cup \{(b_i, b_{i+1}) \mid 1 \leq i < y\},$$
$$U = \{a_i \mid 1 \leq i \leq x\} \cup \{b_2\} \cup \{b_{3i} \mid 1 \leq i \leq c\},$$
$$U^* = \{b_{3i+1} \mid 0 \leq i \leq c\} \ .$$

See fig. 1 for an example of G. Note that U is a k-MDS and U^* is a 1-MDS with respect to G. Adding a new vertex v and a new edge (b_y, v) to G will now invalidate U as a dominating set, while U^* still is a 1-MDS. While it is possible to create a new dominating set U' from U by fixing it locally, i.e., only within c hops of vertex b_y, at least one additional vertex will have to be added: $|U'| \geq |U| + 1$. Since U^* stayed the same, this entails an increased approximation factor k' for U':

$$k' = \frac{|U'|}{|U^*|} = \frac{x + c + 2}{c + 1} = \frac{\lfloor (k-1)(c+1) \rfloor + c + 2}{c + 1} > \frac{(k-1)(c+1) + c + 1}{c + 1} = k$$

Since $k' > k$, k-MDS^{+v_1} is not local for $1 \leq k \leq \frac{n+1}{c} - 2$, and hence $o(n)$-MDS^{+v_1} is not local.

The proofs for the non-locality of $o(n)\text{-}MDS^{-v_1}$, $o(n)\text{-}MDS^{+e}$ and $o(n)\text{-}MDS^{-e}$ are analogous.

4.4 Maximal Independent Set

In this section we will discuss fixing maximal independent sets. Kutten and Peleg [17] already showed that MIS^{+e}, MIS^{-e}, MIS^{+v_1} and MIS^{-v_1} can be fixed in constant time by running a transformed MIS algorithm. We will nevertheless still provide a set of simple proofs for those graph changes. Additionally, we will show that MIS^{+v_*} and MIS^{-v_*} are locally fixable as well.

We say a vertex is *covered* if it is part of the MIS or has a neighbor in the MIS. Note that all vertices in a graph being covered is a sufficient condition for an MIS to be maximal.

We assume that every MIS node knows its 2-hop-neighborhood and is made aware of changes to it (this can be achieved by flooding a message for 2 hops each time a change occurs). This is necessary to allow each MIS node to compute a last will (see Section 2.3), which contains which of its neighbors should enter the MIS in case of a "crash" to ensure retaining a valid MIS.

To compute its last will an MIS node computes the subset of its direct neighbors which are only covered by itself and then computes an MIS on the subgraph of only these neighbors and the edges between them. The nodes of the subgraph's MIS are then chosen to become MIS nodes of the actual graph should the node the last will is for fail. Note that this computation does not require any further messages to be exchanged. Hence, updating the last wills only adds $O(1)$ time to the fixing procedures for each graph change.

Below we will detail the actions which need to be taken in the cases of edge addition and removal of a node with any number of edges. The actions to be taken for the other graph changes are trivial and can be found the in the full version.

Lemma 8. MIS^{+e} *is local.*

Proof. The MIS can be fixed by doing the following: when an edge $e = (v, u)$ is added and both $v \in MIS$ and $u \in MIS$, pick one of v and u (for instance, whichever has the lower identifier), remove it from the MIS and add those nodes to the MIS which are designated in its last will.

If $v \notin MIS$ or $u \notin MIS$ the MIS remains valid: both nodes directly affected by the change are still covered by either being in the MIS themselves or having a neighbor in the MIS (we know this because we had a valid MIS prior to the edge insertion), and independence is still warranted since not both nodes are in the MIS.

Lemma 9. MIS^{-v_*} *is local.*

Proof. The case where the removed node is not part of the MIS is trivial – the remaining MIS on the remaining nodes is still valid. In the following we will consider the other case.

Without the node performing a "sign-off" (i.e., participating in the fixing before actually leaving) or an adequate preparation (such as a last will) it is not possible to salvage the MIS in constant time. To see this just imagine an arbitrarily complex graph where one node is connected to every other node. If an MIS is formed by that node alone, its unprepared removal would require computing a new MIS on the whole remaining graph which is known to take at least $\Omega(\sqrt{\log n})$ time.

Luckily, we stated that every MIS node deposits a last will at each of its neighbors stating which nodes should enter the MIS. This way every node can decide in constant time whether it should join the MIS.

4.5   Maximal Matching

Maximal matchings can be fixed locally as well. Two individual graph changes are discussed below.

We say an edge a *blocks* another edge b with respect to a matching M if the edges share a vertex and $a \in M$ and $b \notin M$. A matching being maximal is equivalent to every edge either being part of the matching or being blocked.

Lemma 10. MM^{-e} *is local.*

Proof. An edge being removed potentially allows for two edges to be added in turn: one at each of the vertices of the edge. Both vertices can identify and choose an unblocked edge adjacent to them to join the matching in constant time, which restores maximality.

Lemma 11. MM^{+v_*} *is local.*

Proof. Of the new edges at most one can become part of the matching, because they all share a vertex. No existing edge can become part of the matching through this change or the matching would not have been maximal before the change. Therefore, by picking any of the new edges which are not blocked (if there are any) and adding the picked edge to the matching, we can obtain a valid maximal matching again.

4.6   Spanning Trees

In this section we will discuss spanning trees which do not necessarily have minimum weight. We will not consider graph changes which cause the graph to become disconnected.

Lemma 12. ST^{-e} *and* ST^{-v_*} *are not local.*

Proof. Consider a spanning tree on a ring graph: it consists of all the graph's edges except for one at some vertex i. Removing vertex $(i + \lfloor \frac{n}{2} \rfloor) \bmod n$, or an edge adjacent to it, requires the edge at vertex i to be added to the spanning tree. For this to happen messages must be sent across up to $\lfloor \frac{n}{2} \rfloor \in \Omega(n)$ links. Hence, ST^{-e} and ST^{-v_*} cannot be fixed locally.

4.7 Minimum Spanning Trees

In this section we will discuss minimum spanning trees. We will not consider graph changes which cause the graph to become disconnected.

Lemma 13. MST^{-e} and MST^{-v*} are not local.

The proof for Lemma 13 follows that of Lemma 12.

Lemma 14. $MST^{w \to w'}$ is not local.

Proof. Consider a minimum spanning tree on a ring graph where every edge has weight 1: it consists of all the graph's edges except for one at some vertex i. Increasing the weight of an edge adjacent to vertex $(i + \lfloor \frac{n}{2} \rfloor) \bmod n$ by any amount requires the edge at vertex i to be added to the minimum spanning tree. For this to happen messages must be sent across up to $\lfloor \frac{n}{2} \rfloor \in \Omega(n)$ links. Hence, $MST^{w \to w'}$ cannot be fixed locally.

Lemma 15. MST^{+e} and MST^{+v*} are not local.

Proof. Consider a minimum spanning tree on a path graph where every edge has weight 1 except for the edge between vertices $\lfloor \frac{n}{2} \rfloor$ and $\lfloor \frac{n}{2} \rfloor + 1$ which has weight 2: it consists of all the graph's edges. Adding an edge between vertices 0 and $n - 1$ with weight 1, or adding a vertex with two edges of weight 1 to vertices 0 and $n - 1$ of the original graph, requires the edge with weight 2 to be removed from the minimum spanning tree. For this to happen messages must be sent across up to $\lfloor \frac{n}{2} \rfloor \in \Omega(n)$ links. Hence, MST^{+e} and MST^{+v*} cannot be fixed locally.

4.8 Shortest Paths Trees

In this section we will discuss shortest paths trees. We will not consider graph changes which cause the graph to become disconnected or which remove the root of the SPT.

Lemma 16. SPT^{-e} and SPT^{-v*} are not local.

The proof for Lemma 16 follows that of Lemma 12. Which node the SPT is rooted in is irrelevant for this proof.

Lemma 17. $SPT^{w \to w'}$ is not local.

Proof. Consider a SPT rooted in node 0 on a ring graph where every edge has weight 1: it consists of all the graph's edges except for one adjacent to node $\lfloor \frac{n}{2} \rfloor$. Increasing the weight of the edge $(0, 1)$ to n requires the missing edge to be inserted into the spanning tree replacing edge $(0, 1)$. For this to happen messages must be sent across up to $\lfloor \frac{n}{2} \rfloor \in \Omega(n)$ links. Hence, $SPT^{w \to w'}$ cannot be fixed locally.

4.9 Maximum Flow

In this section we will discuss maximum flows. We will not consider graph changes which cause source and sink to become to become parts of different graph components or which remove source or sink.

Lemma 18. $Flow^{w \to w'}$, $Flow^{-e}$ and $Flow^{-v*}$ are not local.

Proof. Consider a ring graph where all edge weights are 1 and which has an additional vertex v_{source} which is only attached to vertex $\lfloor \frac{n}{2} \rfloor$ over an edge with weight 1. Let v_{source} be the flow's source and vertex 0 be the flow's sink. All maximum flows on this graph have a strength of 1 and are divided into two parts which travel over vertices $\{0, 1, \ldots, \lfloor \frac{n}{2} \rfloor\}$ and over vertices $\{0, \lfloor \frac{n}{2} \rfloor, \ldots, n-1\}$ respectively.

Decreasing the weight of either edge $(0, 1)$ or edge $(n-1, 0)$ below the strength of the part of the flow on that respective side will require the flow across all edges to be changed (save for the edge adjacent to v_{source}). The same may be caused by removing or removing the adjacent non-sink vertex of either edge $(0, 1)$ or edge $(n-1, 0)$. Hence, $Flow^{w \to w'}$, $Flow^{-e}$ and $Flow^{-v*}$ cannot be fixed locally.

Lemma 19. $Flow^{+e}$ and $Flow^{+v*}$ are not local.

Proof. Consider a path graph where all edge weights are 1. Let vertex 0 be the source of the flow and let vertex $\lfloor \frac{n}{2} \rfloor$ be the sink of the flow. Any maximum flow only uses the edges $\{(a-1, a) \mid 0 < a \leq \lfloor \frac{n}{2} \rfloor\}$.

Adding an edge between vertices 0 and $n-1$, or adding a vertex with two edges of weight 1 to vertices 0 and $n-1$ of the original graph, requires any maximum flow on the resulting graph to use *all* edges. Hence, $Flow^{+e}$ and $Flow^{+v*}$ cannot be fixed locally.

4.10 Leader Election

In this section we will discuss the problem of fixing a leader election. Note that we do not require any node but the leader itself to know who the leader is. The sole requirement is that there is exactly one leader at any time. We will not consider graph changes which cause the graph to become disconnected.

This problem is particularly interesting, because computing it initially takes $\Omega(D)$ rounds [3], while fixing requires little to no effort and can always be done in constant time.

Lemma 20. $Leader^{+e}$, $Leader^{-e}$, $Leader^{+v_1}$, $Leader^{-v_1}$, $Leader^{+v*}$ and $Leader^{-v*}$ are local.

Proof. In all cases where merely edges or non-leader nodes get added or removed, we do not need to change the leader node. This takes constant time.

To cover cases in which the leader node gets deleted, we will make use of the "last will" technique again (see Section 2.3). The leader node has at all times exactly one last will deployed at one of its neighbors, stating that that node should become the leader should the leader node be deleted. However, the node should not become a leader if merely the edge to the leader is deleted; in that case it should scrap the last will. Should the last will node be deleted or should its edge to the leader node be deleted, the leader node will issue a new last will. These operations ensure that there is always exactly one leader after any graph change and also take constant time.

References

1. Afek, Y., Kutten, S., Yung, M.: Memory-efficient self stabilizing protocols for general networks. In: van Leeuwen, J., Santoro, N. (eds.) WDAG 1990. LNCS, vol. 486, pp. 15–28. Springer, Heidelberg (1991)
2. Alon, N., Babai, L., Itai, A.: A fast and simple randomized parallel algorithm for the maximal independent set problem. Journal of Algorithms 7(4), 567–583 (1986)
3. Awerbuch, B.: Optimal distributed algorithms for minimum weight spanning tree, counting, leader election, and related problems. In: Proceedings of the Nineteenth Annual ACM Symposium on Theory of Computing, pp. 230–240. ACM (1987)
4. Awerbuch, B., Sipser, M.: Dynamic networks are as fast as static networks. In: Proc. of 29th Annual Symposium on Foundations of Computer Science (FOCS), pp. 206–219. IEEE (1988)
5. Awerbuch, B., Varghese, G.: Distributed program checking: a paradigm for building self-stabilizing distributed protocols. In: Proc. of 32nd Annual Symposium on Foundations of Computer Science (FOCS), pp. 258–267. IEEE (1991)
6. Dijkstra, E.W.: Self-stabilization in spite of distributed control. Manuscript EWD391 (October 1973)
7. Dijkstra, E.W.: Self-stabilizing systems in spite of distributed control. Communications of the ACM 17(11), 643–644 (1974)
8. Dolev, S.: Self-stabilization. The MIT Press (2000)
9. Elkin, M.: Unconditional lower bounds on the time-approximation tradeoffs for the distributed minimum spanning tree problem. In: Proc. of the 36th ACM Symposium on Theory of Computing (STOC), Chicago, USA, pp. 331–340 (2004)
10. Frischknecht, S., Holzer, S., Wattenhofer, R.: Networks cannot compute their diameter in sublinear time. In: Proc. of the 23rd Annual ACM-SIAM Symposium on Discrete Algorithms (SODA), pp. 1150–1162. SIAM (2012)
11. Hayes, T.P., Saia, J., Trehan, A.: The forgiving graph: a distributed data structure for low stretch under adversarial attack. Distributed Computing 25(4), 261–278 (2012)
12. Israeli, A., Itai, A.: A fast and simple randomized parallel algorithm for maximal matching. Information Processing Letters 22(2), 77–80 (1986)
13. Kuhn, F., Moscibroda, T., Wattenhofer, R.: What cannot be computed locally! In: Proc. of the 23rd ACM Symposium on the Principles of Distributed Computing (PODC), pp. 300–309. ACM (2004)
14. Kuhn, F., Moscibroda, T., Wattenhofer, R.: Local computation: Lower and upper bounds. CoRR, abs/1011.5470 (2010)
15. Kuhn, F., Oshman, R.: Dynamic networks: Models and algorithms. ACM SIGACT News 42(1), 82–96 (2011)
16. Kuhn, F., Wattenhofer, R.: Constant-Time Distributed Dominating Set Approximation. Springer Journal for Distributed Computing 17(4) (May 2005)

17. Kutten, S., Peleg, D.: Tight fault locality. In: Proc. of 36th Annual Symposium on Foundations of Computer Science (FOCS), pp. 704–713. IEEE (1995)
18. Lenzen, C., Suomela, J., Wattenhofer, R.: Local algorithms: Self-stabilization on speed. In: Proc. of 11th International Symposium on Stabilization, Safety, and Security of Distributed Systems (SSS), Lyon, France (November 2009)
19. Linial, N.: Locality in distributed graph algorithms. SIAM Journal on Computing 21(1), 193–201 (1992)
20. Lotker, Z., Patt-Shamir, B., Pettie, S.: Improved distributed approximate matching. In: Proc. of the 20th Annual Symposium on Parallelism in Algorithms and Architectures (SPAA), pp. 129–136. ACM (2008)
21. Lotker, Z., Patt-Shamir, B., Rosen, A.: Distributed approximate matching. In: Proc. of the 26th Annual ACM Symposium on Principles of Distributed Computing (PODC), Portland, Oregon, USA, pp. 167–174. ACM (2007)
22. Luby, M.: A simple parallel algorithm for the maximal independent set problem. SIAM Journal on Computing 15, 1036–1053 (1986)
23. Naor, M., Stockmeyer, L.: What can be computed locally? In: Proc. of the 25th Annual ACM Symposium on Theory of Computing (STOC), San Diego, California, USA, pp. 184–193. ACM (1993)
24. Pandurangan, G., Trehan, A.: Xheal: localized self-healing using expanders. In: Proceedings of the 30th Annual ACM SIGACT-SIGOPS Symposium on Principles of Distributed Computing, pp. 301–310. ACM (2011)
25. Peleg, D., Rubinovich, V.: A near-tight lower bound on the time complexity of distributed minimum-weight spanning tree construction. SIAM Journal on Computing 30(5), 1427–1442 (2001)
26. Sarma, A.D., Holzer, S., Kor, L., Korman, A., Nanongkai, D., Pandurangan, G., Peleg, D., Wattenhofer, R.: Distributed verification and hardness of distributed approximation. Arxiv preprint arXiv:1011.3049 (2010)
27. Suomela, J.: Survey of local algorithms. ACM Computing Surveys (2011)
28. Wattenhofer, M., Wattenhofer, R.: Distributed weighted matching. In: Guerraoui, R. (ed.) DISC 2004. LNCS, vol. 3274, pp. 335–348. Springer, Heidelberg (2004)

A Skiplist-Based Concurrent Priority Queue with Minimal Memory Contention*

Jonatan Lindén and Bengt Jonsson

Uppsala University, Department of Information Technology
P.O. Box 337, SE-751 05 Uppsala, Sweden
{jonatan.linden,bengt}@it.uu.se

Abstract. Priority queues are fundamental to many multiprocessor applications. Several priority queue algorithms based on skiplists have been proposed, as skiplists allow concurrent accesses to different parts of the data structure in a simple way. However, for priority queues on multiprocessors, an inherent bottleneck is the operation that deletes the minimal element. We present a linearizable, lock-free, concurrent priority queue algorithm, based on skiplists, which minimizes the contention for shared memory that is caused by the DELETEMIN operation. The main idea is to minimize the number of global updates to shared memory that are performed in one DELETEMIN. In comparison with other skiplist-based priority queue algorithms, our algorithm achieves a 30 - 80% improvement.

Keywords: Concurrent Data Structures, Priority Queue, Lock-free, Non-blocking, Skiplist.

1 Introduction

Priority queues are of fundamental importance in many multiprocessor applications, ranging from operating system schedulers, over discrete event simulators, to numerical algorithms. A priority queue is an abstract data type, containing a set of key-value pairs. The keys are ordered, and typically interpreted as priorities. It supports two operations: INSERT of a given key-value pair, and DELETEMIN, which removes the pair with the smallest key and returns its value. Traditionally, priority queues have been implemented on top of heap or tree data structures, e.g., [10]. However, for priority queues that are accessed by large numbers of concurrent processor cores, skiplists [15] are an increasingly popular basis. A major reason is that skiplists allow concurrent accesses to different parts of the data structure in a simple way. Several lock-free concurrent skiplist implementations have been proposed [3,4,16].

The performance of skiplist-based data structures can scale well when concurrent threads access different parts of the data structure. However, priority

---

* This work was supported in part by the Swedish Foundation for Strategic Research through the CoDeR-MP project as well as the Swedish Research Council within the UPMARC Linnaeus centre of Excellence.

R. Baldoni, N. Nisse, and M. van Steen (Eds.): OPODIS 2013, LNCS 8304, pp. 206–220, 2013.

queues offer the particular challenge that all concurrent DELETEMIN operations try to remove the same element (viz. the element with the smallest key). This makes DELETEMIN the obvious bottleneck for scaling to large numbers of cores. In existing skiplist-based concurrent priority queues [8,11,17], the deletion of an element proceeds in two phases: first, the node is logically deleted by setting a delete flag in the node; second, the node is physically deleted by moving pointers in adjacent node(s). Both the logical and the physical deletion involve at least one global update operation, which must either be protected by a lock, or use atomic primitives such as Compare-and-Swap (CAS). Each CAS is expensive in itself, but it also incurs other costs, viz. (i) concurrent CAS operations to the same memory cell cause overhead due to contention, since they must be serialized and all but one will fail, and (ii) any other write or read to the same memory location (more precisely, the same cache line) by another core must be serialized by the coherence protocol, thus generating overhead for inter-core communication. In our experimentation, we have found that the global update operations in the DELETEMIN operation are the bottleneck that limits scalability of priority queues. To increase scalability, one should therefore devise an implementation that minimizes the number of such updates.

In this paper, we present a new linearizable, lock-free, concurrent priority queue algorithm, which is based on skiplists. The main advantage of our algorithm is that almost all DELETEMIN operations are performed using only a single global update to shared memory. Our algorithm achieves this by not performing any physical deletion of nodes in connection with logical deletion. Instead, our algorithm performs physical deletion in batches when the number of logically deleted nodes exceeds a given threshold. Each batch deletion is performed by simply moving a few pointers in the sentinel head node of the list, so that they point past logically deleted nodes, thus making them unreachable. Thus only one CAS per DELETEMIN operation is required (for logical deletion). To enable this batch deletion, we have developed a novel technique to maintain that logically deleted nodes always form a prefix of the skiplist.

Since logically deleted nodes are not immediately physically deleted, this implies that subsequent operations may have to perform a larger number of read operations while traversing the list. However, these reads will be cheap, so that the algorithm overall performs significantly better than previous algorithms, where these reads would conflict with concurrent physical deletions, i.e., writes.

The absence of physical deletion makes our algorithm rather simple in comparison to other lock-free concurrent skiplist algorithms. It is furthermore linearizable: in the paper, we present a high-level proof, and we report on a verification effort using the SPIN model checker, which we have used to verify linearizability by extensive state-space exploration [9,19].

We have compared the performance of our algorithm to two skiplist-based priority queue algorithms, each of which employs one of the currently existing DELETEMIN techniques: (i) a lock-free adaptation of Lotan and Shavit's non-linearizable priority queue [11], which is similar to the algorithm by Herlihy and Shavit [8], and (ii) an algorithm which uses the same procedure for DELETEMIN as the algorithm by Sundell and Tsigas [17]. Our algorithm achieves a performance improvement of 30 - 80 % in relation to the compared algorithms, on a limited set of benchmarks. The implementation of the algorithm, and the SPIN model, are both available at user.it.uu.se/~jonli208/opodis2013.

Furthermore, by comparing our algorithm to a specifically designed micro-benchmark, we show that for many cores, it is entirely limited by the logical deletion mechanism in DELETEMIN.

In summary, this paper shows a technique for removing scalability bottlenecks in concurrent data structures that are caused by concurrent memory accesses to the same memory location. We show that it is crucial to minimize the number of concurrent global updates, and present a novel internal representation of a skiplist that allows to use only a single global update per DELETEMIN operation. We hope that this work will inspire analogous work to develop other concurrent data structures, e.g., concurrent heaps, that minimize conflicts due to concurrent updates.

The paper is organized as follows. In Section 2, we give an overview of related work. In Section 3, we present the main new ideas of our algorithm. In Section 4, we present our algorithm in detail, and prove its correctness in Section 5. The performance evaluation is shown in Section 6. Section 7 contains conclusions.

2   Related Work

Skiplists were first proposed for concurrent data structures by Pugh [14,15], one reason being that they easily allow concurrent modification of different parts of the list. Using skiplists for priority queues was first proposed by Lotan and Shavit [11]. They separated logical and physical deletion, as described in the previous section, which allowed physical deletion of different nodes to proceed in parallel. This still incurs costs of type (ii) (i.e., serializing CAS operations with other accesses to the same memory cell), since many other threads are simultaneously accessing the list. By adding a timestamping mechanism, their algorithm was made linearizable. A lock-free adaptation of this algorithm was presented by Herlihy and Shavit [8], who also observed the contention caused by concurrent physical deletions.

Sundell and Tsigas [17] were first to present a lock-free implementation of a skiplist-based priority queue. Their DELETEMIN operation performs logical deletion followed by physical deletion. To achieve linearizability, only a single logically deleted node at the lowest level of the skiplist is allowed at any point in time. Any subsequent thread observing a logically deleted node will help complete the physical deletion. This may reduce latency in some cases, but suffers in general a high cost of both type (i) and (ii). Their skiplist further has back pointers to speed up deletion of elements, and employs back-off to reduce contention of the helping scheme.

Crain et al. [1] propose a technique to reduce memory contention by deferring the physical insertion and removal of logically deleted nodes to a point in the execution when contention is low. In the context of skiplist-based priority queues, this would reduce contention between conflicting global updates, but not reduce the *number* of needed updates, as is done in our algorithm. Thus, they do not reduce the contention between updates and concurrent reads.

Hendler et al. [6] present a methodology, called flat-combining, which reduces contention in arbitrary concurrent data structures (exemplified using a priority queue) that employ coarse-grained locking. In their approach, contended operations are combined into bigger operations, that are handled by the thread currently owning unique access to the structure. Their methodology is orthogonal to ours.

3   Overview of Main Ideas

In this section, we informally motivate and describe our algorithmic invention.

Today's multicore processors are typically equipped with a non-uniform memory architecture, and a cache coherence system, which provides a coherent view of memory to the processor cores. Whenever a core updates a shared memory location, the cache system must first invalidate copies in caches at cores that have previously accessed this location, and afterwards propagate the update to caches in cores that subsequently access the location. The effect is that updates (i.e., writes) cause high latencies if they update memory locations that are accessed by other cores. Thus, a limiting factor for scalability of concurrent data structure is the number of global updates that must be performed to locations that are concurrently accessed by many cores.

Let us now discuss the overhead caused by DELETEMIN operations, and how our algorithm reduces it. Skiplists are search structures consisting of hierarchically ordered linked lists, with a probabilistic guarantee of being balanced. The lowest-level list is an ordered list of all stored elements. Higher-level lists serve as shortcuts into lower-level lists, achieving logarithmic search time. Fig. 1a shows a skiplist with 3 elements, having keys 1, 2, and 5. There are sentinel head and tail nodes at the beginning and end. In existing lock-free skiplists [4,8,17], a node is deleted by first logically deleting it, by setting a delete flag in it. Thereafter it is physically deleted, by moving pointers in adjacent node(s). In Fig. 1a, nodes 1 and 2 have been logically deleted by setting their delete flags, indicated by black dots. In order to physically remove them, all pointers in the head node and node 1 must be moved. Both logical and physical deletion of a node thus require global update operations, typically CASes. During this deletion, many other threads will read the pointers in the first nodes of the list, thus incurring large costs by invalidations and update propagation in the cache coherence system.

In our new algorithm, we implement the DELETEMIN operation *without* performing physical deletion of nodes, in the sense that nodes are never unlinked from the list. Physical deletion is performed in batches, i.e., instead of performing a physical deletion after each logical deletion, we update the pointers in the head node to remove a large number of nodes from the list at a time. The reduction in physical deletion implies that concurrent DELETEMIN operations may have to perform a larger number of read operations when traversing the

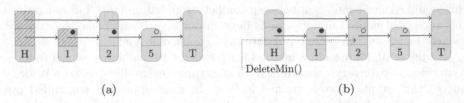

(a)   (b)

Fig. 1. To the left, memory positions affected by physical deletion of node 1 and 2 in skiplist. To the right, ongoing deletion of node 5 in the new algorithm.

list to find the node to be deleted. However, due to the microarchitecture of today's processors, the cost of these reads, relative to the latencies incurred by an increased number of global writes (e.g., CAS), will be very cheap. A read of a non-modified memory position can be up to 30 times faster than that of a modified memory position, in a multi-socket system [13].

A prerequisite for our scheme is that the logically deleted nodes always form a prefix of the list. This is not the case in existing skiplist-based priority queue algorithms. We have therefore added two mechanisms to our algorithm, whose combined effect is the essence of our algorithmic invention.

1. The delete flag, which signals logical deletion of a node is colocated with the pointer of the *preceding* node (for example in the least-order bit, as detailed by Harris [5]), and not in the deleted node itself.
2. The list contains always at least one logically deleted node (or has never contained such a node).

The first mechanism prevents insertions in front of logically deleted nodes, and the second mechanism guarantees that insertions close to the first non-deleted node are safe. Without the second, INSERT would no longer be guaranteed to be correct, as a consequence of the first mechanism.

Fig. 1b shows how the list in Fig. 1a is represented in our algorithm. Note how the delete flags for nodes 1 and 2 are located in the preceding nodes. A typical DELETEMIN operation only needs to traverse the lowest level of the skiplist, and set the delete flag of the first node in which it is not already set.

To illustrate how our location of the delete flag prevents insertions in front of logically deleted nodes, Fig. 2 depicts a situation in which nodes 1 and 2 are deleted, and two concurrent threads are active: a DELETEMIN operation is about to set the delete flag in node 2, and an INSERT(3) operation is about to move the lowest level pointer of node 2 to point to the new node 3. Two outcomes are possible. If INSERT(3) succeeds first, resulting in the situation at bottom right, the DELETEMIN operation will just proceed to set the delete bit on the new pointer pointing to the new node 3, which is then logically deleted. If DELETEMIN succeeds first, resulting in the situation at top right, the CAS of the INSERT(3) will fail, thus preventing insertion in front of the logically deleted node 5, since the pointer has changed, and INSERT will thus have to restart.

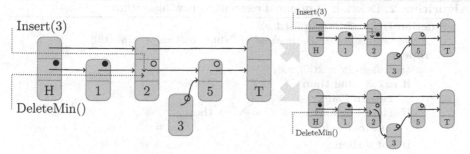

Fig. 2. Concurrent DeleteMin and Insert operation in the new algorithm

4   The Algorithm

In this section, we describe our algorithm in detail. A skiplist stores a set of key-value pairs, ordered according to the keys. Internally, it consists of a set of hierarchically ordered linked lists. The lowest-level list is a complete ordered list of all stored key-value pairs, which also defines the logical state of the skiplist. Higher level linked lists are sublists of the lists at lower levels, and serve as shortcuts into lower levels. Thus, a skiplist with only one level is simply an ordinary linked list. We assume unique keys, but duplicates could be handled as detailed in previous work [4,17].

A node in the skiplist is a structure, as described in Algorithm 1, which contains a value, a key, and an array of next pointers, which contains one pointer for

Algorithm 1. Node and skiplist structures

```
 1  Structure node_t:
 2      value_t value
 3      key_t key
 4      bool d
 5      bool inserting
 6      node_t *next[]

 7  Structure skiplist_t:
 8      node_t *head
 9      node_t *tail
10      integer nlevels
```

each level in which the node participates. Each node also contains a delete flag d, which is true if the successor of the node is deleted from the logical state of the list. In the actual implementation, the delete flag is stored together with the lowest-level next pointer, in its least-order bit, so that atomic operations can be performed on the combination of the lowest-level next pointer and the delete flag. In our pseudocode, we use the notation $\langle x.\text{next}[0], x.\text{d}\rangle$ to denote the combination of a pointer $x.\text{next}[0]$ and a delete flag $x.\text{d}$, which can be stored together in one word. Each node also has a flag **inserting**, which is true until INSERT has completed the insertion.

The DELETEMIN **Operation.** The DELETEMIN operation is shown in Algorithm 2. In the main **repeat-until** loop, it traverses the lowest-level linked list from its head, by following next pointers (which are read at line 4) until it finds a node, whose delete flag (checked at line 9) has not yet been set. At line 5, it is checked whether this next pointer points to the dummy tail node, in which case DELETEMIN returns EMPTY. Otherwise, DELETEMIN attempts to set the delete flag using a CAS instruction. If the CAS fails, then either the delete flag

---

Algorithm 2. Deletion of minimal element in new algorithm

```
 1  function DELETEMIN(skiplist_t q)
 2      x ← q.head, offset ← 0, newhead ← NULL, oldhead ← x.next[0]
 3      repeat
 4          ⟨nxt, d⟩ ← ⟨x.next[0], x.d⟩
 5          if nxt = q.tail then                // If queue is empty, return.
 6              return EMPTY
 7          if x.inserting and newhead = NULL then
 8              newhead ← x                     // Head may not surpass pending insert.
 9          if not d then                       // If succ. of x not deleted.
10              if not CAS(&⟨x.next[0], x.d⟩, ⟨nxt, 0⟩, ⟨nxt, 1⟩) then  //Set x's del. bit.
11                  continue                    // CAS failed. Retry the loop, still at x.
12          offset ← offset +1
13          x ← x.next[0]                       // Traverse list to next node.
14      until not d                             // If delete bit of x set, traverse.
15      v ← x.value                             // Exclusive access to node x, save value.
16      if offset < BOUNDOFFSET then return v
17      if newhead = NULL then newhead ← x
18      if CAS(&⟨q.head.next[0], q.head.d⟩, ⟨oldhead, 1⟩, ⟨newhead, 1⟩) then
19          RESTRUCTURE(q)                      // Update head's upper level pointers.
20          cur ← oldhead
21          while cur ≠ newhead do              // Mark segment for memory reclamation.
22              nxt ← cur.next[0]
23              MARKRECYCLE(cur)
24              cur ← nxt
25      return v
```

---

has already been set by some other DELETEMIN operation, or an insertion has completed between x and nxt, and the loop is retried. Otherwise, it has successfully deleted the node, and may safely read its value (at line 15). After the traversal, it checks whether the prefix of logically deleted nodes (measured by variable *offset*) has now become longer than the threshold BOUNDOFFSET. If so, DELETEMIN tries to update the next[0] pointer of q.head to the value of *newhead*, using a CAS instruction; it can be proven that all nodes that precede the node pointed to by *newhead* are already logically deleted. It should be noted that *newhead* will never go past any node with inserting set to true (i.e., in the process of being inserted), as ensured by the lines 7 and 17. If the CAS is successful, this means that DELETEMIN has succeeded to physically remove a prefix of the lowest-level list. If the CAS is unsuccessful, then some other DELETEMIN operation has started to perform the physical deletion, and the operation returns. In the successful case, the operation must proceed to update the higher level pointers, which is done in the RESTRUCTURE operation, shown in Algorithm 3. After the completion of the RESTRUCTURE operation, the DELETEMIN operation proceeds to mark the nodes between the observed first node, *oldhead*, and the *newhead* node, as ready for recycling (lines 21 - 24).

The RESTRUCTURE operation updates the pointers of q.head, except for level 0, starting from the highest level. At each level, the state of the head's pointer at the current level is first recorded (at lines 3 and 12). Thereafter, that level is traversed until encountering the first node not having a deleted successor, at lines 8 - 9. The head node's next pointer at that level is then updated by means of a CAS, at line 10. If another thread is concurrently modifying the pointers, the CAS operation may fail. The same procedure will then be repeated for the same level.

Algorithm 3. RESTRUCTURE operation

1 **function** RESTRUCTURE(skiplist_t q)
2 $\quad$ $i \leftarrow q$.nlevels -1
3 $\quad$ $h \leftarrow q$.head.next[i], $cur \leftarrow h$
4 $\quad$ **while** $i > 0$ **do**
5 $\quad\quad$ **if** $cur = q$.tail **then**
6 $\quad\quad\quad$ $i \leftarrow i - 1$
7 $\quad\quad\quad$ **continue**
8 $\quad\quad$ **while** cur.d **do**
9 $\quad\quad\quad$ $cur \leftarrow cur$.next[i]
10 $\quad\quad$ $success \leftarrow$ CAS(&q.head.next[i], h, cur)
11 $\quad\quad$ **if** $success$ **then** $i \leftarrow i - 1$
12 $\quad\quad$ $h \leftarrow q$.head.next[i]
13 $\quad\quad$ **if not** $success$ **then** $cur \leftarrow h$

Algorithm 4. Insertion of node with priority k

1 **function** INSERT(skiplist_t q, key_t k, value_t v)
2 $\quad$ $height \leftarrow$ RANDOM(1, q.nlevels), $new \leftarrow$ ALLOCNODE($height$)
3 $\quad$ new.key $\leftarrow k$, new.value $\leftarrow v$, new.d $\leftarrow 0$, new.inserting $\leftarrow 1$
4 $\quad$ **repeat**
5 $\quad\quad$ ($preds$, $succs$, $skew$) $\leftarrow$ LOCATEPREDS(q, k)
6 $\quad\quad$ new.next[0] $\leftarrow succs$[0] $\quad\quad$ //Prepare new to be inserted.
7 $\quad$ **until** CAS(&$\langle preds$[0].next[0], $preds$[0].d$\rangle$, $\langle succs$[0], 0$\rangle$, $\langle new, 0\rangle$)
8 $\quad$ $i \leftarrow 1$
9 $\quad$ **while** $i < height$ **and not** $skew$ **do** $\quad$ //Insert node at higher levels.
10 $\quad\quad$ new.next[i] $\leftarrow succs$[i] $\quad\quad$ //Set next pointer of new node.
11 $\quad\quad$ **if** new.d **then** $\quad\quad$ //new already deleted, finish.
12 $\quad\quad\quad$ **break**
13 $\quad\quad$ **if** CAS(&$preds$[i].next[i], $succs$[i], new) **then**
14 $\quad\quad\quad$ $i \leftarrow i + 1$ $\quad\quad$ //If success, ascend to next level.
15 $\quad\quad$ **else**
16 $\quad\quad\quad$ ($preds$, $succs$, $skew$) $\leftarrow$ LOCATEPREDS(q, k)
17 $\quad\quad\quad$ **if** $succs$[0] $\neq new$ **then break** $\quad$ //New has been deleted.
18 $\quad$ new.inserting $\leftarrow 0$ $\quad\quad$ //Allow batch deletion past this node.

The INSERT Operation. The INSERT operation is similar to a typical concurrent insert operation in skiplist algorithms [4,8,17]. The main difference is that the logically deleted prefix has to be taken into account. The operation (Algorithm 4) works as follows. The new node is first initialized at lines 2 - 3. Thereafter, a search for the predecessor nodes, at each level, of the position where the new key value is to be inserted, is performed by the operation LOCATEPREDS (at line 5). The LOCATEPREDS operation itself is shown in Algorithm 5.

Once the candidate predecessors and successors have been located, the new node is linked in at the lowest level of the skiplist, by updating the lowest-level predecessor's next pointer to point to *new* using CAS, at line 7. Note that this pointer also contains the delete flag: this implies that if meanwhile *succs*[0] has been deleted, the delete bit in *preds*[0].next[0] has been modified, and the CAS fails. If the CAS is successful, then INSERT proceeds with insertion of the *new* node at higher levels. This is done bottom up, so that a node is visible on all lower levels before it is visible on a higher level. As a consequence, it is always possible to descend from a level $i + 1$ list to a level i list.

The insertion at higher levels first sets the *new* node's next pointers to point to the candidate successor, at line 10. Thereafter, it checks whether the successor of the *new* node has been deleted since *new* was inserted, at line 11. If so, this means that *new* is deleted as well, in which case the insertion is considered completed. Otherwise, the level i insertion is attempted at line 13. If successful, the insertion ascends to the next level. If unsuccessful, predecessors and successors will be recorded anew (line 16), and the insertion procedure will be repeated for level i. If at any point *skew* is true, as reported by LOCATEPREDS, the insertion at higher level is aborted. When INSERT is done, completion of the insertion is signaled to other threads by setting inserting to 0, at line 18.

The LOCATEPREDS operation (Algorithm 5), locates the predecessors of a new node that is to be inserted. Starting with the highest-level list, each list is traversed, until a node with a greater key is found, or until either *a*) a node not having its delete flag set is found *and* the level, i, is greater than 0, or *b*) a non-deleted node is found at the lowest level (lines 5 to 9). When such a node is found, the search descends to the level below, or is completed. While traversing level i, if at some point *succs*[i+1] is ob-

---

Algorithm 5. LOCATEPREDS operation

1 **function** LOCATEPREDS(skiplist_t q,key_t k)
2      $i \leftarrow q$.nlevels -1,*skew* $\leftarrow 0$, *pred* $\leftarrow q$.head
3   **while** $i \geq 0$ **do**
4          $\langle cur, d \rangle \leftarrow \langle pred.\text{next}[i], pred.d \rangle$
5   **while** *cur*.key $< k$ **or** *cur*.d **or**
             (d **and** $i = 0$) **do**
6   **if** $i < q$.nlevels -1 **and**
                 $cur = succs[i + 1]$ **then**
7   *skew* $\leftarrow 1$
8   *pred* $\leftarrow cur$
9              $\langle cur, d \rangle \leftarrow \langle pred.\text{next}[i], pred.d \rangle$
10   *preds*[i] $\leftarrow$ *pred*
11   *succs*[i] $\leftarrow$ *cur*
12          $i \leftarrow i - 1$
13   **return** $(preds, succs, skew)$

---

served, this is signaled to the INSERT operation by means of the *skew* variable, at line 7. This situation occurs when the insertion point is close to the first non-deleted node of the queue.

Memory Management. The memory of deleted nodes have to be safely deallocated. This is particularly difficult for non-blocking algorithms, since a thread may be observing outdated parts of the shared state. We use Keir Fraser's epoch based reclamation (EBR) [4] to handle memory reclamation. In short terms, EBR works as follows. Each thread signals when it enters and leaves an operation that accesses the skiplist. After a node has been marked for memory reclamation by some thread, the node will be deallocated (or reused) only when all threads that may possibly have a (direct or indirect) reference to it have returned. Since in our algorithm, a node is marked for recycling only when it can no longer be reached from any pointer of form q.head.next[i], we guarantee that any thread that enters after the marking, cannot reach a node that is recycled.

One property of EBR is that the memory reclamation is not truly non-blocking, but in general it is very fast. Since we focus on global performance, we do not consider this a problem. If it is important to have non-blocking memory reclamation, there are other standard solutions, such as hazard pointers [12] or reference counting [18] for memory reclamation.

5   Correctness and Linearizability

In this section, we establish that our algorithm is a linearizable [7] implementation of a priority queue. We first establish a sequence of invariants of the algorithm. Thereafter, we prove that the algorithm is a linearizable implementation of a priority queue. The linearizability of the algorithm has also been verified using the SPIN model checker, and the model is available at the companion website of the paper. In this SPIN model, we use the same approach as Vechev et al. [19], i.e., whenever the model reaches a linearization point, the update is verified against the state of a sequential model.

In the rest of this section, we will outline a correctness proof for our algorithm. Let us consider the (global) state of the algorithm at some point during its execution. By a *node*, we mean a node of type node_t that has been allocated at some previous point in time. A *head-pointed node* is a node that is pointed to by a head pointer (i.e., a pointer of the form q.head.next[i]). A *live node* is a node that is reachable by a sequence of 0 or more pointers of form next[i] from a head-pointed node. A *deleted node* is a node that is pointed to by a lowest-level pointer (of form next[0]) from a node, which has its delete flag set to 1. A *recycled* node is a node that has been marked for recycling by MARKRECYCLE.

Lowest-Level Invariants. We first establish a set of invariants that characterize the lowest-level list. For each invariant, we provide a short proof sketch which motivates why it is preserved by our algorithm.

A) The set of live nodes and the next[0]-pointers form a linear singly linked list, terminated by the node q.tail.
Proof Sketch: Since nodes are not unlinked from the list, the invariant follows by noting that an insertion into the lowest-level list (at line 8 of INSERT) always inserts new nodes between two contiguous ones.
B) The set of live deleted nodes form a strict prefix of the live nodes in the lowest-level list.
Proof Sketch: The statements that may affect this invariant are: (i) DELETEMIN, line 10, which logically deletes x.next[0]: since x is either deleted or the

head node, DELETEMIN may only extend an existing prefix of deleted nodes, and (ii) INSERT, line 8, which inserts a node: LOCATEPREDS and the semantics of CAS guarantee that then $preds[0].next[0]$ does not have its delete flag set, i.e., the node following the inserted one is not deleted.

C) An inserted node n such that $n.$inserting is set, is live.

Proof: By observing the statements at line 14 - 15 of DELETEMIN, together with the CAS at line 18, we see that the $q.$head.next$[0]$ will not point past n.

D) The non-live nodes are partitioned into sets of nodes, such that for each set, either *a*) all nodes in the set have been marked for recycling, or *b*) a thread is allotted to the set, which currently executes lines 21 - 24 of DELETEMIN, and is in the process of marking all nodes in the set for recycling.

Proof: Nodes are made non-live at line 18 of DELETEMIN, together with the RESTRUCTURE, at line 19. If that CAS statement succeeds, then the thread gains exclusive access to the segment of non-live nodes between the nodes pointed to by local variables *oldhead* and *newhead*. The segments are disjoint since *oldhead* of a DELETEMIN invocation must equal *newhead* of a preceding DELETEMIN invocation.

E) Only non-live nodes are marked for recycling.

Proof: Follows from invariant D, and by noting that when DELETEMIN reaches line 24, then RESTRUCTURE guarantees that $q.$head.next$[i]$, for any i, is reachable from the node pointed to by *newhead*.

Higher Level Invariants. We then establish invariants for the higher-level lists.

F) When LOCATEPREDS returns, then, for any i, the node pointed to by $preds[i]$ is reachable from the node pointed to by $preds[i+1]$, via a sequence of next$[i]$ pointers. Conversely, for any i, the node pointed to by $succs[i+1]$ is reachable from the node pointed to by $succs[i]$, or $skew = 1$.

Proof: The reachability follows by the mechanism for traversing nodes at lines 5-9 in LOCATEPREDS. That $skew = 1$, follows by the test at line 6 in LOCATEPREDS, since at that point, the future node pointed to by $succs[i]$ will be reachable from *cur*.

G) Whenever n is a live node, for any $i > 0$, the node pointed to by $n.$next$[i]$ is reachable from n by a non-empty sequence of next$[i-1]$ pointers.

Proof: The critical statements in INSERT are (i) line 7, which inserts a new node at level 0, between two existing nodes, and (ii) line 13, which is correct for the same reason as in the preceding case if the pointers concern level i, and if the pointers concern level $i + 1$, then by invariant F, the inserted node is reachable from $preds[i+1]$ by a non-empty sequence of next$[i]$ pointers.

Linearizability. We can now establish that our algorithm is a linearizable implementation of a priority queue. Recall that a priority queue can be abstractly specified as having a state which is an ordered sequence of (key,value)-pairs, and two operations: DELETEMIN, which removes the pair with the smallest key value, and INSERT, which inserts a (key,value)-pair at the appropriate position.

To establish that our algorithm is a linearizable implementation, we first define the *abstract state* of our implementation, at any point in time, to be the sequence of (key,value)-pairs in non-deleted live nodes, connected by next[0] pointers. To prove linearizability, we must then for each operation specify a *linearization point*, i.e., a precise points in time at which the implemented operation affects the abstract state. These are as follows.

DELETEMIN: The linearization point is at line 10 of DELETEMIN in the case that the operation succeeds: clearly the abstract state is changed in the appropriate way. In the case of an unsuccessful DELETEMIN, it linearizes at line 4, if the statement assigns q.tail to nxt (this is discovered at line 5).

INSERT: The operation is linearized when the new node is inserted at line 7 of INSERT: clearly the new node is inserted at the appropriate place. Note that INSERT cannot fail: it will retry until successful.

6   Performance Evaluation

In this section, we evaluate the performance of our new algorithm in comparison with two lock-free skiplist-based priority queues, representative of two different DELETEMIN approaches. We also relate its performance to the limits imposed by the necessity of performing one global update per DELETEMIN operation.

The compared algorithms are all implemented on top of Keir Fraser's [4] skiplist implementation, an open source state of the art implementation. Since the same base implementation is used for all compared algorithms, the performance differences between the implementations are directly related to the algorithms themselves, and hence, the DELETEMIN operations. The two compared algorithms are respectively based on:

- *SaT* – Sundell and Tsigas' [17] linearizable algorithm, in which only the first node of the lowest- level list in the skiplist may be logically deleted, at any given moment. If a subsequent thread observes a logically deleted node, it will first help with completing the physical deletion. The algorithm implemented here is a simplified version of their algorithm, neither using back link pointers nor exponential back-off, but it uses the same DELETEMIN operation.
- *HaS* – Herlihy and Shavit's [8] non-linearizable lock-free adaptation of Lotan and Shavit's [11] algorithm, in which DELETEMIN operations are distributed over a prefix of the queue, using atomically updated delete flags. Physical deletion is initiated directly after the logical deletion. Insertions are allowed to occur between logically deleted nodes.

The algorithms are evaluated on two types of synthetic microbenchmarks, and as a part of one real application, URDME [2]. The benchmarks are:

- *Uniform* – Each thread randomly chooses to perform either an INSERT or a DELETEMIN operation. Inserted keys are uniformly distributed. This is the de facto standard when evaluating concurrent data structures [5,11,17].

- *DES* – The second synthetic benchmark is set up to represent a discrete event simulator (DES) workload, in which the priorities represent the time of future events. Hence the key values are increasing: each deleted key generates new key values that are increased by an exponentially distributed offset.
- *URDME* – URDME [2] is a stochastic DES framework for reaction-diffusion models. In this benchmark, the priority queue is used as the event queue in URDME. A model with a queue length of 1000 has been simulated.

The experiments were performed on two machines, (i) a 4-socket Intel Xeon E5-4650 machine, of which each socket has 8 cores and a 20 MB shared L3-cache, and (ii) a 2-socket AMD Opteron 6220 machine, of which each socket has 8 cores and a 16 MB shared L3-cache. In addition, every two cores share a L2-cache. The compiler used was GCC 4.7.2 respectively 4.4.7, at O3 optimization level. Each benchmark was run 5 times, for 10 seconds each time, of which the average throughput is presented. The number of threads varied between 1 to 32. The threshold for updating the head node, BOUNDOFFSET, was chosen in accordance to the maximum latency of the setup, with a higher value for a larger number of threads, ranging from 2 to 200. Threads were pinned to cores, to reduce the variance of the results and to make cache coherence effects visible.

The scaling of the different priority queues is shown in Fig. 3. We see that in general, the new algorithm is between 30 - 80% faster. The largest performance improvement is seen in the *DES* benchmark, on the Intel machine. The performance improvement of the *Uniform* benchmark is slightly smaller. We note that when using uniformly distributed keys, a majority of the inserts occur close to the head, and as a consequence the CAS instructions are more prone to fail, causing retrials of the operations. The penalty of the retry-search loop is slightly higher in our case, because of the traversal of the prefix of logically deleted nodes.

We note a steep drop in performance going from 8 to 9 threads on the Intel machine. This is a direct effect of threads being pinned to cores: up to 8 threads all threads share the L3-cache. At 9 threads or more, communication is done across sockets, with increased latency as a consequence. Likewise, for the AMD machine: every 2 threads share an L2-cache, outside which communication is more expensive.

Performance Limitations. We investigated the scalability bottleneck of the algorithm. For this goal, we devised a microbenchmark in which n threads access a data structure consisting of n nodes in a circular linked list. Each thread k traverses $n - 1$ nodes, and then perform a CAS modifying the kth node. Thus, each node is modified by only one thread, but the modification is read by all other threads. This behavior is intended to represent the read-write synchronization of the DELETEMIN operations. This corresponds to traversing the prefix of deleted nodes, then updating the first non-deleted node using CAS. Of the nodes in the deleted prefix, we can expect one modified node per other thread running, if we assume fair scheduling.

When n reaches 32 threads, the microbenchmark achieves roughly 2.6 million operations per second. This coincides with the maximal throughput of the new

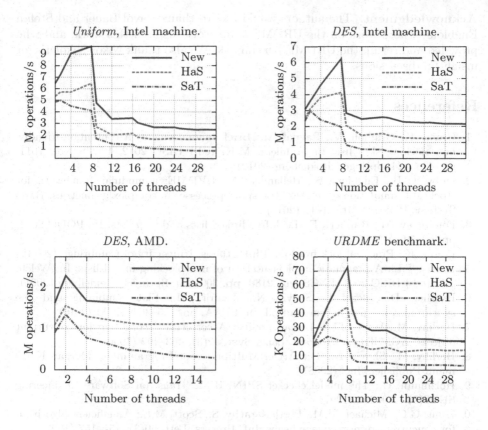

Fig. 3. Throughput of algorithms for the benchmarks

algorithm in *Uniform*. We then conclude that the priority queue is entirely limited by the DELETEMIN operation, and that, to achieve better performance, one would have to read less than one modified memory position per other thread and operation, on average.

7   Conclusion

We have presented a new linearizable skiplist-based priority queue algorithm, which achieves better performance than existing such algorithms, mainly due to reduced contention for shared memory locations. The absence of direct physical deletions makes our algorithm relatively simple to implement, and to verify correct. In addition, a simple benchmark indicates that the scalability of our algorithm is entirely limited by the logical deletion part of the DELETEMIN operation. We believe that similar ideas for improved performance can be applied to other concurrent data structures, whose scalability is limited by contention for shared memory locations.

Acknowledgment. The authors would like to thank Pavol Bauer and Stefan Engblom, whose work on the URDME framework inspired this work, and who provided and set up the URDME benchmark. We also thank Nikos Nikoleris for insightful discussions.

References

1. Crain, T., Gramoli, V., Raynal, M.: Brief announcement: a contention-friendly, non-blocking skip list. In: Aguilera, M.K. (ed.) DISC 2012. LNCS, vol. 7611, pp. 423–424. Springer, Heidelberg (2012)
2. Drawert, B., Engblom, S., Hellander, A.: URDME: a modular framework for stochastic simulation of reaction-transport processes in complex geometries. BMC Systems Biology 6(76), 1–17 (2012)
3. Fomitchev, M., Ruppert, E.: Lock-free linked lists and skip lists. In: PODC 2004, pp. 50–59. ACM (2004)
4. Fraser, K.: Practical lock freedom. Ph.D. thesis, University of Cambridge (2003)
5. Harris, T.L.: A pragmatic implementation of non-blocking linked-lists. In: Welch, J.L. (ed.) DISC 2001. LNCS, vol. 2180, pp. 300–314. Springer, Heidelberg (2001)
6. Hendler, D., Incze, I., Shavit, N., Tzafrir, M.: Flat combining and the synchronization-parallelism tradeoff. In: SPAA, pp. 355–364. ACM (2010)
7. Herlihy, M., Wing, J.M.: Linearizability: A correctness condition for concurrent objects. ACM Trans. Program. Lang. Syst. 12(3), 463–492 (1990)
8. Herlihy, M., Shavit, N.: The Art of Multiprocessor Programming. Morgan Kaufmann Publishers Inc. (2008)
9. Holzmann, G.: The model checker SPIN. IEEE Trans. on Software Engineering SE-23(5), 279–295 (1997)
10. Hunt, G.C., Michael, M.M., Parthasarathy, S., Scott, M.L.: An efficient algorithm for concurrent priority queue heaps. Inf. Process. Lett. 60(3), 151–157 (1996)
11. Lotan, I., Shavit., N.: Skiplist-based concurrent priority queues. In: IPDPS, pp. 263–268. IEEE (2000)
12. Michael, M.M.: Hazard pointers: Safe memory reclamation for lock-free objects. IEEE Trans. Parallel Distrib. Syst. 15(6), 491–504 (2004)
13. Molka, D., Hackenberg, D., Schone, R., Muller, M.: Memory performance and cache coherency effects on an intel nehalem multiprocessor system. In: PACT 2009, pp. 261–270. ACM (2009)
14. Pugh, W.: Concurrent maintenance of skip lists. Tech. Rep. CS-TR-2222, Dept. of Computer Science, University of Maryland, College Park (1990)
15. Pugh, W.: Skip lists: a probabilistic alternative to balanced trees. Commun. ACM 33(6), 668–676 (1990)
16. Sundell, H., Tsigas, P.: Scalable and lock-free concurrent dictionaries. In: SAC 2004, pp. 1438–1445. ACM (2004)
17. Sundell, H., Tsigas, P.: Fast and lock-free concurrent priority queues for multi-thread systems. J. Parallel Distrib. Comput. 65(5), 609–627 (2005)
18. Valois, J.D.: Lock-free linked lists using compare-and-swap. In: PODC 1995, pp. 214–222. ACM (1995)
19. Vechev, M., Yahav, E., Yorsh, G.: Experience with model checking linearizability. In: Păsăreanu, C.S. (ed.) Model Checking Software. LNCS, vol. 5578, pp. 261–278. Springer, Heidelberg (2009)

VirtuCast: Multicast and Aggregation with In-Network Processing

An Exact Single-Commodity Algorithm

Matthias Rost and Stefan Schmid

Telekom Innovation Laboratories (T-Labs) & TU Berlin, Germany
{mrost,stefan}@net.t-labs.tu-berlin.de

Abstract. As the Internet becomes more virtualized and software-defined, new functionality is introduced in the network core: the distributed resources available in ISP central offices, universal nodes, or datacenter middleboxes can be used to process (e.g., filter, aggregate or duplicate) data. Based on this new networking paradigm, we formulate the Constrained Virtual Steiner Arborescence Problem (CVSAP) which asks for optimal locations to perform in-network processing, in order to jointly minimize processing costs and network traffic while respecting link and node capacities.

We prove that CVSAP cannot be approximated (unless $NP \subseteq P$), and accordingly, develop the exact algorithm VirtuCast to compute optimal solutions to CVSAP. VirtuCast consists of: (1) a compact single-commodity flow Integer Programming (IP) formulation; (2) a flow decomposition algorithm to reconstruct individual routes from the IP solution. The compactness of the IP formulation allows for computing lower bounds even on large instances quickly, speeding up the algorithm significantly. We rigorously prove VirtuCast's correctness and show its applicability to solve realistically sized instances close to optimality.

Keywords: Network Virtualization, Network Functions Virtualization, Multicast, In-Network Aggregation, Data-Center, Middleboxes, ISP, Integer Programming.

1 Introduction

Multicast and aggregation are two fundamental functionalities offered by many communication networks. In order to efficiently distribute content (e.g., live TV) to multiple receivers, a multicast solution should duplicate the content as close to the receivers as possible. Analogously, in aggregation applications such as distributed network monitoring, data may be filtered or aggregated along the path to the observer, to avoid redundant transmissions over physical links. Efficient multicasting and aggregation is a mature research field, and many important theoretical and practical results have been obtained over the last decades. Applications range from IPTV [14] over sensor networks [10,11] to fiber-optical transport [14].

This paper is motivated by the virtualization trend in today's Internet, and in particular by network (functions) virtualization [9] and software-defined networking, e.g.,

R. Baldoni, N. Nisse, and M. van Steen (Eds.): OPODIS 2013, LNCS 8304, pp. 221–235, 2013.

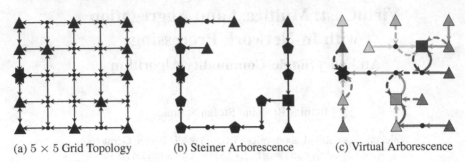

(a) 5 × 5 Grid Topology   (b) Steiner Arborescence   (c) Virtual Arborescence

Fig. 1. An aggregation example on a 5×5 grid. Terminals are depicted as triangles and the receiver as a star. The terminals must establish a route towards the receiver, while multiple data streams may be aggregated by activated processing locations. Such processing locations are pictured as squares or, in case that an active processing location is collocated with a terminal, pentagons. In Figure (c), equally colored and dashed edges represent logical connections (routes), originating at the node with the same color.

OpenFlow [23]. In virtualized environments, resources can be allocated or leased flexibly at the locations where they are most useful or cost-effective: computational and storage resources available at middleboxes in datacenters [5], in universal nodes, or in distributed (micro-)datacenters in the wide-area network [7,20,29] can be used for in-network processing, e.g., to reduce traffic during the MapReduce shuffle phase [6]. Such distributed resource networks open new opportunities on how services can be deployed. Especially in the context of aggregation and multicasting a new degree of freedom arises: the sites (i.e., the number and locations) used for the data processing, becomes subject to optimization.

This paper initiates the study of how to efficiently allocate in-network processing functionality in order to jointly minimize network traffic and computational resources. Importantly, for many of these problem variants, classic Steiner Tree models [28] are no longer applicable [19]. Accordingly, we coin our problem the *Constrained Virtual Steiner Arborescence Problem (CVSAP)*, as the goal is to install a set of processing nodes and to connect all terminals via them to a single root.

Example. To illustrate our model, consider the aggregation example depicted in Figure 1. The terminals must connect to the single receiver (root), while processing functionality can be placed on nodes to aggregate any number of incoming data flows into a single one. Assuming no costs for placing processing functionality, the problem reduces to the Steiner Arborescence Problem: the optimal solution, depicted in Figure 1b, uses 16 edges and 9 processing locations, i.e. nodes where data flows are merged. However, assuming unit edge costs and activation costs of 5 for processing locations, this solution is suboptimal. Figure 1c depicts a solution which only uses 2 processing locations and 26 edges overall: terminals in the first column directly connect to the receiver, while the remaining terminals use one of the two processing nodes. Note that we allow for nested processing of flows: the upper processing node forwards its aggregation result to the lower processing node, from where the result is then forwarded to the receiver.

Contribution. We introduce the *Constrained Virtual Steiner Arborescence Problem (CVSAP)* which captures the trade-off between traffic optimization benefits and in-network processing costs arising in virtualized environments, and which also generalizes many classic in-network processing problems related to multicasting and aggregation. We prove that CVSAP cannot be approximated unless $NP \subseteq P$ holds and therefore focus on obtaining provably good solutions for CVSAP in non-polynomial time. To this end we introduce the algorithm VirtuCast, which is based on Integer Programming (IP) and allows to obtain optimal solutions. The advantage of VirtuCast lies in the fact that even for large problem instances, when optimal solutions cannot be computed in reasonable time, our approach bounds the gap to optimality as lower bounds are computed on the fly.

VirtuCast consists of two components: a single-commodity IP formulation which can be solved by branch-and-cut methods and a decomposition algorithm to construct the routing scheme. Our IP formulation not only uses a smaller number of variables compared to alternative multi-commodity IP formulations, but also yields good linear relaxations in practice, speeding up the branch-and-bound algorithm (see [26] for an in-depth discussion).

Our main contribution is the constructive proof that any solution to our IP formulation can be decomposed to yield a valid routing scheme connecting all terminals via processing nodes to the root. This is intriguing, as the single-commodity flow in the network is not restricted to directed acyclic graphs (cf. Figure 1c). In fact, as already shown in [19], forbidding directed acyclic graphs (DAGs) may yield suboptimal solutions only. Rather, we allow for the iterative processing of flows, such that processing nodes may be connected to other processing nodes.

To implement VirtuCast, we have developed a branch-and-cut framework, including a primal heuristic. Due to space constraints, we refer the interested reader to our technical report [26] for a detailed discussion of the implementation as well as for the full computational evaluation.

Overview. We formally introduce CVSAP and show its inapproximability in Section 2. We continue by presenting our VirtuCast algorithm in Section 3. In Section 4 we shortly outline the results of our computational study. We conclude this paper with summarizing related work in Section 5.

2   The Constrained Virtual Steiner Arborescence Problem

The Constrained Virtual Steiner Arborescence Problem (CVSAP) generalizes several in-network processing problems related with multicasting and aggregation of data where processing locations can be *chosen* to reduce traffic. As using (or leasing) in-network processing capabilities comes at a certain cost (e.g., the corresponding resources cannot be used by other applications), there is a trade-off between additional processing and traffic reduction. In contrast to the classic Steiner Tree Problems [28], our model distinguishes between nodes that merely relay traffic and nodes that may actively *process flows*. Informally, the task is to construct a minimal cost spanning arborescence on the set of active processing nodes, sender(s) and receiver(s), such that edges in the arborescence correspond to paths in the original graph. As edges in the arborescence represent logical links (i.e. routes) between nodes, we refer to the problem as *Virtual* Steiner Arborescence Problem. Based on the notion of virtual edges, the underlying paths may overlap and may use both the nodes and edges in the original graph multiple times (cf. Figure 1c). We naturally adopt the notion of *Steiner nodes* in our model, and refer to processing nodes contained in the virtual arborescence as *active Steiner nodes*. As will be discussed at the end of this section, the multicast case can be easily reduced to the aggregation case. Hence, in the following we only introduce the variant of CVSAP in which data flows are directed towards the root. The following notations will be used throughout this paper.

Notation. In a directed graph $G = (V_G, E_G)$ we denote by $\mathcal{P}_G$ the set of all simple, directed paths in G. Given a set of simple paths $\mathcal{P}$, we denote by $\mathcal{P}[e]$ the subset of paths of $\mathcal{P}$ that contains edge e. We use the notation $P = \langle v_1, v_2, \ldots, v_n \rangle$ to denote the directed path P of length $|P| = n$ where $P_i \triangleq v_i \in V_G$ for $1 \leq i \leq n$ and $(v_i, v_{i+1}) \in E_G$ for $1 \leq i < n$. We denote the set of outgoing and incoming edges, restricted on a subset $F \subseteq E_G$ in G, by $\delta_F^+(v) = \{(v, u) \in F\}$ and $\delta_F^-(v) = \{(u, v) \in F\}$ for $v \in V_G$. We naturally extend this definition to sets: $\delta_F^+(W) = \{(v, u) \in F | v \in W, u \notin W\}$ and $\delta_F^-(W) = \{(u, v) \in F | v \in W, u \notin W\}$ respectively. We abridge $f((y, z))$ to $f(y, z)$ for functions defined on tuples.

Formal Problem Statement. We model the physical infrastructure as capacitated, directed network $G = (V_G, E_G, c_E, u_E)$, where $u_E : E_G \to \mathbb{N}$ defines integral capacities and $c_E : E_G \to \mathbb{R}^+$ defines real-valued, positive costs on the edges. On top of this network, we define an abstract request $R_G = (r, S, T, u_r, c_S, u_S)$, where $T \subseteq V_G$ defines the set of terminals that need to be connected with the root $r \in V_G \setminus T$, for which an integral capacity $u_r \in \mathbb{N}$ is given. The set $S \subseteq V_G \setminus (\{r\} \cup T)$ denotes the set of possible *Steiner sites*, i.e. nodes at which processing nodes may be *activated*. Such Steiner sites are attributed with a positive cost $c_S : S \to \mathbb{R}^+$ that is incurred upon using it, and an integral capacity $u_S : S \to \mathbb{N}$. It should be noted that we require the sets S and T to be disjoint for terminological reasons. A node $v \in S \cup T$ can easily be modeled by introducing a new node $v_T \in T$ and letting $v \in S$ such that v_T is only connected to v with $c_E(v_T, v) = 0$ and $u_E(v_T, v) = 1$.

In the aggregation scenario considered henceforth the terminals hold data that needs to be forwarded to the root (the single receiver) while data may be aggregated at active Steiner nodes. The capacities on the Steiner sites (and on the root) limit the number of flows that can be *actively* processed: any number of incoming flows less than or equal to $u_S(s)$ can be merged into a single flow by $s \in S$ upon activation. To model both routing decisions and paths taken, we introduce the concept of Virtual Arborescences:

Definition 1 (Virtual Arborescence). *Given a directed graph $G = (V_G, E_G)$ and a root $r \in V_G$, a Virtual Arborescence (VA) on G is defined as $\mathcal{T}_G = (V_T, E_T, r, \pi)$ where $\{r\} \subseteq V_T \subseteq V_G$, $E_T \subseteq V_T \times V_T$, r is the root and $\pi : E_T \to \mathcal{P}_G$ maps each edge in the virtual arborescence on a simple directed path $P \in \mathcal{P}_G$ such that*

(VA-1) $(V_\mathcal{T}, E_\mathcal{T}, r)$ *is an arborescence rooted at* r *with edges directed towards* r,

(VA-2) *for all* $(u, v) \in E_\mathcal{T}$ *the directed path* $\pi(u, v)$ *connects* u *to* v *in* G. $\qquad\square$

A link $(v, w) \in E_\mathcal{T}$ represents a logical connection between nodes v and w while the function $\pi(v, w) = P$ defines the route taken to establish this link: in Figure 1c equally colored and dashed paths represent edges of the Virtual Arborescence. Note that the directed path P must, *pursuant* to the orientation (v, w) of the logical link in the arborescence, start with v and end at w. Using the concept of Virtual Arborescences, we can concisely state the problem we are attending to.

Definition 2 (Constrained Virtual Steiner Arborescence Problem). *Given a directed capacitated network* $G = (V_G, E_G, c_E, u_E)$ *and a request* $R_G = (r, S, T, u_r, c_S, u_S)$ *as above, the* Constrained Virtual Steiner Arborescence Problem (CVSAP) *asks for a minimal cost Virtual Arborescence* $\mathcal{T}_G = (V_\mathcal{T}, E_\mathcal{T}, r, \pi)$ *satisfying the following conditions:*

(CVSAP-1) *$\{r\} \cup T \subseteq V_\mathcal{T}$ and $V_\mathcal{T} \subseteq \{r\} \cup S \cup T$,*

(CVSAP-2) *for all terminals $t \in T$ holds $\delta_{E_\mathcal{T}}^+(t) = 1$,*

(CVSAP-3) *for the root $\delta_{E_\mathcal{T}}^-(r) \leq u_r$ holds,*

(CVSAP-4) *for all activated Steiner sites $s \in S \cap V_\mathcal{T}$ holds $\delta_{E_\mathcal{T}}^-(s) \leq u_S(s)$ and*

(CVSAP-5) *for all edges $e \in E_G$ holds $|(\pi(E_\mathcal{T}))[e]| \leq u_E(e)$.*

Any VA $\mathcal{T}_G$ satisfying CVSAP-1 - CVSAP-5 is said to be a feasible solution. The cost of a Virtual Arborescence is defined to be

$$C_{\text{CVSAP}}(\mathcal{T}_G) = \sum_{e \in E_G} c_E(e) \cdot |(\pi(E_\mathcal{T}))[e]| + \sum_{s \in S \cap V_\mathcal{T}} c_S(s),$$

where $|(\pi(E_\mathcal{T}))[e]|$ is the number of times an edge is used in different paths. $\qquad\square$

In the above definition, CVSAP-1 states that terminals and the root must be included in $V_\mathcal{T}$, whereas non Steiner sites are excluded. We identify $V_\mathcal{T} \setminus (\{r\} \cup T)$ with the set of active Steiner nodes. Condition CVSAP-2 states that terminals must be leaves in $\mathcal{T}_G$ and CVSAP-3 and CVSAP-4 enforce degree constraints in $\mathcal{T}_G$. The term $\pi(E_\mathcal{T})$ in Condition CVSAP-5 determines the set of all used paths and consequently $\pi(E_\mathcal{T})[e]$ yields the set of paths that use $e \in E_\mathcal{T}$. As π is injective and maps on simple paths, Condition CVSAP-5 enforces that edge capacities are not violated.

The following theorem motivates our approach in Section 3, namely to search for provably good solutions in non-polynomial time.

Theorem 1. *Checking whether a feasible solution for CVSAP exists is NP-complete. Thus, unless $NP \subseteq P$ holds, there cannot exist an (approximation) algorithm yielding a feasible solution in polynomial time.*

Proof. We give a reduction on the decision variant of set cover. Let U denote the universe of elements and let $\mathcal{S} \subseteq 2^U$ denote a family of sets covering U. To check whether a set cover using at most k many sets exists, we construct the following CVSAP instance. We introduce a terminal t_u for each element $u \in U$ and a Steiner site s_S for each $S \in \mathcal{S}$. A terminal t_u is connected by a directed link to each Steiner site s_S iff. $u \in S$. Each Steiner site s_S is connected to the root r. We set the capacity of the root to k and capacities of Steiner sites to $|U|$. It is easy to check that there exists a feasible solution to this CVSAP instance iff. there exists a set cover of less than k elements. □

Similarly to the above definitions, CVSAP can be defined for multicasting applications in which the task is to distribute a single data item from the root (single sender) to terminals (receivers) via processing nodes (with routing capability) that may duplicate the data and route it to several different destinations. To obtain a formal definition for this scenario, edges in the VA must be oriented away from the root and $\delta^-(\cdot)$ must be replaced by $\delta^+(\cdot)$ and vice versa in Definition 2. Subject to this slight adaption, the root and active Steiner nodes can reproduce an incoming stream, such that terminals must receive this stream. By essentially reversing the direction of edges, the multicasting version of CVSAP can be reduced to the aggregation version presented above.

3   VirtuCast Algorithm

In this section we present the Algorithm VirtuCast to solve CVSAP. VirtuCast first computes a solution for a single-commodity flow Integer Programming formulation and then constructs the corresponding Virtual Arborescence. Even though our IP formulation can be used to compute the optimal solution for any CVSAP instance, feasible solutions to our IP formulation already yield feasible solutions to CVSAP. This allows to derive near-optimal solutions *during* the solution process. Our single-commodity approach improves dramatically upon naive multi-commodity flow formulations and enables us to solve realistically sized instances in the first place (see [26] for a discussion).

3.1   IP Formulation

Our IP (see IP-CVSAP) is based on an *extended graph* containing a single super source o^+ and two distinct super sinks o_S^- and o_r^- (see Definition 3). While o_r^- may only receive flow from the root r, all possible Steiner sites $s \in S$ are connected to o_S^-. Distinguishing between these two super sinks is necessary, as we will require activated Steiner nodes to not *absorb all* incoming flow, but forward at least one unit of flow towards o_r^-, which will ensure connectivity.

Definition 3 (Extended Graph). *Given a directed network* $G = (V_G, E_G, c_E, u_E)$ *and a request* $R_G = (r, S, T, u_r, c_S, u_S)$ *as introduced in Section 2, we define the extended graph* $G_{\text{ext}} = (V_{\text{ext}}, E_{\text{ext}})$ *as follows*

(EXT-1) $V_{\text{ext}} \triangleq V_G \cup \{o^+, o_S^-, o_r^-\}$,

(EXT-2) $E_{\text{ext}} \triangleq E_G \cup \{(r, o_r^-)\} \cup E_{\text{ext}}^{S^-} \cup E_{\text{ext}}^{S^+} \cup E_{\text{ext}}^{T^+}$,

where $E_{\text{ext}}^{S^-} \triangleq S \times \{o_S^-\}$, $E_{\text{ext}}^{S^+} \triangleq \{o^+\} \times S$ *and* $E_{\text{ext}}^{T^+} \triangleq \{o^+\} \times T$. *We define* $E_{\text{ext}}^R \triangleq E_{\text{ext}} \setminus E_{\text{ext}}^{S^-}$, *such that edges towards* o_S^- *are excluded in* E_{ext}^R. □

Further Notation. To clearly distinguish between variables and constants, we typeset constants in bold font: instead of referring to c_E, c_S and u_E, u_r, u_S we use $\mathbf{c}_y$ and $\mathbf{u}_y$, where y may either refer to an edge or a Steiner site. Similarly, we use $\mathbf{u}_y$ where y may either refer to an edge, the root or a Steiner site. We abbreviate $\sum_{y \in Y} f_y$ by $f(Y)$. We use $Y + y$ to denote $Y \cup \{y\}$ and $Y - y$ to denote $Y \setminus \{y\}$ for a set Y and a singleton y. For $f \in \mathbb{Z}_{\geq 0}^{E_{ext}}$ we define the flow-carrying subgraph $G_{ext}^f \triangleq (V_{ext}^f, V_{ext}^f)$ with $V_{ext}^f \triangleq V_{ext}$ and $V_{ext}^f \triangleq \{e | e \in E_{ext} \wedge f(e) \geq 1\}$.

The IP Model. The IP formulation IP-CVSAP uses an integral single-commodity flow. We define flow variables $f_e \in \mathbb{Z}_{\geq 0}$ for each edge $e \in E_{ext}$ in the extended graph (see IP-11). As we use an aggregated flow formulation, that does not model routing decisions explicitly, we show in Section 3.2 how this single-commodity flow can be decomposed into paths for constructing an actual solution for CVSAP.

The binary variable $x_s \in \{0, 1\}$ (see IP-10) decides, whether a Steiner site $s \in S$ is activated. By Constraint IP-8, each terminal $t \in T$ is forced to send a single unit of flow, as flow conservation is enforced on all original nodes $v \in V_G$ (see IP-1). Therefore, all flow originating at o^+ must be forwarded to one of the super sinks o_r^- or o_S^-, while not violating link capacities (see IP-7).

As the definition of CVSAP requires that each terminal $t \in T$ establishes a path to r, we need to enforce connectivity; otherwise active Steiner nodes would simply absorb flow by directing it towards o_S^-. To prohibit this, we adopt well-known *Connectivity Inequalities* IP-2 [18] and *Directed Steiner Cuts* IP-3* [16]. Our connectivity inequalities (see IP-2) state that each set of nodes containing a Steiner site $s \in S$ must emit at

---

Integer Program IP-CVSAP

minimize $\quad C_{IP}(x, f) = \sum_{e \in E_G} \mathbf{c}_e f_e + \sum_{s \in S} \mathbf{c}_s x_s$ $\hfill$ (IP-OBJ)

subject to $\quad f(\delta_{E_{ext}}^+(v)) = f(\delta_{E_{ext}}^-(v)) \hfill \forall v \in V_G \quad$ (IP-1)

$$f(\delta_{E_{ext}^R}^+(W)) \geq x_s \hfill \forall W \subseteq V_G, s \in W \cap S \neq \emptyset \quad \text{(IP-2)}$$

$$f(\delta_{E_{ext}^R}^+(W)) \geq 1 \hfill \forall W \subseteq V_G, T \cap W \neq \emptyset \quad \text{(IP-3*)}$$

$$f_e \geq x_s \hfill \forall e = (s, o_S^-) \in E_{ext}^{S^-} \quad \text{(IP-4*)}$$

$$f_e \leq \mathbf{u}_s x_s \hfill \forall e = (s, o_S^-) \in E_{ext}^{S^-} \quad \text{(IP-5)}$$

$$f_{(r, o_r^-)} \leq \mathbf{u}_r \hfill \text{(IP-6)}$$

$$f_e \leq \mathbf{u}_e \hfill \forall e \in E_G \quad \text{(IP-7)}$$

$$f_e = 1 \hfill \forall e \in E_{ext}^{T^+} \quad \text{(IP-8)}$$

$$f_e = x_s \hfill \forall e = (o^+, s) \in E_{ext}^{S^+} \quad \text{(IP-9)}$$

$$x_s \in \{0, 1\} \hfill \forall s \in S \quad \text{(IP-10)}$$

$$f_e \in \mathbb{Z}_{\geq 0} \hfill \forall e \in E_{ext} \quad \text{(IP-11)}$$

---

least one unit of flow in E_{ext}^R, if s is activated. As E_{ext}^R does not contain edges towards o_S^-, this constraint therefore enforces that there exists a path in G_{ext}^f from each activated Steiner node s to the root r. Analogously, Constraint IP-3* enforces that there exists a path from each terminal $t \in T$ towards r in G_{ext}^f. The directed Steiner cuts constitute valid inequalities which are implied by IP-1 and IP-2 (see [26] for the proof). However, these cuts can strengthen the model by improving the LP-relaxation during the branch-and-cut process. As discussed in [26], including these constraints substantially improved the quality of lower bounds in our computational evaluation. As they are not needed for proving the correctness and could technically be removed, we mark them with a * (star).

As a Steiner node $s \in S$ is activated iff. $x_s = 1$, Constraint IP-9 requires activated Steiner nodes to receive one unit of flow while being able to maximally absorb u_s many units of flow by forwarding it to o_S^- (see IP-5). Furthermore, by IP-5 inactive Steiner sites may not absorb flow at all. The Constraint IP-4* requires active Steiner nodes to at least absorb one unit of flow. This is a valid inequality, as activating a Steiner site $s \in S$ incurs a non-negative cost. We introduce this constraint here, as it specifies a condition needed in the proof of correctness later on.

Constraint IP-6 defines an upper bound on the amount of flow that the root may receive and the objective function IP-OBJ mirrors the CVSAP cost function (see Definition 2). We denote with $\mathcal{F}_{IP} = \{(x, f) \in \{0, 1\}^S \times \mathbb{Z}_{\geq 0}^{E_{ext}} | \text{IP-1 - IP-11}\}$ the set of feasible solutions to IP-CVSAP.

3.2 Decomposition Algorithm

Given a feasible solution $(\hat{x}, \hat{f}) \in \mathcal{F}_{IP}$ for IP-CVSAP, Algorithm Decompose constructs a feasible solution $\hat{\mathcal{T}}_G \in \mathcal{F}_{CVSAP}$ for CVSAP. Similarly to well-known algorithms for computing flow decompositions for simple s-t flows (see e.g. [2]), our algorithm iteratively deconstructs the flow into paths from the super source o^+ to one of the super sinks o_S^- or o_r^- and reduces flow along the found paths to yield a solution to a subproblem. However, as IP-CVSAP does not pose a simple flow problem, we constantly need to ensure that Connectivity Inequalities IP-2 hold after removing flow in $G_{ext}^{\hat{f}}$. We first present Algorithm Decompose in more detail and then prove its correctness.

Synopsis of Algorithm. Algorithm Decompose constructs a feasible VA $\hat{\mathcal{T}}_G$ given a solution $(\hat{x}, \hat{f}) \in \mathcal{F}_{IP}$. In Line 2, $\hat{\mathcal{T}}_G$ is initialized without any edges but containing all the nodes the final solution will consist of, namely the root r, the terminals T and the activated Steiner nodes $\{s \in S | x_s = 1\}$.

Unconnected terminals in $\hat{T}$ are connected iteratively. For an unconnected terminal $t \in \hat{T}$ the path generation procedure from Line 6 to 14 computes a path P from o^+ via t to o_S^- or o_r^-. If the path P terminates in o_r^- then t is connected to r. Otherwise, if P terminates in o_S^-, then the second last node of P is an active Steiner node and t is connected to it (see Line 18). During the path generation procedure the flow variables $\hat{f}$ are decremented. If the second last node of P was indeed an active Steiner node $s \in \hat{S}$ and s does not forward any flow towards o_S^- anymore, s itself is added to the set of unconnected terminals (see Line 16). Note that in Line 18 the (virtual) edge $(t, P_{|P|-1})$

Algorithm Decompose

Input : Network $G = (V_G, E_G, c_E, u_E)$, Request $R_G = (r, S, T, u_r, c_S, u_S)$,
    Solution $(\hat{x}, \hat{f}) \in \mathcal{F}_{\text{IP}}$ to IP-CVSAP

Output: Feasible Virtual Arborescence $\hat{\mathcal{T}}_G$ for CVSAP

1 set $\hat{S} \triangleq \{s \in S | x_s \geq 1\}$ and $\hat{T} \triangleq T$

2 set $\hat{\mathcal{T}}_G \triangleq (\hat{V}_T, \hat{E}_T, r, \hat{\pi})$ where $\hat{V}_T \triangleq \{r\} \cup \hat{S} \cup \hat{T}$, $\hat{E}_T \triangleq \emptyset$ and $\hat{\pi} : \hat{E}_T \to \mathcal{P}_G$

3 **while** $\hat{T} \neq \emptyset$ **do**

4 $\quad$ **let** $t \in \hat{T}$ and $\hat{T} \leftarrow \hat{T} - t$

5 $\quad$ **choose** $P \triangleq \langle o^+, t, \ldots, o_r^- \rangle \in G_{\text{ext}}^{\hat{f}}$

6 $\quad$ **for** $j = 1$ **to** $|P| - 1$ **do**

7 $\quad\quad$ set $\hat{f}(P_j, P_{j+1}) \leftarrow \hat{f}(P_j, P_{j+1}) - 1$

8 $\quad\quad$ **if** Constraint IP-2 is violated *with respect to* $\hat{f}$ and $\hat{S}$ **then**

9 $\quad\quad\quad$ **choose** $W \subseteq V_G$ such that $W \cap \hat{S} \neq \emptyset$ and $\hat{f}(\delta_{E_{\text{ext}}^R}^+(W)) = 0$

10 $\quad\quad\quad$ **choose** $P' \triangleq \langle P_j, \ldots, o_S^- \rangle \in G_{\text{ext}}^{\hat{f}}$ such that $P_i \in W$ for $1 \leq i < m$

11 $\quad\quad\quad$ set $\hat{f}(P_j, P_{j+1}) \leftarrow \hat{f}(P_j, P_{j+1}) + 1$ and $\hat{f}(P_1', P_2') \leftarrow \hat{f}(P_1', P_2') - 1$

12 $\quad\quad\quad$ set $P \leftarrow \langle P_1, \ldots, P_{j-1}, P_j = P_1', P_2', \ldots, P_m' \rangle$

13 $\quad\quad$ **end**

14 $\quad$ **end**

15 $\quad$ **if** $P_{|P|} = o_S^-$ and $\hat{f}(P_{|P|-1}, P_{|P|}) = 0$ **then**

16 $\quad\quad$ set $\hat{S} \leftarrow \hat{S} - P_{|P|-1}$ and $\hat{x}(P_{|P|-1}) \leftarrow 0$ and $\hat{T} \leftarrow \hat{T} + P_{|P|-1}$

17 $\quad$ **end**

18 $\quad$ set $\hat{E}_T \leftarrow \hat{E}_T + (t, P_{|P|-1})$ and $\hat{\pi}(t, P_{|P|-1}) \triangleq \texttt{simplify}(\langle P_2, \ldots, P_{|P|-1} \rangle)$

19 **end**

---

is added to $\hat{E}_T$ and $\hat{\pi}(t, P_{|P|-1})$ is set accordingly to the truncated path P, where any cycles are removed (function $\texttt{simplify}$).

Proof of Correctness. We will now prove the correctness of Algorithm Decompose, thereby showing that IP-CVSAP can be used to compute (optimal) solutions to CVSAP. Our proof relies on an inductive argument similar to the one used for proving the existence of flow decompositions (see [2]): we assume that all constraints of IP-CVSAP hold and show that for any terminal $t \in T$ a path towards the root or to an active Steiner node can be constructed, such that decrementing the flow along the path by one unit does again yield a feasible solution to IP-CVSAP, in which t has been removed from the set of terminals (see Theorem 2 below). During the course of this induction, the well-definedness of the **choose** operations is shown. As the complete proof is included in [26], we allow us to mainly sketch the proofs.

Theorem 2. *Assuming that the constraints of Decompose hold with respect to $\hat{S}, \hat{T}, \hat{f}, \hat{x}$ before executing Line 4, then the constraints of Decompose will also hold in Line 18 with respect to the then reduced problem $\hat{S}, \hat{T}, \hat{f}, \hat{x}$.*

To prove the above theorem, we use the following Lemmas 1 through 3 of which we only prove the essential third one; the proofs for Lemmas 1 and 2 are included in [26]. Lemma 1 shows the well-definedness of choosing the path in Line 5 and is easy to check. Lemma 2 states that flow conservation (IP-1) holds during the execution of Decompose except at node P_{j+1} at which the outgoing flow exceeds the incoming flow by exactly one unit.

Lemma 1. *Assuming that IP-1 and IP-2 hold, there exists a path $P = \langle o^+, t, \dots, o_r^- \rangle \in G_{\text{ext}}^{\hat{f}}$ in Line 5.*

Lemma 2. *Assuming that IP-1 has held in Line 5, $f(\delta_{E_{\text{ext}}}^+(v)) - f(\delta_{E_{\text{ext}}}^-(v)) = \delta_{v,P_{j+1}}$ holds for all $v \in V_G$ during construction of P (Lines 8-13), where $\delta_{x,y} \in \{0,1\}$ and $\delta_{x,y} = 1$ iff. $x = y$.*

Lemma 3. *Assuming that connectivity inequalities IP-2 have held before executing Line 7, these inequalities will hold again at Line 13.*

Proof Sketch. We only need to consider the case in which the Constraint IP-2 was violated after executing Line 7. Assume therefore that IP-2 is violated in Line 8. The **choose** operation in Line 9 is well-defined, as IP-2 is violated. Let $W \subseteq V_G$ be any violated set with $\hat{S} \cap W \neq \emptyset$. Our proof relies on the following four statements:

(a) P_j is contained in W while P_{j+1} is not contained in W.
(b) $\hat{f}(P_j, P_{j+1}) = 0$ holds in Lines 9-10.
(c) Before flow reduction in Line 7, there existed a path
$$P'' = \langle s, \dots, P_j, P_{j+1}, \dots, o_r^- \rangle \in G_{\text{ext}}^{\hat{f}} \text{ for } s \in \hat{S} \cap W.$$
(d) There exists a path $P' = \langle P_j, \dots, o_S^- \rangle$ with $P_i' \in W$ for $1 \leq i < |P'|$ in $G_{\text{ext}}^{\hat{f}}$ after reduction of flow.

To see that statement (a) holds, consider the following. Before the reduction of flow on (P_j, P_{j+1}) all inequalities IP-2 held. For the expression $\hat{f}(\delta_{E_{\text{ext}}}^+(W)) = 0$ to hold after reduction of flow, the edge (P_j, P_{j+1}) must be contained in $\delta_{E_{\text{ext}}}^+(W)$.

The correctness of (b) directly follows from (a), as $(P_j, P_{j+1}) \in \delta_{E_{\text{ext}}}^+(W)$ holds.

We now prove statement (c). As connectivity inequalities IP-2 have held *before* the flow reduction in Line 7, for each activated Steiner node $s \in \hat{S} \cap W$ there existed a path from s to o_r^- in $G_{\text{ext}}^{\hat{f}}$. By (b), (P_j, P_{j+1}) was the only edge in $G_{\text{ext}}^{\hat{f}}$ leaving W before reduction of flow. Therefore a path as claimed in (c) must have existed before reduction of flow.

By statement (c), the prefix $\langle s, \dots, P_j \rangle$ of path P'' still exists in $G_{\text{ext}}^{\hat{f}}$ after reduction of flow. This implies that P_j is reached by a positive amount of flow. By Lemma 2 flow conservation holds for all nodes $w \in W$, since by (a) P_{j+1} is not included in W. As o_r^- is not included in W, there must exist a path $P' = \langle P_j, \dots, o_S^- \rangle \in G_{\text{ext}}^{\hat{f}}$ after reduction of flow with $P_i' \in W$ for $1 \leq i < |P'|$. This proves the fourth statement (d) and shows that the **choose** operation in Line 10 is well-defined.

To see that the main statement of this lemma holds, consider the case that after Line 11 any connectivity inequality of IP-2 is violated. Let $W' \subseteq V_G$ with $W' \cap \hat{S} \neq \emptyset$ be a violated node set such that $\hat{f}(\delta_{E_{\text{ext}}}^+(W')) = 0$ holds. By the same argument as used for proving statement (a), it is easy to see that $P_1' \in W'$ and $P_2' \notin W'$ must hold. However, by statement three, after having reverted the flow reduction along (P_j, P_{j+1}), the path $\langle P_j, P_{j+1}, \ldots, o_r^- \rangle$ was re-established in $G_{\text{ext}}^{\hat{f}}$. As flow along any of the edges contained in this path is greater or equal to one, W' cannot possibly violate IP-2. Therefore all Connectivity Inequalities IP-2 hold.   □

Using the above lemmas, we now outline the proof of Theorem 2. By Lemmas 1 and 3 the algorithm is well-defined. Lemma 2 implies that flow preservation holds at Line 18 as node P_{j+1} is one of the super sinks. Lemma 3 directly ensures that connectivity constraints IP-2 hold. As capacity related constraints trivially hold as flow was only decreased, it only remains to check that placing a former active Steiner node into the set of terminals in Line 16 does not violate the terminal related constraint IP-8. This however is easy to check as constraint IP-9 ensured that this node received one unit of flow from the super source.

Using Theorem 2 it is easy to check that Algorithm Decompose terminates: Since no constraint is violated during the execution of the path generation and as flow is only reduced, the inner loop must eventually terminate

Theorem 3. *Algorithm Decompose terminates.*

We can now turn to proving that Algorithm Decompose indeed constructs a feasible solution for CVSAP. As the proof is of a rather technical nature, we again only sketch the proof and refer the interested reader to [26] for the complete argument.

Theorem 4. *Algorithm Decompose constructs a feasible solution $\hat{\mathcal{T}}_G \in \mathcal{F}_{\text{CVSAP}}$ for CVSAP given a solution $(\hat{x}, \hat{f}) \in \mathcal{F}_{\text{IP}}$. Additionally, $C_{\text{CVSAP}}(\hat{\mathcal{T}}_G) \leq C_{\text{IP}}(\hat{x}, \hat{f})$ holds.*

By Theorem 3 the algorithm terminates such that we only need to check feasibility of the solution. First note that, as the VA $\hat{\mathcal{T}}_G$ is constructed using only resources accounted for in IP-CVSAP, $C_{\text{CVSAP}}(\hat{\mathcal{T}}_G) \leq C_{\text{IP}}(\hat{x}, \hat{f})$ must hold. Clearly, as capacity constraints of CVSAP are modeled explicitly in IP-CVSAP, it must only be checked whether indeed $\hat{\mathcal{T}}_G$ is a Virtual Arborescence. For proving that, first note that at the end of executing Decompose both sets $\hat{S}$ and $\hat{T}$ are empty and therefore all terminals and active Steiner nodes have been connected. While this holds by definition of the outer loop for $\hat{T}$, proving $\hat{S} = \emptyset$ requires the following argument. Assume that $\hat{S} \neq \emptyset$ while $\hat{T} = \emptyset$. By constraint IP-9 each active Steiner node $s \in \hat{S}$ receives one unit of flow from the super source. On the other hand constraint IP-4* safeguards that each active Steiner node absorbs at least a single unit of flow. This implies that no flow reaches o_r^-, violateing Constraint IP-2. Lastly, to check that $\hat{\mathcal{T}}_G$ defines an arborescence note that the order in which terminals (or former active Steiner sites) are removed from $\hat{T}$ defines a topological order on $(\hat{V}_T, \hat{E}_T, r)$.

To prove that formulation IP-CVSAP indeed computes optimal solutions, we need the following technical lemma showing that each solution to CVSAP can be mapped on a solution of IP-CVSAP with equal cost. As this mapping is straightforward, we refer the reader to [26] for the construction and only state the following lemma.

Lemma 4. *Given a network $G = (V_G, E_G, c_E, u_E)$, a request $R_G = (r, S, T, u_r, c_S, u_S)$ and a feasible solution $\hat{\mathcal{T}}_G = (\hat{V}_T, \hat{E}_T, r, \hat{\pi})$ to the corresponding CVSAP. There exists a solution $(\hat{x}, \hat{f}) \in \mathcal{F}_{\text{IP}}$ with $C_{\text{CVSAP}}(\hat{\mathcal{T}}_G) = C_{\text{IP}}(\hat{x}, \hat{f})$.*

The above lemma fills the last gap in our proof to show that algorithm VirtuCast, which first computes an optimal solution to IP-CVSAP and then constructs a corresponding Virtual Arborescence using Decompose, solves CVSAP to optimality.

Theorem 5. *Algorithm VirtuCast solves CVSAP to optimality.*

We conclude this section with stating that each **choose** operation in Algorithm Decompose and checking whether connectivity inequalities IP-2 hold can be implemented using depth-first search. Implementing Decompose in this way and assuming that an optimal solution for IP-CVSAP is given and that G does not contain zero-cost cycles, we can bound the runtime of Decompose by $\mathcal{O}\left(|V_G|^2 \cdot |E_G| \cdot (|V_G| + |E_G|)\right)$ [26].

4 Computational Evaluation

We have implemented VirtuCast using SCIP [1] as underlying branch-and-cut framework. As the separation procedures employed to enforce Constraints IP-2 and IP-3* are well-known [16], we only shortly outline the results of our computational evaluation. A detailed discussion of all our results can be found in [26]. Furthermore, our solver as well as the test instances are obtainable from [25]. All our experiments were conducted on machines equipped with an 8-core Intel Xeon L5420 processor running at 2.5 Ghz and 16 GB RAM.

In our computational evaluation we consider two complementary graph topologies with varying sizes: symmetric $n \times n$ grid graphs and ISP topologies generated by IGen [24]. We report only on results obtained for the largest topology sizes, namely on a 20×20 grid and an IGen topology with 3200 nodes (further refered to by IGen.3200). The IGen.3200 topology is created by populating a world map with 3200 nodes, applying a local clustering and then connecting these clusters, yielding 19410 edges.

For each of the both test sets we generated 25 instances according to the following parameters. The receiver as well as the Steiner sites and the terminals are picked uniformly at random. For the grid instances we selected 80 Steiner sites and 100 terminals. For the IGen.3200 topology we chose 400 Steiner sites and 600 terminals. Common to both test sets, we set the edge capacities to 3 and the capacity of Steiner sites and the root to 5. On the grid topology, we set edge costs to 1 and activation costs for Steiner sites to 20. For IGen.3200 instances, edge costs are defined by the euclidean distance and activation costs are distributed uniformly according to $\mu(c_E) \cdot \mathcal{U}(25, 75)$, where $\mu(c_E)$ denotes the average edge length.

Figure 2 shows the objective gap, i.e. the relative quality guarantee, over time for both test sets, consisting of 25 instances each. Independent of the test set, the objective gap stabilizes after one hour of computation. For the highly symmetric grid instances as well as for the IGen.3200 instances, a median gap of less than 4% is achieved. As documented in [26], the lower bound improves by less than 2% for IGen.3200 instances and by less than 12% for grid instances. Hence, the lower bounds obtained initially are already reasonably accurate and the progress of the objective gap (cf. Figure 2) is driven by the quality of the solutions found. Based on this observation, we have implemented a primal heuristic to generate feasible solutions based on the linear relaxations during the branch-and-bound search (see [26]).

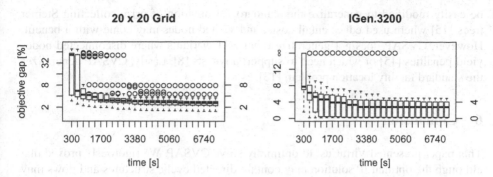

Fig. 2. Objective gap over time for the 20×20 grid and IGen.3200 test sets. Note the logarithmic y-axis for the grid instances. A gap of ∞ indicates that no primal solution has been found.

5 Related Work

The CVSAP problem differs from many models studied in the context of IPTV [14], sensor networks [10,11], fiber-optical transport [14], or Active Networking [3], to just name a few, in that the number and placement of processing locations is subject to optimization as well. The problem is complicated further by the fact that the communication between sender and receiver may be processed *repeatedly* within the network. The result from [19] on multi-constrained multicast routing also applies to CVSAP: any algorithm limited to (directed) acyclic graphs cannot solve the problem in general. Generally, while there exist many heuristic and approximate algorithms for related problem variants, we are the first to consider exact solutions.

The two closest models to CVSAP are studied in [21] and [27]. While [27] already showed the applicability of selecting only a few processing nodes for multicasting, no concise formalization is given and the described heuristic does not provide performance guarantees. In a series of publications, Oliviera and Pardalos consider the Flow Streaming Cache Placement Problem (FSCPP) [21]. Unfortunately, their FSCPP definition is inherently flawed as it does not guarantee connectivity (see [26] for a discussion). Interestingly, the authors also provide a correct approximation algorithm, which however only considers the rather weak model which ignores traffic.

Other Related Problems and Algorithms. The CVSAP is related to several classic problems. For example, CVSAP generalizes the *light-tree* concepts [4] in the sense that "light splitting" locations can be chosen depending on the repeatedly processed traffic; our approach can directly be used to optimally solve the light-tree problem. In the context of wave-length assignment, Park et al. [22] show that a small number of virtual splitters can be sufficient for efficient multicasting. Our formalism and the notion of hierarchy is based on the paper by Molnar [19] who studies the structure of the so-called multi-constrained multicast routing problem. However, unlike in CVSAP, an edge may be only used once in the solution. If the cost of in-network processing is zero and all nodes are possible Steiner sites, the CVSAP boils down to the classic Steiner Tree Problem [12] and its degree-bounded variants [17]. A closer look shows that CVSAP can be easily modified to generalize the standard formulation of prize-collecting Steiner trees [15] where used edges entail costs, and visited nodes may come with a benefit. However, CVSAP does not generalize other STP variants where disconnected nodes yield penalties [15] or which need to support anycasts [8]. Lastly, CVSAP generalizes the standard facility location problem [13].

6   Conclusion

This paper presented VirtuCast to optimally solve CVSAP. We rigorously proved that although the optimal IP solution may contain directed cyclic structures and flows may be merged repeatedly, there exists an algorithm to decompose the solution into individual routes. Using VirtuCast, we solved realistically sized instances to within 4% of optimality. Since CVSAP is related to several classical optimization problems, we believe that our approach is of interest beyond the specific model studied here.

An interesting direction for future research regards the design of approximation algorithms as an efficient alternative to the rigorous optimization approach proposed in this paper. While in its general form CVSAP cannot be approximated, we believe that there exist good approximate solutions, e.g., for uncapacitated variants or bi-criteria models where capacities may be violated slightly.

Acknowledgement. We would like to thank Marten Schönherr from Deutsche Telekom. This research was supported by the EU projects BigFoot and UNIFY.

References

1. Achterberg, T.: SCIP: Solving Constraint Integer Programs. Mathematical Programming Computation 1(1), 1–41 (2009)
2. Ahuja, R.K., Magnanti, T.L., Orlin, J.B.: Network Flows: Theory, Algorithms and Applications. Prentice Hall (1993)
3. Banchs, A., Effelsberg, W., Tschudin, C., Turau, V.: Multicasting Multimedia Streams with Active Networks. In: Proc. Local Computer Network Conference (LCN). IEEE (1998)
4. Cai, Z., Lin, G., Xue, G.: Improved Approximation Algorithms for the Capacitated Multicast Routing Problem. In: Wang, L. (ed.) COCOON 2005. LNCS, vol. 3595, pp. 136–145. Springer, Heidelberg (2005)
5. Costa, P., Donnelly, A., Rowstron, A., Shea, G.O.: Camdoop: Exploiting In-network Aggregation for Big Data Applications. In: Proc. USENIX Symposium on Networked Systems Design and Implementation (NSDI) (2012)
6. Costa, P., Migliavacca, M., Pietzuch, P., Wolf, A.L.: NaaS: Network-as-a-Service in the Cloud. In: Proc. USENIX Hot-ICE Workshop (2012)
7. Cranor, C., Johnson, T., Spataschek, O., Shkapenyuk, V.: Gigascope: A Stream Database for Network Applications. In: Proc. ACM SIGMOD International Conference on Management of Data, pp. 647–651 (2003)
8. Demaine, E.D., Hajiaghayi, M., Klein, P.N.: Node-weighted steiner tree and group steiner tree in planar graphs. In: Albers, S., Marchetti-Spaccamela, A., Matias, Y., Nikoletseas, S., Thomas, W. (eds.) ICALP 2009, Part I. LNCS, vol. 5555, pp. 328–340. Springer, Heidelberg (2009)

9. European Telecommunications Standards Institute. Network Functions Virtualisation - Introductory White Paper. SDN and OpenFlow World Congress, Darmstadt-Germany (2012)
10. Eyal, I., Keidar, I., Patterson, S., Rom, R.: In-Network Analytics for Ubiquitous Sensing. In: Afek, Y. (ed.) DISC 2013. LNCS, vol. 8205, pp. 507–521. Springer, Heidelberg (2013)
11. Fasolo, E., Rossi, M., Widmer, J., Zorzi, M.: In-Network Aggregation Techniques for Wireless Sensor Networks: A Survey. IEEE Wireless Communications 14, 70–87 (2007)
12. Goemans, M.X., Myung, Y.-S.: A catalog of Steiner tree formulations. Networks 23(1), 19–28 (1993)
13. Gollowitzer, S., Ljubić, I.: MIP models for Connected Facility Location: A theoretical and computational study. Computers & Operations Research 38(2), 435–449 (2011)
14. Hermsmeyer, C., Hernandez-Valencia, E., Stoll, D., Tamm, O.: Ethernet aggregation and core network models for effcient and reliable IPTV services. Bell Labs Technical Journal 12(1), 57–76 (2007)
15. Johnson, D.S., Minkoff, M., Phillips, S.: The prize collecting Steiner tree problem: theory and practice. In: Proc. 11th Annual ACM-SIAM Symposium on Discrete Algorithms, pp. 760–769. Society for Industrial and Applied Mathematics (2000)
16. Koch, T., Martin, A.: Solving Steiner tree problems in graphs to optimality. Networks 32(3), 207–232 (1998)
17. Lee, Y., Lu, L., Qiu, Y., Glover, F.: Strong formulations and cutting planes for designing digital data service networks. Telecommunication Systems 2(1), 261–274 (1993)
18. Lucena, A., Resende, M.G.: Strong lower bounds for the prize collecting Steiner problem in graphs. Discrete Applied Mathematics 141(1), 277–294 (2004)
19. Molnár, M.: Hierarchies to Solve Constrained Connected Spanning Problems. Technical Report lrimm-00619806, University Montpellier 2, LIRMM (2011)
20. Narayana, S., Jiang, W., Rexford, J., Chiang, M.: Joint Server Selection and Routing for Geo-Replicated Services. In: Proc. Workshop on Distributed Cloud Computing (DCC) (2013)
21. Oliveira, C., Pardalos, P.: Streaming Cache Placement. In: Mathematical Aspects of Network Routing Optimization. Springer Optimization and Its Applications, pp. 117–133. Springer, New York (2011)
22. Park, J.-W., Lim, H., Kim, J.: Virtual-node-based multicast routing and wavelength assignment in sparse-splitting optical networks. Photonic Network Communications 19(2), 182–191 (2010)
23. Qazi, Z., Tu, C.-C., Chiang, L., Miao, R., Sekar, V., Yu, M.: SIMPLE-fying Middlebox Policy Enforcement Using SDN. In: Proc. ACM SIGCOMM (2013)
24. Quoitin, B., Van den Schrieck, V., Franois, P., Bonaventure, O.: IGen: Generation of router-level Internet topologies through network design heuristics. In: Proc. 21st International Teletraffic Congress (ITC), pp. 1–8 (2009)
25. Rost, M., Schmid, S.: CVSAP-Project Website (2013),
   http://www.net.t-labs.tu-berlin.de/~stefan/cvsap.html
26. Rost, M., Schmid, S.: The Constrained Virtual Steiner Arborescence Problem: Formal Definition, Single-Commodity Integer Programming Formulation and Computational Evaluation. Technical report, arXiv: 1310.0346 (2013)
27. Shi, S.: A Proposal for A Scalable Internet Multicast Architecture. Technical Report WUCS-01-03, Washington University (2001)
28. Voß, S.: Steiner Tree Problems in Telecommunications. In: Handbook of optimization in telecommunications, ch. 18. Spinger Science + Business Media, New York (2006)
29. Zhang, Z., Zhang, M., Greenberg, A., Hu, Y.C., Mahajan, R., Christian, B.: Optimizing cost and performance in online service provider networks. In: Proc. 7th USENIX Conference on Networked Systems Design and Implementation (NSDI) (2010)

Mobile Byzantine Agreement
on Arbitrary Network*

Toru Sasaki, Yukiko Yamauchi, Shuji Kijima, and Masafumi Yamashita

Graduate School of Information Science and Electrical Engineering, Kyushu
University, Japan, 744, Motooka, Nishi, Fukuoka, Japan
{toru.sasaki,yamauchi,kijima,mak}@inf.kyushu-u.ac.jp

Abstract. The mobile Byzantine agreement problem on *general* network is investigated for the first time. We first show that the problem is unsolvable on any network with the order n and the vertex connectivity d, if $n \leq 6t$ or $d \leq 4t$, where t is an upper bound on the number of faulty processes. Assuming full synchronization and the existence of a permanently non-faulty process, we next propose two t-resilient mobile Byzantine agreement algorithms for some families of not fully connected networks. They are optimal on some networks, in the sense that they correctly work if $n > 6t$ and $d > 4t$.

Keywords: agreement problem, Byzantine fault, distributed network, distributed algorithm, mobile Byzantine agreement problem.

1   Introduction

Reaching an agreement among processes on a distributed network is a fundamental distributed problem, and becomes formidable when processes are subject to Byzantine faults. The problem was originally formulated by Pease et al. [13] in 1980, and a huge number of researches followed this pioneering work. A few of the earliest ones, which are closely related to this paper, are [6–8, 10]. They, in particular, showed that Byzantine agreement is possible only on synchronous networks [8], and there is a t-resilient Byzantine agreement algorithm on a synchronous network with the order n and the vertex connectivity d, if and only if $n > 3t$ and $d > 2t$ [6]. They all assumed that Byzantine fault is permanent and a faulty process never recover.

Transient Byzantine faults were investigated by Garay [9], in the context of the *mobile Byzantine agreement problem*. He investigated a distributed network which includes a malicious mobile agent who controls its host process and forces it to behave as an adversary to the algorithm. The process is faulty as long as the malicious agent is there, but can recover when it leaves. (A malicious agent may stay on a process forever to cause a permanent Byzantine fault.) The mobile Byzantine agreement problem differs from the self-stabilizing agreement

* This work is supported in part by JSPS KAKENHI (No. 22300004, No. 24650008, No. 23700019, and No. 24106005).

R. Baldoni, N. Nisse, and M. van Steen (Eds.): OPODIS 2013, LNCS 8304, pp. 236–250, 2013.

problem in that, in the former, an algorithm is requested to achieve and maintain an agreement, in the presence of possibly infinite number of transient faults plus some finite number of permanent faults. Provided that 1) the network is synchronous, 2) it is fully connected, 3) there are at most t malicious agents in a round, and 4) there is a process which is non-faulty forever, Garay presented a t-resilient mobile Byzantine algorithm, assuming $n > 6t$.[1] Burhman et al. [4] slightly changed the fault model and assumed that only a message can carry a malicious agent from a process to another. Under this model, which is more generous to the algorithm, they could provide a t-resilient mobile Byzantine algorithm, when $n > 3t$.[2] As for the Garay's original model, Banu et al. [1] proposed a t-resilient mobile Byzantine algorithms provided $n > 4t$, but they did not discuss its necessity. Other moving failures were also investigated in [2, 3, 14, 15].

The mobile Byzantine agreement problem has been investigated only on a complete network in the literature, as far as we know. In this paper, we investigate it on *general* networks to obtain results similar to those in [6].

Compared with the Garay's model, our system model, which we present in Section 2, is slightly more generous to an adversary, and (we think) is more natural. We assume that a round consists of the send, receive and (local) computation steps, and all messages being sent in the next round are prepared in the computation step of this round. We then show that there is no t-resilient mobile Byzantine algorithm, if either $n \leq 6t$ or $d \leq 4t$ holds, where d is the vertex connectivity of the network. As for the positive side, let $\mathcal{G}(\alpha, \beta)$ be the family of networks such that for any pair of processes, there are at least α disjoint paths connecting them, whose lengths are at most β. Then we show that there is a t-resilient mobile Byzantine algorithm on any network in $\mathcal{G}(\alpha, \beta)$, if both $n > 6t$ and $\alpha > 2\beta t$ hold. The algorithm is optimal when $\alpha = d$ and $\beta = 2$. We also propose another mobile Byzantine agreement algorithm that works on a network which is the k-th power of a graph. The algorithm is also optimal on some networks.

To construct these algorithms, we take the same approach as in [6]; we combine an algorithm to realize reliable communication on an arbitrary network with a mobile Byzantine algorithm MBA in [1] for a complete network.

The paper is organized as follows: Section 2 formalizes the mobile Byzantine agreement problem and describes mobile Byzantine agreement algorithm MBA proposed in [1]. Section 3 derives a necessary condition for a network to have a t-resilient mobile Byzantine agreement algorithm, and Section 4 presents two mobile Byzantine agreement algorithms. We then conclude the paper by presenting some open problems in Section 5.

---

[1] The number of rounds necessary for a malicious agent to move from a process to another is called the roaming pace, and is denoted by ρ. This algorithm works for $\rho = 1$. Ref. [9] includes another algorithm for $\rho = 2$, provided $n > 4t$. We assume $\rho = 1$, unless otherwise stated.

[2] The algorithm is optimal, since the Byzantine algorithm is unsolvable if $n \leq 3t$ [13].

2   Preliminaries

2.1   System Model

Consider a distributed network consisting of processes, some pairs of which can directly communicate with each other through a bidirectional communication link between them. We model the network by a simple, connected, undirected graph $G = (\Pi, E)$, where vertex set Π represents the set of processes and edge set E the set of communication links. We use graph and network, vertex and process, and edge and communication link, interchangeably. Graph G may not be complete, and some pair of processes not in E cannot directly exchange messages. By $N_i = \{j \in \Pi : (i, j) \in E\} \cup \{i\}$, we denote the *closed* set of the i's neighbors. Then i can send a message to, and receive a message from a process $j \in N_i$.[3] Let d be the (vertex) connectivity of G, which is the minimum number of vertices necessary to remove to disconnect G. Without loss of generality, we assume that $\Pi = \{1, 2, \ldots, n\}$ and regard $i \in \Pi$ as the ID of process i. We also assume that every process knows G, as a part of its initial information.

In this paper, we consider a *synchronous* network, whose execution consists of a sequence of rounds. A round consists of the send, receive and computation steps. In round $r = 1, 2, \ldots$, every process i sends a message m_{ij}^r to each process $j \in N_i$ and receives a message m_{ji}^r from each $j \in N_i$, where m_{ij}^r was computed in the computation step of the previous round $r - 1$. Process i then computes its new local state s_i^r and messages m_{ij}^{r+1} for its neighbors $j \in N_i$, for the next round $r + 1$, using a given deterministic algorithm A.

Round 0 is a special round representing the initialization; in round 0, every process i receives initial data and prepares an initial local state s_i^0. It also constructs messages m_{ij}^1 for each of the neighbors $j \in N_i$. This paper investigates a (local) algorithm A to compute local state s_i^r and messages m_{ij}^r in each round r on every process i, in such a way that the network executing A can solve the mobile Byzantine problem, which we will define shortly.

2.2   Fault Model: Mobile Byzantine Faults

In this paper, every communication link in the network is reliable, and a message is never lost, duplicated, and corrupted. More clearly, when a process i sends a message m to a process $j \in N_i$ in a round r, then ID i is automatically attached to m, and m and i are correctly received by j in the same round r, if j is nonfaulty. Although i can send a corrupted m when i is faulty, j always receives the correct ID of the sender i.

Some process i may be unreliable on the other hand. It may suffer from a *Byzantine* fault, and independently of its algorithm A, it can take any malicious actions by corrupting local memories, as well as by skipping the send and/or

---

[3] We regard that every process can send a message to itself, to simplify descriptions.

receive steps. We make two assumptions. First, despite the definition, we can assume that a faulty process never skip the send and receive steps, without loss of generality, as those omissions do not help it. Second, we assume that ID i, algorithm A and its program counter (i.e., the current round number) are stored in the anti-tamper memory, and are not subjected to corruption, like transient faults in the literature of self-stabilization, although a faulty process can arbitrarily change the messages waiting for being sent, those received from other processes, and of course the contents of local variables.

Unlike the conventional Byzantine faults, faults in this paper are *mobile* as in [9]. Let F_r be the set of faulty processes in round $r = 0, 1, \ldots$. Then F_r may change round by round. We consider that F_r is chosen by a malicious adversary, but assume that for some known constant t, $|F_r| \leq t$ holds for any r.[4] There may be faulty processes even in round 0, i.e., F_0 may not be empty, and they may corrupt the initial local state and messages for round 1.

There is a delicate difference between our and the Garay's models in how a process recovers from a fault. In the Garay's model, when a malicious agent leaves a process i at the end of round r, i *autonomously* notices the fact when round $r + 1$ starts, skips the send step, and learns its algorithm A and the current computation state from the messages received in round $r + 1$. It then starts participating in the on-going computation from round $r + 2$. In our model, on the other hand, i does not notice the fact, but i in round $r+1$ can start executing A, since A is not corrupted (as we assumed in above). The computation however is possibly based on corrupted messages and local state. Intuitively, in our model, recovery from a fault is the responsibility of A, which is a natural requirement when we consider fault tolerant algorithms, but is not, in the Garay's model. We consider the worst behavior of Byzantine faults, including the case where the current computation states of processes are modified without the processes recognizing it.

Finally, we assume that there is a process i_0 which is non-faulty all the time, i.e., there is an i_0 such that $i_0 \notin \cup_r F_r$, since the agreement problem is unsolvable, if there is no permanently non-faulty process [9, 14].

2.3 Mobile Byzantine Agreement Problem

In the mobile Byzantine agreement problem, each process i has an initial value $d_i \in \{0, 1\}$, and we wish to design an algorithm A for each process i to eventually decide on a value v_i satisfying all of the following four properties:

(**Decision**). Every process i which is not faulty in infinite number of rounds eventually irreversibly decides on a value $v_i \in \{0, 1\}$, i.e., there is a round r_0

---

[4] Besides an upper bound t on the number of faulty processes, we do not make any assumption on F_r. That is, we consider the case in which the roaming pace ρ is 1(that is, faults can move every round).

after which process i outputs this very same value v_i as its decision at the end of every round $r \geq r_0$ where i is non-faulty.[5]

(**Agreement**). All processes i which are not faulty in infinite number of rounds decide on the same value $v_i \in \{0, 1\}$.

(**Validity**). If all processes which are not faulty in round 0 have the same initial value v, then for all non-faulty processes i, $v_i = v$.

(**Consistency Maintenance**). Once an agreement is reached among the currently non-faulty processes, it is preserved among the (possibly different) non-faulty processes.

Note that the first three conditions are borrowed from the definition of the Byzantine agreement problem [10]. Unlike a Byzantine agreement algorithm, any mobile Byzantine agreement algorithm never terminate to satisfy Consistency Maintenance property.

2.4   MBA: Banu et al.'s Mobile Byzantine Agreement Algorithm

Banu et al.[1] proposed a mobile Byzantine agreement algorithm MBA for a complete graph, using broadcast as the communication primitive [6]. They showed that if $4t < n$, MBA correctly solves the mobile Byzantine algorithm on a complete graph under the Garay's model. We slightly modify MBA so that it can run in our model. We call this algorithm mMBA and use it as the skeleton of our mobile Byzantine agreement algorithms for general graphs in Section 4.

We present mMBA on process i. It is based on the rotating coordinator paradigm, and consists of infinite number of phases $s = 1, 2, \ldots$, each of which consists of three rounds. In the first round, process i collects other processes' decision values and stores them in PV_i, and sets its decision value v_i to either $v \in \{0, 1\}$ or $\perp$. In the second round, it stores other processes' decision values in SV_i, then if most of them are the same value, say v, i adopts v as a new decision value. Finally, in the third round, i stores other processes's SV's and tries to make all processes have the same decision value by checking the coordinator's SV at the end of this round. The coordinator in phase s is $c_i = s \mod n$, where $c_i = 0$ means $c_i = n$. An agreement is reached when c_i is non-faulty. Since there

---

[5] Specifically, algorithm DP-Byz (we will show in 4.1) guarantees $r_0 = 3\beta n$ and KP-Byz (we will show in 4.2) guarantees $r_0 = 3n(D_G - k + 1)$. Both literature [1, 9] defined Decision property as follows: Every non-faulty process i eventually irreversibly decides on a value $v_i \in \{0, 1\}$. Since the meaning of "non-faulty process" is not clear enough, we correct the definition. To preserve Consistency Maintenance property below, any mobile Byzantine agreement algorithm does not terminate, and a process may become faulty infinitely many times. The variable that stores the decision value thus may be corrupted infinitely many times. After r_0, whenever it is corrupted, the algorithm correct it with the decision value as soon as the process recovers (once deciding it). This is what "irreversibly decides" means. We note that this correction does not change the essence of the problem, and both algorithms [1, 9] solve the mobile Byzantine agreement in the sense of the corrected definition.

[6] MBA is a binary consensus algorithm.

is a process i_0 who is permanently non-faulty, an agreement is always achieved up to (and including) $s = i_0$. Among local variables that mMBA uses, d_i and v_i store the initial and the decision (i.e., output) values. We assume that, when mMBA starts, the initial value has been assigned to d_i in each process i, but d_i may be corrupted in at most t processes.[7] Although every process i is requested to irreversibly decide the value v_i by the definition of agreement problem, we allow i to change v_i to save the number of symbols; v_i stabilizes after phase i_0 and satisfies Consistency Maintenance property forever after phase i_0.

Algorithm 2.1. mMBA on process i

1: $v_i \leftarrow d_i$
2: **for** Phase $s = 1$ **to** ∞ **do**
3:     $PV_i[1..n] \leftarrow [\bot, \ldots, \bot]$;
4:     (Round 1)
5:     send v_i to all processes;
6:     **for all** $j \in \Pi$ **do**
7:         $PV_i[j] \leftarrow v_j$ (if $v_j \notin \{0,1\}$, store $\bot$ instead of v_j);
8:     **if** $w \in \{0,1\}$ occurs at least $n - 2t$ times in $PV_i[1..n]$ **then**
9:         $v_i \leftarrow w$;
10:     **else**
11:         $v_i \leftarrow \bot$;
12:     $SV_i[1..n] \leftarrow [\bot, \ldots, \bot]$;
13:     (Round 2)
14:     send v_i to all processes;
15:     **for all** $j \in \Pi$ **do**
16:         $SV_i[j] \leftarrow v_j$ (if $v_j \notin \{0,1\}$, store $\bot$ instead of v_j);
17:     **if** $w \in \{0,1\}$ occurs more than $2t$ times in $SV_i[1..n]$ **then**
18:         $v_i \leftarrow w$;
19:     **else**
20:         $v_i \leftarrow \bot$;
21:     $EV_i \leftarrow [\bot, \ldots, \bot][\bot, \ldots, \bot]$;
22:     (Round 3)
23:     send $SV_i[1..n]$ to all processes;
24:     **for all** $j \in \Pi$ **do**
25:         $EV_i[j][1..n] \leftarrow SV_j[1..n]$;
26:     RECONSTRUCT(EV_i);
27:     $c_i \leftarrow s \bmod n$;
28:     **if** $w \in \{0,1\}$ occurs more than $2t$ times in $EV_i[c_i][1..n]$ **then**
29:         $cv_i \leftarrow w$;
30:     **else**
31:         $cv_i \leftarrow 0$;
32:     **if** $accept_i = True$ **then**
33:         $v_i \leftarrow cv_i$;

---

[7] This assumption is consistent with our fault model.

Algorithm 2.2. RECONSTRUCT(EV_i) on process i

1: **for all** $j \in \Pi$ **do**
2:     **if** $w \in \{0, 1\}$ occurs more than $n - 2t$ times in column j of EV_i **then**
3:         $SV_i[j] \leftarrow w$;
4:     **else**
5:         $SV_i[j] \leftarrow \perp$;
6: **if** $w \in \{0, 1\}$ occurs more than $4t$ times in $SV_i[1..n]$ **then**
7:     $v_i \leftarrow w$;
8:     $accept_i \leftarrow False$;
9: **else if** $w \in \{0, 1\}$ occurs more than $2t$ times in $SV_i[1..n]$ **then**
10:     $v_i \leftarrow w$;
11:     $accept_i \leftarrow True$;
12: **else**
13:     $v_i \leftarrow \perp$;
14:     $accept_i \leftarrow True$;

Observe that mMBA executes three message exchanges between any two processes in Lines 5–7, 14–16 and 23–25. On a complete graph, a reliable message exchange between two non-faulty processes is obviously realizable just by using the communication primitives, as mMBA does. Suppose that there is a reliable message exchange algorithm between any two non-faulty processes on some non-complete graph. Then we can modify mMBA so that it can correctly work on the graph, by replacing the direct message exchange between two processes i and j in Lines 5–7, 14–16 and 23–25 with a reliable message transmission algorithm for i and j (provided that mMBA correctly works on a complete graph). In Section 4, we propose two algorithms DPT and PerT to implement a reliable communication. As a corollary to Theorem 1 in [1], we have:

Corollary 1. *Provided $n > 6t$, mMBA solves the mobile Byzantine agreement problem on a complete graph under our model.*

Proof. By following the argument in the proof of Theorem 1 in [1], one can show the corollary, since mMBA is a slight modification of MBA.   □

3   Upper Bound on the Number of Faulty Processes

A main technique we use in this section to derive an impossibility result is to reduce the impossibility regarding the static (i.e., conventional) Byzantine agreement problem to that of the mobile Byzantine agreement problem. To this end, we first compare the behavior of any algorithm A on a graph $G = (\Pi, E)$ under the static and mobile Byzantine settings. Let C (resp. $\hat{C}$) be an execution of A on G under the static (resp. mobile) setting, where, in C (resp. $\hat{C}$), a message m_{ij}^r (resp. $\hat{m}_{ij}^r$) is sent from process i to j in round r, and s_i^r (resp. $\hat{s}_i^r$) is the local state of i at the end of round r. Despite that A is deterministic and G is synchronous, C (resp. $\hat{C}$) is not unique, because of the adversary that controls

Byzantine faults. Let F be the set of faulty processes in the static Byzantine setting, and F_r be the one in round r in the mobile Byzantine setting.

Lemma 1. *Suppose that $|F| = 2\max_r\{|F_r|\} = 2t$. Then for any C that terminates in round r_f, there is a $\hat{C}$ that satisfies the following two conditions:*

$$\forall i \in \Pi, \forall j \in N_i, \forall r(0 \le r \le r_f), m_{ij}^r = \hat{m}_{ij}^r, \tag{1}$$

$$\forall i \in \Pi \setminus F, \forall r(0 \le r \le r_f), s_i^r = \hat{s}_i^r. \tag{2}$$

Proof. For any C, we construct $\hat{C}$ that satisfies the conditions. It is worth emphasizing that $\hat{C}$ depends on the adversary, so that a main part of the construction explains F_r in each round r and how the faulty processes in F_r behave. Recall that we have assumed that the contents of m_{ij}^r (resp. $\hat{m}_{ij}^r$) is s_i^{r-1} (resp. $\hat{s}_i^{r-1}$) for any $j \in N_i$, if i is non-faulty in round r.

We arbitrarily partition F (of size $2t$) into two sets F_{odd} and F_{even} with the same size t, and define that $F_r = F_{odd}$ if r is odd, and otherwise if r is even, then $F_r = F_{even}$. By induction on r, we show that a desired $\hat{C}$ is constructible by arranging the behaviors of faulty processes in F_r.

The base case ($r = 0$) is obvious, since $s_i^0 = \hat{s}_i^0$ for all non-faulty processes $i \notin F$ because $F_{even} \subseteq F$.

For induction step, consider round $r \ge 1$, and assume that r is odd. (The case in which r is even can be shown by a symmetric argument.) By induction hypothesis, for all $i \in \Pi$ and $j \in N_i$, $m_{ij}^{r-1} = \hat{m}_{ij}^{r-1}$, and for all $i \in \Pi \setminus F$, $s_i^{r-1} = \hat{s}_i^{r-1}$.

We first show $m_{ij}^r = \hat{m}_{ij}^r$ for all $i \in \Pi$ and $j \in N_i$. If $i \notin F$, that is, if i is non-faulty both in rounds $r - 1$ and r, and in both static and mobile Byzantine settings, then, by induction hypothesis, in round $r - 1$, for each of neighbors $j \in N_i$, A constructed the same message in C and $\hat{C}$, and i sends it to j in round r. Thus $m_{ij}^r = \hat{m}_{ij}^r$ for all $j \in N_i$. If $i \in F_{odd} = F_r$, since i is faulty in round r (in the mobile Byzantine setting), it can send any message to each of neighbors $j \in N_i$, in particular, it can send m_{ij}^r. Finally, if $i \in F_{even}$, then $i \notin F_r$ and i is non-faulty in round r in the mobile Byzantine setting. Since $i \in F$ and i is faulty in round r in the static Byzantine setting, so that m_{ij}^r can be arbitrary. In this case, since i is faulty in round $r - 1$ in the mobile Byzantine setting, i could choose an arbitrary message as $\hat{m}_{ij}^r$ in round $r - 1$. In particular, i could choose m_{ij}^r as $\hat{m}_{ij}^r$.

We go on showing $s_i^r = \hat{s}_i^r$ for any $i \notin F$. If $i \notin F$, then obviously the claim holds, since $s_i^{r-1} = \hat{s}_i^{r-1}$ and $m_{ji}^r = \hat{m}_{ji}^r$ for all $j \in N_i$. $\qquad\square$

Theorem 1. *There is no algorithm to solve the mobile Byzantine agreement problem on any (simple, connected, undirected) graph $G = (\Pi, E)$, if $n \le 6t$ or $d \le 4t$, where $|\Pi| = n$ and d is the vertex connectivity of G.*

Proof. To derive a contradiction, let us assume that there is an algorithm A to solve the mobile Byzantine agreement problem on G, despite that either $n \le 6t$

or $d \leq 4t$ holds, where t is an upper bound on the number of faulty processes in a round r, i.e., $|F_r| \leq t$. Fix any initial value v_i assigned to each process i, and assume that A achieves an agreement up to round r_f. That is, Decision, Agreement and Validity properties hold when round r_f finishes.

Now, on G, we execute A to solve the static Byzantine agreement problem for the same initial values v_i, assuming that $|F| = 2t$, where F is the set of faulty processes (in the static Byzantine setting). Note that $n \leq 3|F|$ and $d \leq 2|F|$. Since there is no algorithm to solve the static Byzantine algorithm when either $n \leq 3|F|$ or $d \leq 2|F|$ holds [6], A of course cannot solve the problem. Let C be an execution that A does not reach the agreement in the static Byzantine setting. It is a contradiction, since there is an execution $\hat{C}$ that does not reach the agreement in the mobile Byzantine setting by Lemma 1.

$\square$

4   Two Mobile Byzantine Agreement Algorithms

In this section, we propose two mobile Byzantine agreement algorithms DP-Byz and KP-Byz. Both of them are based on mMBA, and only the algorithms to realize reliable communication differ. They work on different families of graphs. Both algorithms are shown to be optimal for some graphs, in the sense that they work if $n > 6t$ and $d > 4t$.

4.1   Algorithm DP-Byz

The celebrated Menger's theorem (see, e.g., [5]) says that every graph with vertex connectivity d contains, for any two distinct vertices, d (vertex) disjoint paths connecting them. Let $\mathcal{G}(\alpha, \beta)$ be the class of graphs $G = (\Pi, E)$ such that, for any pair (i, j) of vertices in Π, there are α disjoint paths connecting i and j, whose length (in terms of the number of edges) is at most β. If d is the vertex connectivity of G, then $d \geq \alpha$ by the Menger's theorem. In this subsection, we present an algorithm DP-Byz (Disjoint Path Byzantine), and shows that it solves the mobile Byzantine agreement problem on any graph $G \in \mathcal{G}(\alpha, \beta)$, if an upper bound t on the number of faulty processes satisfies $n > 6t$ and $\alpha > 2\beta t$. Thus DP-Byz is optimal for any graph in $\mathcal{G}(d, 2)$ by Theorem 1.

Dolev [6] combined an algorithm to establish a reliable communication route between any two non-faulty processes with a Byzantine agreement algorithm on a complete graph to solve the Byzantine agreement problem on arbitrary graphs. To establish a reliable communication route, based on the Menger's theorem, a process i sends d copies of a message m to another process j along d vertex disjoint paths connecting i and j. Then j can correctly decide m simply by taking the majority of the received messages, if $d > 2t$.

Algorithm DP-Byz extends this trick to the mobile Byzantine case and applies it to mMBA. In the Dolev's algorithm, $2t + 1$ vertex disjoint paths are sufficient to tolerate t Byzantine faults, since the faults are stationary and a single fault can corrupt at most one path. In the mobile Byzantine setting, on the other hand, one fault may be able to corrupt many paths. Procedure DPT (Disjoint Path Transmission) is an implementation of the trick to guarantee a reliable transmission of a message in the mobile Byzantine setting. We now present DPT.

DPT on process i takes two parameters, a value w_i (which may be a vector) and an array X_i, where w_i is a message that i wishes to send to all processes $j \in \Pi$, and X_i is used to store received messages w_j from all processes $j \in \Pi$. If all processes invoke DPT(w_i, X_i) in round r, in the presence of at most t mobile faulty processes, $X_i[j] = w_j$ holds for each of the non-faulty processes $j \notin (F_{r-1} \cup F_r)$, i.e., those which are non-faulty both in rounds $r - 1$ and r. DPT on i makes use of the α vertex disjoint paths connecting i and j for all $j \in \Pi$, which are guaranteed to exist, since $G \in \mathcal{G}(\alpha, \beta)$. For any $1 \leq k \leq \alpha$, we denote the k-th path by π_{ij}^k, where for any $1 \leq \ell \leq |\pi_{ij}^k|$, $\pi_{ij}^k[\ell]$ is the ℓ-th vertex in π_{ij}^k. Here $|\pi_{ij}^k| - 1$, which is the length of path π_{ij}^k, is at most β, $\pi_{ij}^k[1] = i$, and $\pi_{ij}^k[|\pi_{ij}^k|] = j$. An array W_i is used as a working memory. $W_i[j][k]$ holds the message relayed to i through the k-th path from j. We assume that π_{ij}^k for all i, j, k are available to DPT, although the description of DPT does not explicitly state this fact.

We will take v_i as the actual parameter of w_i. As usual, when DPT is invoked, the value of v_i is copied to w_i, which is a local variable of DPT. Notice that v_i also appears in DPT as a global variable, whose value may change when i becomes faulty while DPT is executed. DPT makes use of v_i to preserve Consistency Maintenance.

Note that in some round r, a process i may send more than one messages to a process $j \in N_i$. In such a case, we assume that these messages are aggregated into a single message before sending them.

Lemma 2. *Suppose that $G \in \mathcal{G}(\alpha, \beta)$. If each process $i \notin F_{r_0-1} \cup F_{r_0}$ has a value w_i and invokes* DPT(w_i, X_i) *on G in a round r_0, then for any $i \notin F_{r_0+\beta}$ and $j \notin F_{r_0-1} \cup F_{r_0}$, $X_i[j] = w_j$ holds in round r_f when $\alpha > 2\beta t$, where $r_f = r_0 + \beta$ is the round that DPT terminates. Furthermore, Consistency Maintenance is preserved between (and including) rounds r_0 and $r_f + \beta$.*

Proof. Suppose that $i \notin F_{r_f}$ and $j \notin F_{r_0-1} \cup F_{r_0}$. Since α paths π_{ji}^k are vertex disjoint, and there are at most βt faulty processes between rounds r_0 and $r_f - 1$,[8] at most βt paths π_{ji}^k are corrupted; in other words, i receives the correct value w_j via at least $\alpha - \beta t > \beta t$ paths in round r_f. Thus taking the majority i can easily obtain the correct w_j.

We should also consider the following case. If many of the paths π_{ji}^k are shorter than β, and if we send w_j to i in round $r_0 + |\pi_{ji}^k| - 1$, which is earlier than r_f, i cannot compute the correct w_j, when i is faulty in round $r_f - 1$. That is, we need to send w_j to i via all paths exactly in round r_f simultaneously. To this end, a process immediately before i in a path π_{ji}^k postpones sending w_j to i until round r_f, if $|\pi_{ji}^k| < \beta$.

---

[8] Observe that a faulty process in round r_0 can corrupt some π_{ji}^k.

Algorithm 4.1. DPT(w_i, X_i) on process i

```
1:  W_i[1..n][1..α] ← [⊥ ··· ⊥][⊥ ··· ⊥];
2:  LAST_i ← ∅
3:  (Round r (1 ≤ r ≤ β − 1))
4:  for r = 1 to β − 1 do
5:      if r = 1 then
6:          for all j ∈ Π do
7:              for all k = 1 to α do
8:                  send (w_i, k, i, j) to π_{ij}^k[2];
9:          for all j ∈ N_i do
10:             send v_i to j;
11:     else
12:         for all (w, k, h, j) ∈ NEXT_i do
13:             send (w, k, h, j) to π_{hj}^k[r + 1];
14:         for all j ∈ N_i do
15:             send v_i to j;
16:     NEXT_i ← ∅
17:     for all messages (w, k, h, j) received do
18:         if (w, k, h, j) arrived from π_{hj}^k[r − 1] then
19:             if |π_{hj}^k| = r + 1 then
20:                 LAST_i ← LAST_i ∪ {(w, k, h, j)};
21:             else
22:                 NEXT_i ← NEXT_i ∪ {(w, k, h, j)};
23:     if v arrived from more than 2t processes then
24:         v_i ← v;
25: (Round β)
26: for all (w, k, h, j) ∈ LAST_i do
27:     send (w, k, h, j) to π_{hj}^k[|π_{hj}^k|];
28: for all messages (w, k, h, i) received do
29:     if (w, k, h, i) arrived from π_{hi}^k[|π_{hi}^k| − 1] then
30:         W_i[h][k] ← w;
31: for all j ∈ Π do
32:     if w occurs in W_i[j][1..α] more than βt times then
33:         X_i[j] ← w;
```

To observe that Consistency Maintenance is preserved after reaching agreement, notice that, in each round, v_i is exchanged among neighbors j in Lines 9–10 (when round r = 1) or in Lines 14–15 ($r \geq 2$), and i corrects v_i with the agreed value v in Line 24, which is the value that occurs more than $2t$ times in the received v_j's. Since incorrect values are at most $2t$, v must be the correct agreed value after reaching agreement. Note that there is always a value v satisfying the condition, since $|N_i| \geq d \geq \alpha > 2\beta t$ and $\beta \geq 2$.   □

Now we propose a mobile Byzantine agreement algorithm DP-Byz.

Algorithm DP-Byz: DP-Byz is exactly the same as mMBA, except that the following three algorithm segments (of mMBA) are replaced with three DPT invocations:

- Replace Lines 5–7 with DPT(v_i, PV_i).
- Replace Lines 14–16 with DPT(v_i, SV_i).
- Replace Lines 23–25 with DPT(SV_i, EV_i).

Theorem 2. *Algorithm* DP-Byz *solves the mobile Byzantine agreement problem on any* $G \in \mathcal{G}(\alpha, \beta)$, *when both* $n > 6t$ *and* $\alpha > 2\beta t$ *hold.*[9]

Proof. Immediate from Lemma 2 and Corollary 1.   □

Corollary 2. *Algorithm* DP-Byz *solves the mobile Byzantine agreement problem on any* $G \in \mathcal{G}(\alpha, 2)$, *when both* $n > 6t$ *and* $\alpha > 4t$ *hold; it is optimal by Theorem 1.*

A complete k-partite graph with partite sets $V_1, V_2, \ldots, V_k$ is a graph $G = (V_1 \cup V_2 \cup \ldots \cup V_k, E)$ such that for any two vertices $v_i \in V_i$ and $v_j \in V_j$ ($i \neq j$), $(v_i, v_j) \in E$. Without loss of generality, we may assume that $|V_1| \leq |V_2| \leq \ldots \leq |V_k|$.

Corollary 3. *Suppose* $k \geq 3$. *Algorithm* DP-Byz *solves the mobile Byzantine agreement problem on any* k-*partite graph* $G = (V_1 \cup V_2 \cup \cdots \cup V_k, E)$, *if* $n > 6t$ *and* $n - |V_k| - |V_{k-1}| + 1 > 4t$.

Proof. Since $k \geq 3$, the vertex connectivity of G is $n - |V_k| - |V_{k-1}| + 1$. On the other hand, for any two vertices, there are at least $n - |V_k| - |V_{k-1}| + 1$ vertex disjoint paths of length 2. Thus DP-Byz can solves the mobile Byzantine agreement problem on complete k-partite graph, if $n > 6t$ and $n - |V_k| - |V_{k-1}| + 1 > 4t$ by Theorem 2.   □

Corollary 4. *Algorithm* DP-Byz *solves the mobile Byzantine agreement problem on any complete bipartite graph* $G = (V_1 \cup V_2, E)$, *if* $|V_1| > 6t$.

Proof. Since there are at least $|V_1|$ vertex disjoint paths of length at most 3 between any pair of vertices from different partites.   □

4.2   Algorithm KP-Byz

The kth power of a connected graph G, denote G^k, is the graph having the same vertex set as G and an edge between any two vertices such that the distance between them is at most k in G. This subsection proposes a mobile Byzantine agreement algorithm KP-Byz on G^k. To this end, we present an algorithm PerT (Permeate Transmission) that is used to realize reliable communication on G^k. As in Subsection 4.1, KP-Byz is constructed from mMBA by replacing the three message exchange sections into invocations of PerT.

We explain the idea behind reliable communication by PerT in G^k, assuming $k > 4t$. For any pair (i, j) of vertices in Π, let $i_0 (= i), i_1, \ldots, i_m (= j)$ be the shortest path in G (not G^k) from i to j, and consider the following message exchanges in G^k to safely carry a value w from i to j. In the first round, i_0

---

[9]   Recall that we have assumed that there is a process which is non-faulty all the time.

simply broadcasts w to all neighbors $N_i^{G^k}$ in G^k. In each of the later rounds, a process h takes the majority of the values received, put it to a variable w_h, and broadcasts it to all neighbors $N_h^{G^k}$. Suppose that i_0 is non-faulty in the initial round r_0 and the round immediately before it. In the first round, by the definition of G^k, in particular, processes $i_0, i_1, \ldots, i_k$ receive w. In the second round, each process i_h ($0 \leq h \leq k$) receives at least $k+1$ values w_p from $p \in N_{i_h}^{G^k}$. Obviously all the values except at most $2t$ values are w, and hence i_h takes w as w_{i_h}, since $k > 4t$. Furthermore, i_{k+1} also receives at least k values from $i_1, i_2, \ldots, i_k$, and i_{k+1} takes w as the value of $w_{i_{k+1}}$. By an easy induction, we can show that every non-faulty process i_x in $\{i_0, i_1, \ldots, i_{k+h}\}$ has w as the value of w_{i_x} in the $(h+1)$-st round, and hence $j(= i_m)$ assigns w to w_j in the $(m - k + 1)$-st round. Since m is at most the diameter D_G of G, PerT can terminate, in the $D_G - k + 1$-st round, with reliably sending w to all processes in Π.

The algorithm description of PerT looks a bit difficult to understand, because of the two reasons. First, it also includes the trick to preserve Consistency Maintenance we used in DPT. Second, it tries to reduce the number of messages by omitting unnecessary message exchanges described in the above explanation. For example, in the second round, i_k does not need to send w_{i_k} to the processes further than i_{k+1} from i_k, and they are thus eliminated in PerT.

PerT plays exactly the same role as DPT on G^k; PerT on process i takes two parameters, a value w_i (which may be a vector) and an array X_i, where w_i is a message that i wishes to send to all processes $j \in \Pi$, and X_i is used to store received messages w_j from all processes $j \in \Pi$. If all processes invoke PerT(w_i, X_i) in round r, in the presence of at most t mobile faulty processes, $X_i[j] = w_j$ for each of the non-faulty processes $j \notin (F_{r-1} \cup F_r)$, i.e., those which are non-faulty both in rounds $r - 1$ and r, when it terminates. Let $L_j^G(\ell)$ be the set of processes in Π whose distance from j in G is at most ℓ, and let D_G be the diameter of G. We assume that G^k, G, $L_j^G(\ell)$ and D_G are all available to PerT. In the description of PerT, $N_i^{G^k}$ is the closed neighbors of i in G^k.

It is easy to show the next lemma by the explanation of the basic idea of PerT.

Lemma 3. *If each process i has a value w_i and invokes* PerT(w_i, X_i) *on G^k in a round r_0, then for any $i \notin F_{r_f}$ and $j \notin F_{r_0-1} \cup F_{r_0}$, $X_i[j] = w_j$ holds in round r_f, when $k > 4t$, where $r_f = r_0 + D_G - k$ is the round that* PerT *terminates, and D_G is the diameter of G. Furthermore, Consistency Maintenance is preserved between (and including) rounds r_0 and r_f.*

Now we propose algorithm KP-Byz.

Algorithm KP-Byz: Algorithm KP-Byz is exactly the same as mMBA, except that the following three algorithm segments (of mMBA) are replaced with three PerT invocations as follows:
 - Replace Lines 5–7 with PerT(v_i, PV_i).
 - Replace Lines 14–16 with PerT(v_i, SV_i).
 - Replace Lines 23–25 with PerT(SV_i, EV_i).

Algorithm 4.2. $\text{PerT}(w_i, X_i)$

1: (Round1)
2: **for all** $j \in N_i^{G^k}$ **do**
3:      send (w_i, i) and v_i to j;
4: **for all** messages (w, j) received **do**
5:      **if** (w, j) arrived from j **then**
6:          $X_i[j] \leftarrow w$;
7: **if** v arrived from more than $2t$ processes **then**
8:      $v_i \leftarrow v$;
9: (Round r $(2 \leq r \leq D_G - k + 1)$)
10: **for** $r = 2$ **to** $D_G - k + 1$ **do**
11:      $W_i[1..n][1..n] \leftarrow [\bot, \cdots, \bot][\bot, \cdots, \bot]$;
12:      **for all** $j \in \Pi$ **do**
13:          **if** $i \in L_j^G(k + r - 2)$ **then**
14:              send $(X_i[j], j)$ to $p \in N_i^{G^k}$;
15:      **for all** $j \in N_i^{G^k}$ **do**
16:          send v_i to j;
17:      **for all** messages (w, j) received **do**
18:          **if** (w, j) arrived from $p \in L_j^G(k + r - 2)$ **then**
19:              $W_i[j][p] \leftarrow w$;
20:      **for all** $j \in \{p | i \in L_p^G(k + r - 1)\}$ **do**
21:          **if** w occurs in $W_i[j][1..n]$ more than $2t$ times **then**
22:              $X_i[j] \leftarrow w$
23:      **if** receive v more than $2t$ times **then**
24:          $v_i \leftarrow v$;

Recall that we have assumed that there is a process which is non-faulty all the time.

Theorem 3. *Algorithm* KP-Byz *solves the mobile Byzantine agreement problem on the k-th power graph G^k of any graph G, when both $n > 6t$ and $k > 4t$ hold.*

Proof. Immediate from Lemma 3 and Corollary 1. □

Corollary 5. *Suppose that G has a "tail" of length more than k (like a lollipop graph and a path graph). Since* KP-Byz *solves the mobile Byzantine agreement problem on the k-th power graph G^k, if $n > 6t$ and $d > 4t$, where d is the vertex connectivity of G^k, it is optimal for G^k.*

Proof. Since the vertex connectivity of G^k is k. □

5 Conclusion

We have considered the mobile Byzantine agreement problem on general network for the first time, and derived a necessary condition for a t-resilient mobile Byzantine algorithm on a network to exist. We then designed two mobile Byzantine agreement algorithms DP-Byz and KP-Byz. DP-Byz works on any graph such that, for any two vertices, there are at least α vertex disjoint paths connecting them, whose lengths are at most β, if $n > 6t$ and $\alpha > 2\beta t$, and KP-Byz, on the other hand, works on the k-th power of a graph G, if $n > 6t$ and $k > 4t$.

As observed, both algorithms are optimal for some graphs, but, unfortunately, we could not find an optimal mobile Byzantine agreement algorithm that works on any graph as long as $n > 6t$ and $k > 4t$ hold, nor a tighter upper bound on t. Those are the main open problems we left. Also, investigating an algorithm on general graph for a roaming pace ρ greater than 1 is an interesting issue. In the context of self-stabilization, networks which may contain both transient and Byzantine faults have been investigated as a problem of the fault containment [11, 12]. It would be a good research direction to relate these two fields.

References

1. Banu, N., Souissi, S., Izumi, T., Wada, K.: An improved Byzantine agreement algorithm for synchronous systems with mobile faults. Int'l J. Computer Applications 43, 21 (2011)
2. Biely, M., Hutle, M.: Consensus when all processes may be Byzantine for some time. In: Guerraoui, R., Petit, F. (eds.) SSS 2009. LNCS, vol. 5873, pp. 120–132. Springer, Heidelberg (2009)
3. Biely, M., Charron, B., Gaillard, A., Hutle, M., Schiper, A., Widder, J.: Tolerating Corrupted Communication. Swiss National Science Foundation, 200021–111701 (2007)
4. Buhrman, H., Garay, J.A., Hoepman, J.: Optimal resiliency against mobile faults. In: Proc. 25th Int'l Symp. Fault-Tolerant Computing (FTCS 1995), pp. 83–88 (1995)
5. Diestel, R.: Graph Theory. Springer (1997)
6. Dolev, D.: The Byzantine generals strike again. J. Algorithms 3, 14–30 (1982)
7. Dolev, D., Strong, H.R.: Authenticated algorithms for Byzantine agreement. SIAM J. Computing 12(4), 656–666 (1983)
8. Fisher, M.J., Lynch, N.A., Paterson, M.S.: Impossibility of distributed consensus with one faulty process. J. ACM 32(2), 374–382 (1985)
9. Garay, J.A.: Reaching (and maintaining) agreement in the presence of mobile faults. In: Tel, G., Vitányi, P.M.B. (eds.) WDAG 1994. LNCS, vol. 857, pp. 253–264. Springer, Heidelberg (1994)
10. Lamport, L., Shostak, R., Pease, M.: The Byzantine generals problem. ACM Trans. Programming Languages and Systems 4, 382–401 (1982)
11. Masuzawa, T., Tixeuil, S.: Bounding the impact of unbounded attacks in stabilization. In: Datta, A.K., Gradinariu, M. (eds.) SSS 2006. LNCS, vol. 4280, pp. 440–453. Springer, Heidelberg (2006)
12. Nesterenko, M., Rora, A.: Tolerance to unbounded Byzantine faults. In: Proc. of 21st IEEE Symposium on Reliable Distributed Systems (SRDS 2002), pp. 22–29 (2002)
13. Pease, M., Shostak, R., Lamport, L.: Reaching agreement in the presence of fault. J. ACM 27(2), 228–234 (1980)
14. Santoro, N., Widmayer, P.: Time is not a healer. In: Cori, R., Monien, B. (eds.) STACS 1989. LNCS, vol. 349, pp. 304–313. Springer, Heidelberg (1989)
15. Schmid, U., Weiss, B., Keidar, I.: Impossibility results and lower bounds for consensus under link failures. SIAM J. Computing 38(5), 1912–1951 (2009)

On Scheduling Algorithms for MapReduce Jobs in Heterogeneous Clouds with Budget Constraints

Yang Wang[1] and Wei Shi[2]

[1] Faculty of Computer Science
University of New Brunswick, Fredericton, Canada
[2] Faculty of Business and Information Technology
University of Ontario Institute of Technology, Ontario, Canada

Abstract. In this paper, we consider task-level scheduling algorithms with respect to budget constraints for a bag of MapReduce jobs on a set of provisioned heterogeneous (virtual) machines in cloud platforms. The heterogeneity is manifested in the popular "pay-as-you-go" charging model where the service machines with different performance would have different service rates. We organize a bag of jobs as a κ-stage workflow and consider the scheduling problem with budget constraints. In particular, given a total monetary budget, by combining a greedy-based local optimal algorithm and dynamic programming techniques, we first propose a global optimal scheduling algorithm to achieve a minimum scheduling length of the workflow in pseudo-polynomial time. Then, we extend the idea in the greedy algorithm to efficient global distribution of the budget among the tasks in different stages for overall scheduling length reduction. Our empirical studies verify the proposed optimal algorithm and show the efficiency of the greedy algorithm to minimize the scheduling length.

1 Introduction

The Cloud, with its abundant on-demand computing resources and elastic charging models, have emerged as a promising platform to address various data processing and task computing problems [7,10]. Also, MapReduce [6], characterized by its remarkable simplicity, fault tolerance, and scalability, is becoming a popular programming framework to automatically parallelize large scale data processing as in web indexing, data mining, and bioinformatics. Since a cloud supports on-demand "massively parallel" applications with loosely coupled computational tasks, it is amenable to the MapReduce framework and thus suitable for diverse MapReduce applications. Therefore, many cloud infrastructure providers have deployed the MapReduce framework on their commercial clouds as one of their infrastructure services (e.g., Amazon Elastic MapReduce).

Given MapReduce is extremely powerful and runs fast for diverse application areas, it is becoming a viable service in the form of *MapReduce as a Service* (MRaaS) for cloud service providers (CSPs). It is typically set up as a kind of

R. Baldoni, N. Nisse, and M. van Steen (Eds.): OPODIS 2013, LNCS 8304, pp. 251–265, 2013.

Software as a Service (SaaS) on the provisioned MapReduce cluster of cloud instances. Clearly, for CSPs to reap the benefits of such a deployment, many challenging problems have to be addressed. However, most current studies focus solely on the system issues pertaining to deployment, such as overcoming the limitations of the cloud infrastructure to build-up the framework [15], evaluating the performance harm from running the framework on virtual machines [8], and other issues in fault tolerance [4], reliability [16], data locality [21], etc.

We are also aware of some recent research tackling the scheduling problem of MapReduce in Clouds [12, 13, 19]. These contributions mainly address the scheduling issues with various concerns placed on dynamic loading [19], energy reduction [13], and network performance [12]. To the best of our knowledge, no one has optimized the scheduling of MapReduce jobs with budget constraints at the task level. In our opinion several factors that may account for this status quo. Specifically, as mentioned above, the MapReduce service, like other basic database and system services, could be provided as an infrastructure service by the cloud infrastructure providers (e.g., Amazon), rather than CSPs. Consequently, it would be charged together with other infrastructure services. Hence, the problem we are proposing to study would be irrelevant. Also, some properties of the MapReduce framework (e.g., automatic fault tolerance with speculative execution [22]) make it difficult for CSPs to track job execution in a reasonable way, thus making scheduling very complex.

Since cloud resources are typically provisioned on demand with a "pay-as-you-go" billing model, cloud-based applications are usually budget driven. Consequently, in practice the cost-effective use of resources to satisfy relevant performance requirements within budget is always a pragmatic concern for CSPs, and solving this problem with respect to MapReduce framework could dramatically exploit the cloud potentials.

A MapReduce job essentially consists of two sets of tasks: map tasks and reduce tasks as shown in Fig. 1. The executions of both sets of tasks are synchronized into a map stage followed by a reduce stage. In the map stage, the entire dataset is partitioned into several smaller chunks in forms of key-value pairs, each chunk being assigned to a map node for partial computation results. The map stage ends up with a set of intermediate key-value pairs on each map node, which are further shuffled based on the intermediate keys into a set of scheduled reduce nodes where the received pairs are aggregated to obtain the final results.

A bag of MapReduce jobs may have multiple stages of MapReduce computation, each stage running either map or reduce tasks in parallel, with enforced synchronization only between them. Therefore, the executions of the jobs can be viewed as a fork&join workflow characterized by multiple synchronized stages, each consisting of a collection of sequential or parallel map/reduce tasks. An example of such a workflow is shown in Fig. 2 which is composed of 4 stages, respectively with 8, 2, 4 and 1 (map or reduce) tasks. These tasks are to be scheduled on different nodes for parallel execution. However, in heterogeneous clouds, different nodes may have different performance and/or configuration

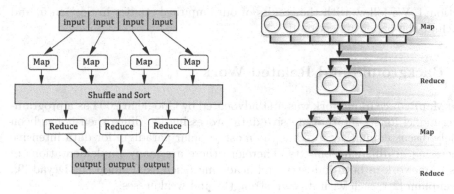

Fig. 1. MapReduce framework   **Fig. 2.** A 4-stage MapReduce workflow

specifications, and thus may have different service rates. Since resources are provisioned on-demand in cloud computing, the CSPs are faced with a general practical problem: how are resources to be selected and utilized for running each task in a cost-effective way? This problem is, in particular, directly relevant to CSPs wanting to compute their MapReduce workloads, especially when the computation budget is fixed.

In this paper, we investigate the problem of scheduling a bag of MapReduce jobs within budget constraints. This bag of MapReduce jobs could be an iterative MapReduce job, a set of independent MapReduce jobs, or a collection of jobs related to some high-level applications such as Hadoop Hive [17]. We address *task-level scheduling*, which is fine grained compared to the frequently-discussed job-level scheduling, where the scheduled unit is a job instead of a task. Specifically, given a fixed amount of budget, we focus on how to efficiently select a machine from a candidate set for each task so that the total scheduling length of the job (aka makespan of the job) is minimum without breaking the budget. This problem is of particular interest to CSPs wanting to deploy MRaaS on heterogeneous cloud instances in a cost-effective way.

To address this problem, we first design an efficient greedy algorithm for computing the minimum execution time with a given budget for each stage and show its optimality with respect to execution time and budget use. Then, with this result we develop a dynamic programming algorithm to achieve a global optimal solution within time of $O(\kappa B^2)$. To overcome the time complexity, we extend the idea in the greedy algorithm to efficient global distribution of the budget among the tasks in different stages for overall scheduling length reduction. Our empirical studies verify the proposed optimal algorithm and show the efficiency of the greedy algorithm to minimize the scheduling length.

The rest of this paper is organized as follows: in Section 2, we introduce some background knowledge regarding the MapReduce framework and survey some related work. Section 3 formulates our problem. The proposed budget-driven algorithms including the optimal and the greedy algorithms are discussed in

Section 4. We follow with the results of our empirical studies in Section 5, and conclude the paper in Section 6.

2   Background and Related Work

The MapReduce framework was first advocated by Google in 2004 as a programming model for its internal massive data processing [5]. Since then it has been widely discussed and accepted as the most popular paradigm for data intensive processing in different contexts. Therefore there are many implementations of this framework in both industry and academia (such as Hadoop [1], Dryad [9], Greenplum [2]), each with its own strengths and weaknesses.

MapReduce is made up of an execution runtime and a distributed file system. The execution runtime is responsible for job scheduling and execution. It is composed of one master node and slave nodes. A distributed file system is used to manage task and data across nodes. When the master receives a submitted job, it first splits the job into a number of map and reduce tasks and then allocates them to the slave nodes, As with most distributed systems, the performance of the task scheduler greatly affects the scheduling length of the job, as well as, in our particular case, the budget consumed.

There exists research on the scheduler of MapReduce aiming at improving its scheduling policies. For instance, Hadoop adopts *speculative task scheduling* to minimize the slowdown in the synchronization phases caused by straggling tasks in a homogeneous environment [1]. To extend this idea to heterogeneous clusters, Zaharia et al. [22] proposed the LATE algorithm. But this algorithm does not consider the phenomenon of dynamic loading, which is common in practice. This limitation was studied by You et al. [19] who proposed a load-aware scheduler. In addition, there are Other work on *power-aware scheduling* [14], *deadline constraint scheduling* [11], and scheduling based on automatic task slot assignments [18]. While these contributions do address different aspects of MapReduce scheduling, they are mostly centred on system performance and do not consider the budget, which is our main focus.

Budget constraints have been considered in studies focusing on scientific workflow scheduling on HPC platforms including the Grid and Cloud [3, 20, 23]. For example, Yu et al. [20] discussed this problem based on service Grids and presented a QoS-based workflow scheduling method to minimize execution cost and yet meet the time constraints imposed by the user. In the same vein, Zeng et al. [23] considered the executions of large scale many-task workflows in Clouds with budget constraints. They proposed *ScaleStar*, a budget-conscious scheduling algorithm to effectively balance execution time with the monetary costs. Now recall that, in the context of this paper, we view the executions of the jobs as a fork&join workflow characterized by multiple synchronized stages, each consisting of a collection of sequential or parallel map/reduce tasks. From this perspective, this abstract fork&join workflow can be viewed as a special case of general workflows. However, our focus is on MapReduce scheduling with budget constraints, rather than on general workflow scheduling. Therefore, the

Table 1. Time-price table of task J_{jl}

$$\begin{bmatrix} t_{jl}^1 \ t_{jl}^2 \ \cdots \ t_{jl}^{m_{jl}} \\ p_{jl}^1 \ p_{jl}^2 \ \cdots \ p_{jl}^{m_{jl}} \end{bmatrix}$$

characteristics of MapReduce framework are fully exploited in the designs of the scheduling algorithms.

3 Problem Formulation

3.1 Workflow Model

We model a bag of MapReduce job as a multi-stage fork&join workflow that consists of κ stages (called a κ-stage job), each stage j having a collection of independent map or reduce tasks, denoted as $J_j = \{J_{j0}, J_{j1}, ..., J_{jn_j}\}$, where $0 \leq j < \kappa$, and $n_j + 1$ is the size of stage j. In a cloud, each map or reduce task may be associated with a set of machines that are provided by a cloud infrastructure provider to run this task, each machine with possibly distinct performance and configuration and thus with different charge rates. More specifically, for Task J_{jl}, $0 \leq j < \kappa$ and $0 \leq l \leq n_j$ the available machines and corresponding prices (service rates) are listed in Table 1, the values could be determined by the VM power and the computational loads of each task, where $t_{jl}^u, 1 \leq u \leq m_{jl}$ represents the time to run task J_{jl} on machine M_u whereas p_{jl}^u represents the corresponding price for using that machine, and m_{jl} is the total number of the machines that can run J_{jl}.

Table 2. Notation frequently used in model and algorithm descriptions

Symbol	Meaning	Symbol	Meaning
κ	the number of stages	m_{jl}	the total number of the machines
J_{ji}	the ith task in stage j	m	the total size of time-price tables of the workflow
J_j	task set in stage j	B_{jl}	the budget used by J_{jl}
n_j	the number of tasks in stage j	B	the total budget for the MapReduce job
n	the total number of tasks in the workflow	$T_{jl}(B_{jl})$	the shortest time to finish J_{jl} given B_{jl}
t_{jl}^u	time to run task J_{jl} on machine M_u	$T_j(B_j)$	the shortest time to finish stage j given B_j
p_{jl}^u	the cost rate for using M_u	$T(B)$	the shortest time to finish the job given B

Without loss of generality, we assume that times have been sorted in increasing order and prices in decreasing order, and furthermore, that both time and price values are unique in their respective sorted sequence. These assumptions are reasonable since given any two machines with same run time for a task, the expensive one should never be selected. Similarly, given any two machines with same price for a task, the slow machine should never be chosen.

For clarity and quick reference, we provide in Table 2 a summary of some symbols frequently used hereafter.

3.2 Budget Constraints

Given budget B_{jl} for task J_{jl}, the shortest time to finish it, denoted as $T_{jl}(B_{jl})$ is defined as

$$T_{jl}(B_{jl}) = t_{jl}^u \qquad p_{jl}^{u+1} < B_{jl} < p_{jl}^{u-1} \tag{1}$$

Obviously, if $B_{jl} < p_{jl}^{m_{jl}}$, $T_{jl}(B_{jl}) = +\infty$.

The time to complete a stage j with budget B_j, denoted as $T_j(B_j)$, is defined as the time consumed when the last task in that stage completes within the given budget:

$$T_j(B_j) = \max_{\sum_{l \in [0,n_j]} B_{jl} \leq B_j} \{T_{jl}(B_{jl})\} \tag{2}$$

In fork&join, a stage cannot start until its immediately preceding stage has terminated. Thus the total makespan under budget B to complete the workflow is defined as the sum of all stages' time. Our goal is to minimize the time within the given budget B.

$$T(B) = \min_{\sum_{j \in [0,\kappa)} B_j \leq B} \sum_{j \in [0,\kappa)} T_j(B_j) \tag{3}$$

Some readers may question the feasibility of this model since the number of stages and the number of tasks in each stage need to be known a prior to the scheduler. But, in reality, it is entirely possible since a) the number of map tasks for a given job is driven by the number of input splits (which is known to the scheduler) and b) the number of reduce tasks can be preset as with all other parameters (e.g., parameter mapred.reduce.tasks in Hadoop). As for the number of stages, it is not always possible to predefine it for MapReduce workflows. This is the main limitation of our model. But under the default FIFO job scheduler, we can treat a set of independent jobs as a single fork&join workflow. Therefore, we believe our model is still representative of most cases in reality.

4   Budget-Driven Algorithms

In this section, we propose our task-level scheduling algorithms for MapReduce workflows with the goals of optimizing Equations (3) under budget constraints. To this end, we first leverage dynamic programming techniques to obtain an optimal solution and then present an efficient greedy algorithm to overcome its inherent complexity.

4.1 Optimization under Budget Constraints

The proposed algorithm should be able to distribute the budget among the stages, and in each stage distributing the assigned budget to each constituent task in an optimal way. To take these effects, we design the algorithm in two steps:

1. Given budget B_j for stage j, distribute the budget to all constituent tasks in such a way that $T_j(B_j)$ is minimum (see Equation (2)). Clearly, the computation for each stage is independent of other stages. Therefore such computations can be treated in parallel using κ machines.
2. Given budget B for a workflow and the results in Equation (2), optimize our goal of Equation (3).

In-Stage Distribution. To address the first step, we develop an optimal local greedy algorithm to distribute budget B_j between the $n_j + 1$ tasks for stage $j, 0 \leq j \leq \kappa - 1$ in such a way that $T_j(B_j)$ is minimized.

The idea of the algorithm is simple. To ensure that all the tasks in stage j have sufficient budget to finish while minimizing $T_j(B_j)$, we first require $B'_j = B_j - \sum_{l \in [0, n_j]} p_{jl}^{m_{jl}} \geq 0$ and then iteratively distribute B'_j in a greedy manner each time to the task whose current execution time determines $T_j(B_j)$ (i.e., the slowest one). This process continues until no sufficient budget is left. Clearly, having considered the structure of this problem, we can easily show its optimality with respect to minimizing the scheduling length within the given budget.

Global Distribution. Given the results of the first step, we now consider the second step by using a dynamic programming recursion to compute the global optimal result. To this end, we use $T(j, r)$ to represent the minimum total time to complete stages indexed from j to κ when budget r is available and use $T_j[n_j, q]$ to store the optimal time computed for stage j given budget q. Then, we have the following recursion ($0 < j \leq \kappa, 0 < r \leq B$):

$$T(j, r) = \begin{cases} \min_{0 < q \leq r} \{T_j(n_j, q) + T(j + 1, r - q)\} & \text{if } j < \kappa \\ T_j(n_j, r) & \text{if } j = \kappa \end{cases} \quad (4)$$

where the optimal solution can be found in $T(1, B)$. The scheduling scheme can be reconstructed from $T(1, B)$ by recursively backtracking the Dynamic Programming (DP) matrix in (4) up to the initial budget distribution at stage κ which can, phase by phase, steer to the final optimal result. To this end, in addition to the time value, we only store the budget q and the index of the previous stage (i.e., $T(j + 1, r - q)$) in each cell of the matrix since, given the budget for each stage, we can simply use the algorithm in the first step to recompute the budget distribution. Based on these descriptions, we can easily have the following results:

Theorem 1. *Given budget B for a κ-stage MapReduce job, each stage j having n_j tasks, Recursion (4) yields an optimal solution to the distribution of budget B to all the κ stages with time complexity $O(\kappa B^2)$ when $T_j(n_j, q), 0 < j \leq \kappa, 0 < q \leq B$ is pre-computed.*

4.2 Efficiency Improvements

In the previous subsection, we briefly introduced an optimal solution to the distribution of a given budget among different stages to minimize the workflow execution time. The time complexity of the proposed algorithm is pseudopolynomial and proportional to the square of the budget, which is fairly high. To address this problem, we now propose a heuristic algorithm, called *Global Greedy Budget* (GGB), which extends the idea of the algorithm in computing $T_j[n_j, B_j]$ (Section 4.1) to the whole multi-stage workflow. More specifically, GGB applies the idea of the algorithm in Section 4.1 with some extensions to the selection of candidate tasks for budget assignments across *all* the stages of the workflow. The pseudo code of GGB is shown in Algorithm 1. Similar to the algorithm in Section 4.1, we also need to ensure the given budget has a lower bound $\sum_{j \in [1,\kappa]} B_j$ where $B'_j = \sum_{l \in [0,n_j]} p_{jl}^{m_{jl}}$ that guarantees the completion of the workflow (Lines 2-3). We also use the three profile variables T_{jl}, B_{jl} and M_{jl} for each task J_{jl} in stage j to record its execution time, assigned budget, and selected machine (Lines 6-12).

Since in each stage, the slowest task determines the stage completion time, we first need to allocate the budget to the slowest task in each stage. After the slowest task is allocated, the second slowest will become the bottleneck. In our heuristic, we must consider this fact. To this end, we first identify the slowest and the second slowest tasks in each stage j, which are indexed by jl and jl', respectively. Then we gather these index pairs in a set L thereby determining which task in L should be allocated budget (Lines 14-18). To measure the quality of a budget investment, we define a *utility value*, v_{jl}^u, for each given task J_{jl}, which is a value assigned to an investment on the basis of anticipated performance:[1]

$$v_{jl}^u = \alpha \beta_j + (1 - \alpha)\beta'_j \tag{5}$$

where $\beta_j = \frac{t_{jl}^u - t_{jl'}^u}{p_{jl}^{u-1} - p_{jl}^u} \geq 0$, $\beta'_j = \frac{t_{jl}^u - t_{jl}^{u-1}}{p_{jl}^{u-1} - p_{jl}^u} \geq 0$, and α is defined as:

$$\alpha = \begin{cases} 1 & \text{if } \sum_{j=1}^{\kappa} \beta_j > 0 \\ 0 & \text{Otherwise} \end{cases} \tag{6}$$

β_j represents time saving on per-budget unit when task J_{jl} is moved from machine u to run on the next faster machine $u - 1$ in stage j ($\beta_j > 0$) while β'_j is used when there are multiple slowest tasks in stage j ($\beta_j = 0$). α is defined to allow β_j to have a higher priority than β'_j in task selection. Put simply, unless for $\forall j \in [1, \kappa], \beta_j = 0$ in which case β'_j is used, we use the value of $\beta_j, j \in [1, \kappa]$ as the criteria to select the allocated tasks.

---

[1] Recall that the sequences of t_{jl}^u and p_{jl}^u are sorted, respectively in Table 1.

Algorithm 1. Global-Greedy-Budget Algorithm (GGB)

1: **procedure** $T(1, B)$ ▷ Dist. B among κ stages
2:      $B' = B - \sum_{j \in [1,\kappa]} B'_j$ ▷ $B'_j = \sum_{l \in [0,n_j]} p_{jl}^{m_{jl}}$
3:      **if** $B' < 0$ **then return** $(+\infty)$
4:      **end if**   ▷ No sufficient budget!
5:   ▷ Initialization
6:      **for** $j \in [1, \kappa]$ **do**   ▷ $O(\sum_{j=1}^{\kappa} n_j) = \#$ of tasks
7:          **for** $J_{jl} \in J_j$ **do**
8:              $T_{jl} \leftarrow t_{jl}^{m_{jl}}$ ▷ record exec. time
9:              $B_{jl} \leftarrow p_{jl}^{m_{jl}}$ ▷ record budget dist.
10:             $M_{jl} \leftarrow m_{jl}$ ▷ record assigned machine index.
11:         **end for**
12:     **end for**
13:     **while** $B' \geq 0$ **do**   ▷ $\leq O(\frac{B}{\min_{1 \leq j \leq \kappa, 0 \leq l \leq n_j} \{\delta_{jl}\}})$
14:         $L \leftarrow \varnothing$
15:         **for** $j \in [1, \kappa]$ **do**   ▷ $O(\sum_{j=1}^{\kappa} \log n_j)$
16:             $< jl, jl' >^* \leftarrow \arg\max_{l \in [0, n_j]} \{T_{jl}(B_{jl})\}$
17:             $L \leftarrow L \cup \{< jl, jl' >^*\}$ ▷ $|L| = \kappa$
18:         **end for**
19:         $V \leftarrow \varnothing$
20:         **for** $< jl, jl' > \in L$ **do**   ▷ $O(\kappa)$
21:             $u \leftarrow M_{jl}$
22:             **if** $u > 1$ **then**
23:                 $< p_{jl}^{u-1}, p_{jl}^u > \leftarrow \textbf{Lookup}(J_{jl}, u - 1, u)$
24:                 $v_{jl}^u \leftarrow \alpha\beta_j + (1 - \alpha)\beta'_j$
25:                 $V \leftarrow V \cup \{v_{jl}^u\}$ ▷ $|V| \leq \kappa$
26:             **end if**
27:         **end for**
28:         **while** $V \neq \varnothing$ **do**   ▷ $O(\kappa \log \kappa)$
29:   ▷ sel. task with max. u.value
30:             $jl^* \leftarrow \arg\max_{v_{jl}^u \in V} \{v_{jl}^u\}$
31:             $u \leftarrow M_{jl^*}$ ▷ Lookup matrix in Table 1
32:             $\delta_{jl^*} \leftarrow p_{jl^*}^{u-1} - p_{jl^*}^u$ ▷ $u > 1$
33:             **if** $B' \geq \delta_{jl^*}$ **then**   ▷ reduce J_{jl^*}'s time
34:                 $B' \leftarrow B' - \delta_{jl^*}$
35:                 $B_{jl^*} \leftarrow B_{jl^*} + \delta_{jl^*}$
36:                 $T_{jl^*} \leftarrow t_{jl^*}^{u-1}$
37:                 $M_{jl^*} \leftarrow u - 1$
38:                 **break**   ▷ restart from scratch
39:             **else**
40:                 $V \leftarrow V \setminus \{v_{jl^*}^u\}$ ▷ select the next one in V
41:             **end if**
42:         **end while**
43:         **if** $V = \varnothing$ **then**
44:             **return**   ▷ $B_j = \sum_{l \in [0, n_j]} B_{jl}$
45:         **end if**
46:     **end while**
47: **end procedure**

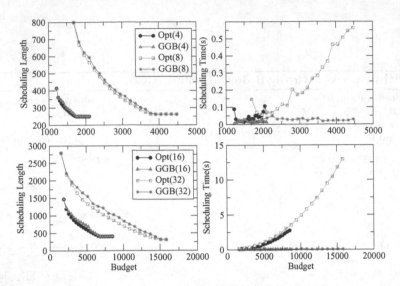

Fig. 3. Impact of time-price table (TP) size on the scheduling length (e.g., makespan of the job in terms of indivisible time units) and the scheduling time (Stage:8, Task: ≤ 20/each stage, and the numbers in the brackets represent the different TP table sizes)

In the algorithm, all the values of the tasks in L are collected into a set V (Lines 19-28). We note that the tasks running on machine $u = 1$ in each stage have no definition of this value since they are already running on the fastest machine under the given budget (and thus no further improvement is available).

Given set V, we can iterate over it to select the task in V that has the largest utility value, indexed by jl^*, to be allocated budget for minimizing the stage computation time (Lines 29-30). We fist obtain the machine u to which the selected task is currently mapped and then compute the extra monetary cost δ_{jl^*} if the task is moved from u to the next faster machine $u - 1$ (Lines 31-32). If the leftover budget B' is insufficient, the selected task will not be considered and removed from V (Line 40). In the next step, a task in a different stage will be selected for budget allocation (given each stage has at most one task in V). This process will be continued until either the leftover budget B' is sufficient for a selected task or V becomes empty. In the former case, δ_{jl^*} will be deducted from B' and added to the select task. At the same time, other profile information related to this allocation is also updated (Lines 33-37). After this, the algorithm exits from the loop and repeats the computation of L (Line 13) since L has been changed due to this allocation. In the latter case, when V becomes empty, the algorithm returns directly, indicating that the final results of the budget distribution and the associated execution time of each tasks in each stage are available as recorded in the corresponding profile variables.

Theorem 2. *The time complexity of GGB is not greater than $O(B(n+\kappa \log \kappa))$. In particular, when $n \geq \kappa \log \kappa$, the complexity of GGB is upper bounded by $O(nB)$.*

Proof. The time complexity of this algorithm is largely determined by the nested loops (Lines 13-42). Since each allocation of budget B' is at least $\min\limits_{1 \leq j \leq \kappa, 0 \leq l \leq n_j} \{\delta_{jl}\}$, the algorithm has at most $O(\frac{B}{\min\{\delta_{jl}\}}), 1 \leq j \leq \kappa, 0 \leq l \leq n_j$ iterations at Line 13. On the other hand, if some advanced data structure such as a *priority queue* is used to optimize the search process, the algorithm can achieve a time complexity of $O(\sum_{j=1}^{\kappa} \log n_j)$ at Line 15 and $O(\kappa \log \kappa)$ at Line 29. Therefore, the overall time complexity can be written as

$$O(n + \frac{B}{\min\{\delta_{jl}\}}(\sum_{j=1}^{\kappa} \log n_j + \kappa \log \kappa)) < O(B(n + \kappa \log \kappa)) \qquad (7)$$

where $\delta_{jl} = p_{jl}^{u-1} - p_{jl}^{u}, 1 \leq j \leq \kappa, 0 \leq l \leq n_j$ and $n = \sum_{j=1}^{\kappa} n_j$ the total number of tasks in the workflow. Here, we leverage the fact that $\log n < n$. Obviously, when $n \geq \kappa \log \kappa$, which is reasonable in multi-stage MapReduce jobs, we obtain a time complexity of $O(nB)$.

5   Empirical Studies

To verify and evaluate the proposed algorithms and study their performance behaviours in reality, we developed a *Budget Distribution Solver* (BDS) in Java that efficiently implements the algorithms for the specified scheduling problem in MapReduce. Since the monetary cost is our primary interest, in BSD we did not consider some properties and features of the network platforms. Rather, we focus on the factors closely related to our research goal. In practical, how efficient the algorithms are in minimizing the scheduling lengths of the workflow subject to different budget constraints are our concern.

The BDS accepts as an input a bag of MapReduce jobs that are organized as a multi-stage fork&join workflow by the scheduler at run-time. Each task of the job is associated with a time-price table, which is pre-defined by the cloud providers. As a consequence, the BDS can be configured with several parameters, including those described time-price tables, the number of tasks in each stage and the total number of stages in the workflow. Since there is no well-accepted model to specify these parameters, we assume them to be automatically generated in a uniform distribution where the task execution time and the corresponding prices in particular are varied in the ranges of [1, 12.5*table_size] and [1, 10*table_size], respectively. As intuitively, with the table size being increased, the scheduler has more choices to select the candidate machines to execute a task. On the other hand, in each experiment we allow the budget resources to be increased from its lower bound to upper bound and thereby comparing the scheduling lengths and the scheduling time of the proposed algorithms with respect to different configuration parameters. Here, the lower and upper bound are defined to be the minimal and maximal budget resources, respectively, that can be used to complete the workflow.

All the experiments are conducted by comparing the proposed GGB algorithm with the optimal algorithm Opt and the numerical results are obtained from a Ubuntu 12.04 platform having a hardware configuration of 3392.183 MHz processors, with a total of 8 processors activated, each with 8192K cache.

5.1   Impact of Time-Price Table Size

We first evaluate the impact of the time-price table size on the total scheduling length of the workflow with respect to different budget constraints. To this end, we fix a 8-stage workflow with at most 20 tasks in each stage. The size of the time-price table associated with each task varies from 4, 8, 16 to 32.

The results of the GGB algorithm compared with those of the optimal algorithm are shown in Fig. 3. While the budget increases, for all sizes of the tables, the scheduling lengths decrease super-linearly. These results are interesting also difficulty to make from algorithm analysis alone. We attribute these results to the fact that the opportunities of reducing the execution time of each stage are super-linearly increased with the budget growth, especially for those large size workflows. This phenomenon implies that the *performance/cost* ratio increases if cloud users are willing to pay more for MapReduce computation. This figure also shows that the performance of GGB is very close to the optimal algorithm, but its scheduling time is significantly less than that of the optimal algorithm

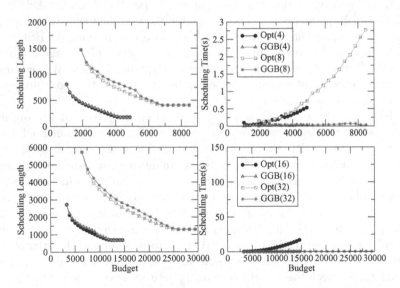

Fig. 4. Impact of the number of stages on the total scheduling length (e.g., makespan of the job in terms of indivisible time units) and scheduling time (Task: ≤ 20, Table Size ≤ 16, and the numbers in the brackets represent the different number of stages)

(quadratic in its time complexity). These results demonstrate how effective and efficient the proposed GGB algorithm is to achieve the best performance for MapReduce workflows subject to different budget constraints.

5.2   Impact of Workflow Size

In this set of experiments, we evaluate the performance changes with respect to different workflow sizes when the budget resources for each workflow are increased from the lower bound to the upper bound as we defined before. To this end, we fix the maximum number of tasks in the MapReduce workflow to 20 in each stage, and each task is associated with a time-price table with a maximum size of 16. We vary the number of stages from $4, 8, 16$ to 32, and observe the performance and scheduling time changes in Fig. 4. From this figure, we can see that all the algorithms exhibit the same performance patterns with those we observed when the impact of the table size is considered. These results are expected as both the number of stages and the size of tables are linearly correlated with the total workloads in the computation. This observation can be also made when the number of tasks in each stage is changed.

6   Conclusions

In this paper, we studied the scheduling of a bag of MapReduce jobs with budget constraints on a set of (virtual) machines in Clouds. To this end, we first presented a parallel optimal algorithm to address the constraints within pseudo polynomial time. The algorithm is based on dynamic programming techniques and integrates an in-stage local greedy algorithm to achieve the global optimality.

To further improve the efficiency, we then developed a global greedy algorithm GGB that extends the idea of the local greedy algorithm to the distribution of the budget among the tasks across different stages of the workflow while minimizing the scheduling length as a goal. The performance of the proposed algorithms were evaluated by empirical studies. The results show that the GGB is close to the optimal results in terms of the scheduling length but entails much lower time overhead.

References

1. Apache Software Foundation. Hadoop, http://hadoop.apache.org/core
2. Greenplum HD, http://www.greenplum.com
3. Caron, E., Desprez, F., Muresan, A., Suter, F.: Budget constrained resource allocation for non-deterministic workflows on an iaas cloud. In: Xiang, Y., Stojmenovic, I., Apduhan, B.O., Wang, G., Nakano, K., Zomaya, A. (eds.) ICA3PP 2012, Part I. LNCS, vol. 7439, pp. 186–201. Springer, Heidelberg (2012)
4. Correia, M., Costa, P., Pasin, M., Bessani, A., Ramos, F., Verissimo, P.: On the feasibility of byzantine fault-tolerant mapreduce in clouds-of-clouds. In: 2012 IEEE 31st Symposium on Reliable Distributed Systems (SRDS), pp. 448–453 (2012)

5. Dean, J., Ghemawat, S.: Mapreduce: simplified data processing on large clusters. In: Proceedings of the 6th Conference on Symposium on Opearting Systems Design & Implementation, OSDI 2004, vol. 6, p. 10 (2004)
6. Dean, J., Ghemawat, S.: Mapreduce: simplified data processing on large clusters. Commun. ACM 51(1), 107–113 (2008)
7. Hoffa, C., Mehta, G., Freeman, T., Deelman, E., Keahey, K., Berriman, B., Good, J.: On the use of cloud computing for scientific workflows. In: IEEE Fourth International Conference on eScience, eScience 2008, pp. 640–645 (December 2008)
8. Ibrahim, S., Jin, H., Lu, L., Qi, L., Wu, S., Shi, X.: Evaluating mapreduce on virtual machines: The hadoop case. In: Jaatun, M.G., Zhao, G., Rong, C. (eds.) Cloud Computing 2009. LNCS, vol. 5931, pp. 519–528. Springer, Heidelberg (2009)
9. Isard, M., Budiu, M., Yu, Y., Birrell, A., Fetterly, D.: Dryad: distributed data-parallel programs from sequential building blocks. In: Proceedings of the 2nd ACM SIGOPS/EuroSys European Conference on Computer Systems 2007, EuroSys 2007, pp. 59–72 (2007)
10. Juve, G., Deelman, E., Berriman, G.B., Berman, B.P., Maechling, P.: An evaluation of the cost and performance of scientific workflows on amazon ec2. J. Grid Comput. 10(1), 5–21 (2012)
11. Kc, K., Anyanwu, K.: Scheduling hadoop jobs to meet deadlines. In: 2010 IEEE Second International Conference on Cloud Computing Technology and Science, CloudCom, pp. 388–392 (2010)
12. Kondikoppa, P., Chiu, C.-H., Cui, C., Xue, L., Park, S.-J.: Network-aware scheduling of mapreduce framework ondistributed clusters over high speed networks. In: Proceedings of the 2012 Workshop on Cloud Services, Federation, and the 8th Open Cirrus Summit, FederatedClouds 2012, pp. 39–44 (2012)
13. Li, Y., Zhang, H., Kim, K.H.: A power-aware scheduling of mapreduce applications in the cloud. In: 2011 IEEE Ninth International Conference on Dependable, Autonomic and Secure Computing (DASC), pp. 613–620 (2011)
14. Li, Y., Zhang, H., Kim, K.H.: A power-aware scheduling of mapreduce applications in the cloud. In: 2011 IEEE Ninth International Conference on Dependable, Autonomic and Secure Computing (DASC), pp. 613–620 (2011)
15. Liu, H., Orban, D.: Cloud mapreduce: A mapreduce implementation on top of a cloud operating system. In: 2011 11th IEEE/ACM International Symposium on Cluster, Cloud and Grid Computing (CCGrid), pp. 464–474 (2011)
16. Marozzo, F., Talia, D., Trunfio, P.: Enabling reliable mapreduce applications in dynamic cloud infrastructures. ERCIM News 2010(83), 44–45 (2010)
17. Thusoo, A., Sarma, J., Jain, N., Shao, Z., Chakka, P., Zhang, N., Antony, S., Liu, H., Murthy, R.: Hive - a petabyte scale data warehouse using hadoop. In: 2010 IEEE 26th International Conference on Data Engineering (ICDE), pp. 996–1005 (2010)
18. Wang, K., Tan, B., Shi, J., Yang, B.: Automatic task slots assignment in hadoop mapreduce. In: Proceedings of the 1st Workshop on Architectures and Systems for Big Data, ASBD 2011, pp. 24–29 (2011)
19. You, H.-H., Yang, C.-C., Huang, J.-L.: A load-aware scheduler for mapreduce framework in heterogeneous cloud environments. In: Proceedings of the 2011 ACM Symposium on Applied Computing, SAC 2011, pp. 127–132 (2011)
20. Yu, J., Buyya, R.: Scheduling scientific workflow applications with deadline and budget constraints using genetic algorithms. Sci. Program 14(3,4), 217–230 (2006)

21. Zaharia, M., Borthakur, D., Sen Sarma, J., Elmeleegy, K., Shenker, S., Stoica, I.: Delay scheduling: a simple technique for achieving locality and fairness in cluster scheduling. In: Proceedings of the 5th European Conference on Computer Systems, pp. 265–278 (2010)
22. Zaharia, M., Konwinski, A., Joseph, A.D., Katz, R., Stoica, I.: Improving mapreduce performance in heterogeneous environments. In: Proceedings of the 8th USENIX Conference on Operating Systems Design and Implementation, OSDI 2008, pp. 29–42 (2008)
23. Zeng, L., Veeravalli, B., Li, X.: Scalestar: Budget conscious scheduling precedence-constrained many-task workflow applications in cloud. In: Proceedings of the 2012 IEEE 26th International Conference on Advanced Information Networking and Applications, AINA 2012, pp. 534–541 (2012)

Fast and Scalable Queue-Based Resource Allocation Lock on Shared-Memory Multiprocessors

Deli Zhang, Brendan Lynch, and Damian Dechev

Department of EECS, University of Central Florida
Orlando, FL 32816, USA
{de-li.zhang,brendan.lynch}@knights.ucf.edu, dechev@eecs.ucf.edu

Abstract. We present a fast and scalable lock algorithm for shared-memory multiprocessors addressing the resource allocation problem. In this problem, threads compete for k shared resources where a thread may request an arbitrary number $1 \leq h \leq k$ of resources at the same time. The challenge is for each thread to acquire exclusive access to desired resources while preventing deadlock or starvation. Many existing approaches solve this problem in a distributed system, but the explicit message passing paradigm they adopt is not optimal for shared-memory. Other applicable methods, like two-phase locking and resource hierarchy, suffer from performance degradation under heavy contention, while lacking a desirable fairness guarantee. This work describes the first multi-resource lock algorithm that guarantees the strongest first-in, first-out (FIFO) fairness. Our methodology is based on a non-blocking queue where competing threads spin on previous conflicting resource requests. In our experimental evaluation we compared the overhead and scalability of our lock to the best available alternative approaches using a micro-benchmark. As contention increases, our multi-resource lock obtains an average of ten times speedup over the alternatives including GNU C++'s lock method and Boost's lock function.

1 Introduction

Improving the scalability of resource allocation algorithms on shared-memory multiprocessors is of practical importance due to the trend of developing many-core chips. The performance of parallel applications on a shared-memory multiprocessor is often limited by contention for shared resources, creating the need for efficient synchronization methods. In particular, the limitations of the synchronization techniques used in existing database systems leads to poor scalability and reduced throughput on modern multicore machines [16]. For example, when running on a machine with 32 hardware threads, Berkeley DB spends over 80% of the execution time in its Test-and-Test-and-Set lock [16].

Mutual exclusion locks eliminate race conditions by limiting concurrency and enforcing sequential access to shared resources. Comparing to more intricate approaches like lock-free synchronization [15] and software transactional memory [14], mutual exclusion locks introduce sequential ordering that eases the

R. Baldoni, N. Nisse, and M. van Steen (Eds.): OPODIS 2013, LNCS 8304, pp. 266–280, 2013.

reasoning about correctness. Despite the popular use of mutual exclusion locks, one requires extreme caution when using multiple mutual exclusion locks together. In a system with several shared resources, threads often need more than just one resource to complete certain tasks, and assigning one mutual exclusion lock to one resource is common practice. Without coordination between locks this can produce undesirable effects such as deadlock, livelock and decrease in performance.

Consider two clerks, *Joe* and *Doe*, transferring money between two bank accounts C_1 and C_2, where the accounts are exclusive shared resources and the clerks are two contending threads. To prevent conflicting access, a lock is associated with each bank account. The clerks need to acquire both locks before transferring the money. The problem is that mutual exclusion locks cannot be composed, meaning that acquiring multiple locks inappropriately may lead to deadlock. For example, when *Joe* locks the account C_1 then he attempts to lock C_2. In the meantime, *Doe* has acquired the lock on C_2 and waits for the lock on C_1. In general, one seeks to allocate multiple resources among contending threads that guarantees forward system progress, which is known as the resource allocation problem [12]. Two pervasive solutions, namely resource hierarchy [9] and two-phase locking [10], prevent the occurrence of deadlocks but do not respect the fairness among threads and their performance degrades as the level of contention increases. Nevertheless, both the GNU C++ library[1] and the Boost library[2] adopt the two-phase locking mechanism as a means to avoid deadlocks.

In this paper, we propose the first FIFO (first-in, first-out) multi-resource lock algorithm for solving the resource allocation problem on shared-memory multiprocessors. Given k resources, instead of having k separate locks for each one, we employ a non-blocking queue as the centralized manager. Each element in the queue is a resource request bitset[3] of length k with each bit representing the state of one resource. The manager accepts the resource requests in a first-come, first-served fashion: new requests are enqueued to the tail, and then they progress through the queue in a way that no two conflicting requests can reach the head of the queue at the same time. Using the bitset, we detect resource conflict by matching the correspondent bits. The key algorithmic advantages of our approach include:

1. The FIFO nature of the manager guarantees fair acquisition of locks, while implying starvation-freedom and deadlock-freedom
2. The lock manager has low access overhead and is scalable with the cost of enqueue and dequeue being only a single `compare_and_swap` operation
3. The maximum concurrency is preserved as a thread is blocked only when there are outstanding conflicting resource requests
4. Using a bitset allows an arbitrary number of resources to be tracked with low memory overhead, and does not require atomic access

---

[1] http://gcc.gnu.org
[2] http://www.boost.org
[3] A bitset is a data structure that contains an array of bits.

We evaluate the overhead and scalability of our lock algorithm using a micro-benchmark. We compare our work to the state-of-the-art approaches in the field, which include resource hierarchy locking combined with std::mutex, two-phase locking, such as std::lock and boost::lock, and an extended Test-and-Test-and-Set (TATAS) lock. At low levels of contention, our lock sacrifices performance for fairness resulting in a worst case slowdown of 2 times. As contention increases, it outperforms the two-phase locking methods by a factor of 10 with a worst case speedup of 1.5 to 2 times against the resource hierarchy lock and the extended TATAS lock. Moreover, the timings of our multi-resource lock are significantly more consistent and regular throughout all test scenarios when compared to other approaches.

2   Background

In this section, we briefly review the mutual exclusion problem and its variations, with emphasis on the resource allocation problem and the desirable properties for a solution. We also provide a summary on the lock-free data structures and the atomic primitives used in our algorithm.

2.1   Mutual Exclusion and Resource Allocation

Mutual exclusion algorithms are widely used to construct synchronization primitives like locks, semaphores and monitors. Designing efficient and scalable mutual exclusion algorithms has been extensively studied (Raynal [21] and Anderson [1] provide excellent surveys on this topic). In the classic form of the problem, competing threads are required to enter the critical section one at a time. In the k-mutual exclusion problem [12], k units of an identical shared resource exist so that up to k threads are able to acquire the shared resource at once. Further generalization of k-mutual exclusion gives the h-out-of-k mutual exclusion problem [20], in which a set of k identical resources are shared among threads. Each thread may request any number $1 \leq h \leq k$ of the resources, and the thread remains blocked until all the required resources become available.

We address the resource allocation problem [17] on shared-memory multiprocessors, which extends the h-out-of-k mutual exclusion problem in the sense that the resources are not necessarily identical. The resource allocation problem can also be seen as a generalization to the prominent *Dining Philosophers Problem* (DPP) originally formulated by Dijkstra [9]. It drops the static resource configuration used in the DPP and allows an arbitrary number of resources to be requested from a pool of k resources. The minimal safety and liveness properties for any solution include mutual exclusion and deadlock-freedom [1]. Mutual exclusion means a resource must not be accessed by more than one thread at the same time, while deadlock-freedom guarantees system wide progress. Starvation-freedom, a stronger liveness property than deadlock-freedom, ensures every thread eventually gets the requested resources. In the strongest FIFO ordering, the threads are served in the order they arrive. It is preferable for ensuring starvation-freedom because it enforces strict fairness between contenders [18].

2.2   Atomic Primitives and Synchronization

Atomic primitives are the cornerstones of any synchronization algorithm. compare_and_swap(address, expectedValue, newValue)[4], or CAS for short, always returns the original value at the specified address but only writes newValue to address if the original value matches expectedValue. A slightly different version compare_and_set returns a Boolean value indicating whether the comparison succeeded. In C++ memory model, the use of an atomic operation is accompanied by std::memory_order, which specify how regular memory accesses made by different threads should be ordered around the atomic operation. More specifically, a pair of std::memory_order_acquire and std::memory_order_release requires that when a thread does a atomic load operation with acquire order, prior writes made to other memory locations by the thread that did the release become visible to it. std::memory_order_relaxed, on the hand, poses no ordering constraints.

Algorithm 1. TATAS lock for resource allocation

```
1  typedef uint64 bitset;                    11    }while(!compare_and_set(1, b, b |
2                                                        r));
3  //input l: address of the lock            12  }
4  //input r: request bit mask               13
5  void lock(bitset* l, bitset r){           14  void unlock(bitset* l, bitset r){
6    bitset b;                               15    bitset b;
7    do{                                     16    do{
8      b = *l; //read bits value             17      b = *l;
9      if(b & r) //check for                 18    }while(!compare_and_set(1, b, b &
         confliction                                    ~r));
10       continue; //spin with reads         19  }
```

Given the atomic CAS instruction, it is straightforward to develop simple spin locks. In Algorithm 1 we present an extended TATAS lock that solves the resource allocation problem for a small number of resources. The basic TATAS lock is a spin lock that allows threads to busy-wait on the initial test instruction to reduce bus traffic. The key change we made is to treat the lock integer value as a bit array instead of a Boolean flag. A thread needs to specify the resource requests through a bitset mask when acquiring and releasing the lock. With each bit representing a resource, the bits associated with the desired resources are set to 1 while others remain 0. The request updates the relevant bits in the lock bitset if there is no conflict, otherwise the thread spins. One drawback of this extension is that the total number of resources is limited by the size of integer type because a bitset capable of representing arbitrary number of resources may span across multiple memory words. Updating multiple words atomically is not possible without resorting to multi-word CAS [13], which is not readily available on all platforms.

Non-blocking synchronization, eliminates the use of locks completely. A concurrent object is lock-free if at least one thread makes forward progress in a finite number of steps [15]. It is wait-free if all threads make forward progresses

---

[4] Also known as compare_exchange

in a finite number of steps [7]. Compared to their blocking counterparts, non-blocking objects promise greater scalability and robustness. In this work, we take advantage of a non-blocking queue to increase the scalability and throughput of our lock mechanism.

3 Algorithms

We implement a queue-based multi-resource lock that manages an arbitrary number of exclusive resources on shared-memory architectures. Our highly scalable algorithm controls resource request conflicts by holding all requests in a FIFO queue and allocating resources to the threads that reach the top of the queue. We achieve scalable behavior by representing resource requests as a bitset and employing a non-blocking queue that grants fair acquisition of the locks.

3.1 Handle Locking Request with FIFO Queue

Our conflict management approach is built on an array-based bounded lock-free FIFO queue [15]. The lock-free property is desirable as our lock manager must guarantee deadlock freedom. The FIFO property of the data structure allows for serving threads in their arriving order, implying starvation-freedom for all enqueued requests. We favor an array-based queue over other high performance non-blocking queues because it does not require dynamic memory management. Link-list based queues involve dynamic memory allocation for new nodes, which could lead to significant performance overhead and the ABA problem [19]. With a pre-allocated continuous buffer, our lock algorithm is not prone to the ABA problem and has low runtime overhead by using a single CAS for both enqueue and dequeue operations.

Given a set of resources, each bit in a request bitset is uniquely mapped to one resource. A thread encapsulates a request of multiple resources in one bitset with the correspondent bit of the requested resources set to 1. The multi-resource lock handles requests atomically meaning that a request is fulfilled only if all requested resources are made available, otherwise the thread waits in the queue. This all-or-nothing atomic acquisition allows the maximum number of threads, without conflicting requests, to use the shared resources.

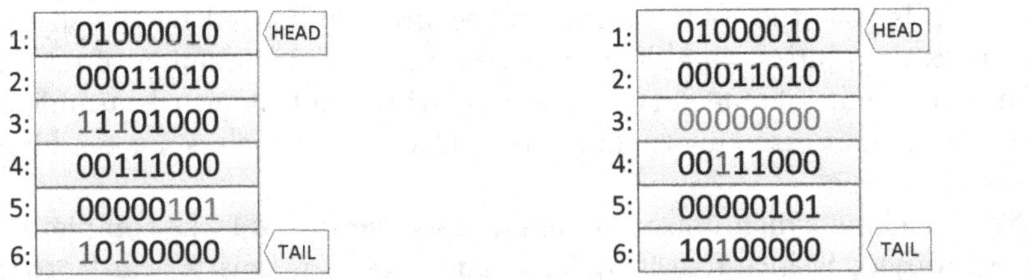

(a) Cell 6 spins on cell 3   (b) Release of cell 3. Cell 6 spins on cell 4

Fig. 1. Atomic lock acquisition process

The length of the bitset is unlimited and can be determined either at runtime as in boost::dynamic_bitset, or at compile time as in std::bitset. Using variable length bitset is also possible to accommodate growing number of total resources at runtime, as long as the resource mapping is maintained. Figure 1a demonstrates this approach. A newly enqueued request is placed at the tail. Starting from the queue head, it compares the bitset value with each request. In the absence of conflict, it moves on to the next one until it reaches itself. Here, the thread on 5th cell successfully acquires all needed resources. The thread on the tail (6th cell) spins on the 3rd request due to conflict. In Figure 1b, the thread on the tail proceeds to spin on the 4th cell when the 3rd request was released.

Algorithm 2. Multi-Resource Lock Data Structures

```
1  #include<bitset.h>                   19   l.buffer = new cell[siz];
2  #include<atomic>                      20   l.mask = siz - 1;
3  using namespace std;                  21   l.head.store(0,
4                                                 memory_order_relaxed);
5  struct cell{                          22   l.tail.store(0,
6     atomic<uint32> seq;                          memory_order_relaxed);
7     bitset bits;                       23   //initialize bits to all 1s
8  }                                     24   for (uint32 i = 0; i < siz; i++)
9  struct mrlock{                                {
10    cell* buffer;                      25      l.buffer[i].bits.set();
11    uint32 mask;                       26      l.buffer[i].seq.store(i,
12    atomic<uint32> head;                          memory_order_relaxed);
13    atomic<uint32> tail;               27   }
14 }                                     28 }
15                                       29
16 //input l: reference to the lock      30 void uninit(mrlock& l){
17 //input siz: suggested buffer size    31   delete[] l.buffer;
18 void init(mrlock& l, uint32 siz){     32 }
```

Algorithm 2 defines the lock manager's class. The cell structure defines one element in the queue, it consists of a bitset that represents a resource request and an atomic *sequence number* that coordinates concurrent access. The mrlock structure contains a cell buffer pointer, the size mask, and the queue head and tail. We use the size mask to apply fast index modulus. In our implementation, the head and tail increase monotonically; we use an index modulus to map them to the correct array position. Expensive modulo operation can be replaced by bitwise AND operation if the buffer size is chosen to be a power of two.

3.2   Acquiring and Releasing Locks

We list the code for lock acquire function in Algorithm 3, which consists of two steps: enqueue and spin. The code from line 7 to 16 outlines a CAS-based loop, with threads competing to update the queue tail on line 13. If the CAS attempt succeeds the thread is granted access to the cell at the tail position, and the tail is advanced by one. The thread then stores its resource request, which is passed to lock function as the variable r, in the cell along with a sequence number. The sequence number serves as a sentinel in our implementation. During the enqueue operation the thread assigns a sequence number to its cell as it enters the queue as seen on line 18. The nature of a bounded queue allows the head and

Algorithm 3. Lock Acquire	**Algorithm 4.** Lock Release

```
1  //input l:  referenc to mrlock
2  //input r:  resource  request
3  //output : the lock handle
4  uint32 lock(mrlock& l, bitset r){
5    cell* c;
6    uint32 pos;
7    for(;;){
8      pos = l.tail.load(
           memory_order_relaxed);
9      c = &l.buffer[pos & l.mask];
10     uint32 seq = c->seq.load(
           memory_order_acquire);
11     int32 dif = (int32)seq - (int32
           )pos;
12     if(dif == 0){
13       if(l.tail.
             compare_exchange_weak(
             pos, pos + 1,
             memory_order_relaxed))
14         break;
15     }
16   }
17   c->bits = r;
18   c->seq.store(pos + 1,
         memory_order_release);
19   uint32 spin = l.head;
20   while(spin != pos){
21     if(pos - l.buffer[spin & l.mask
         ].seq > l.mask || !(l.
         buffer[spin & l.mask].bits
         & r))
22       spin++;
23   }
24   return pos;
25 }
```

```
1  //input l:  reference to mrlock
2  //input h:  the lock handle
3  void unlock(mrlock& l, uint32 h){
4    l.buffer[h&l.mask].bits.reset();
5    uint32 pos = l.head.load(
         memory_order_relaxed);
6    while(l.buffer[pos & l.mask].bits
         == 0){
7      cell* c = &l.buffer[pos & l.
         mask];
8      uint32 seq = c->seq.load(
           memory_order_acquire);
9      int32 dif = (int32)seq - (int32
           )(pos + 1);
10     if(dif == 0){
11       if(l.head.
             compare_exchange_weak(
             pos, pos + 1,
             memory_order_relaxed)){
12         c->bits.set();
13         c->seq.store(pos + l.mask
             + 1,
             memory_order_release
             );
14       }
15     }
16     pos = l.head.load(
           memory_order_relaxed);
17   }
18 }
```

tail pointers to move through a circular array. Dequeue attempts to increment the head pointer towards the current tail, while a successful call to enqueue will increment the tail pointer pulling it away from head. The sequence numbers are initialized on line 26 in Algorithm 2.

Once a thread successfully enqueues its request, it spins in the while loop on line 20 to 23. It traverses the queue beginning at the head. When there is a conflict of resources indicated by the bitset, the thread will spin locally on the conflicting request. Line 21 displays two conditions that allow the thread to advance: 1) the cell the thread is spinning on is free and recycled, meaning the cell is no longer in front of this thread. This condition is detected by the use of sequence numbers; 2) The request in the cell has no conflict, which is tested by bitwise and of the two requests. Once the thread reaches its position in the queue, it is safe to assume the thread has acquired the requested resources. The position of the enqueued request is returned as a handle, which is required when releasing the locks.

The unlock function releases the locks on the requested resources by setting the bitset fields to zero using the lock handle, on line 4 of Algorithm 4. This allows threads waiting for this position to continue traversing the queue.

The removal of the request from the queue is delayed until the request in the head cell is cleared (line 6). If a thread is releasing the lock on the head cell, the releasing operation will perform dequeue and recycle the cell. The thread will also examine and dequeue the cells at the top of the queue until a nonzero bitset is found. The code between lines 6 and 17 outlines a CAS loop that is similar to the enqueue function. The difference is that here threads assist each other with the work of advancing the head pointer. With this release mechanism, threads which finish before becoming the head of the queue do not block the other threads.

4 Related Work

As noted in section 2.1, a substantial body of work addresses the mutual exclusion problem and the generalized resource allocation problem. In this section, we summarize the solutions to the resource allocation problem and related queue-based algorithms. We skip the approaches targeting distributed environments [4,20]. These solutions do not transfer to shared-memory systems because of the drastically different communication characteristics. In distributed environments processes communicate with each other by message passing, while in shared-memory systems communication is done through shared memory objects. We also omit early mutual exclusion algorithms that use more primitive atomic read and write registers [21,1]. As we show in section 2.2, the powerful CAS operation on modern multiprocessors greatly reduces the complexity of mutual exclusion algorithms.

4.1 Resource Allocation Solutions

Assuming each resource is guarded by a mutual exclusion lock, lock acquiring protocols can effectively prevent deadlocks. Resource hierarchy is one protocol given by Dijkstra [9] based on total ordering of the resources. Every thread locks resources in an increasing order of enumeration; if a needed resource is not available the thread holds the acquired locks and waits. Deadlock is not possible because there is no cycle in the resource dependency graph. Lynch [17] proposes a similar solution based on a partial ordering of the resources. Resource hierarchy is simple to implement, and when combined with queue mutex it is the most efficient existing approach. However, total ordering requires prior knowledge of all system resources, and dynamically incorporating new resources is difficult. Besides, FIFO fairness is not guaranteed because the final acquisition of the resources is always determined by the acquisition last lock in this hold-and-wait scheme. Two-phase locking [10] was originally proposed to address concurrency control in databases. At first, threads are allowed to acquire locks but not release them, and in the second phase threads are allowed to release locks without acquisition. For example, a thread tries to lock all needed resources one at a time; if anyone is not available the thread releases all the acquired locks and start over again. When applied to shared-memory systems, it requires a try_lock method that returns immediately instead of blocking the thread when the lock is not available. Two-phase locking is flexible requiring no prior knowledge on resources other than the desired ones, but its performance degrades drastically under contention, because the release-and-wait protocol is vulnerable to failure and retry. Time stamp ordering [5] prevents deadlock by selecting an ordering among the threads. Usually a unique time stamp is assigned to the thread before it starts to lock the resources. Whenever there is a conflict the thread with smaller time stamp wins.

4.2   Queue-Based Algorithms

Fischer et al. [11] describes a simple FIFO queue algorithm for the k-mutual exclusion problem. Awerbuch and Saks [3] proposed the first queuing solution to the resource allocation problem. They treat it as a dynamic job scheduling problem, where each job encapsulates all the resources requested by one process. Newly enqueued jobs progress through the queue if no conflict is detected. Their solution is based on a distributed environment in which the enqueue and dequeue operation are done via message communication. Due to this limitation, they need to assume no two jobs are submitted concurrently. Spin locks such as the TATAS lock shown in Algorithm 1 induce significant contention on large machines, leading to irregular timings. Queue-based spin locks eliminate these problems by making sure that each thread spins on a different memory location [22]. Anderson [2] embeds the queue in a Boolean array, the size of which equals the number of threads. Each thread determines its unique spin position by drawing a ticket. When relinquishing the lock, the thread resets the Boolean flag on the next slot to notify the waiting thread. The MCS lock [18] designed by Scott et al., employs a linked list with pointers from each thread to its successor. The CLH lock by Craig et al. [6] also employs a linked list but with pointers from each thread to its predecessor. A Recent queue lock based on *flat-combining* synchronization [8] exhibits superior scalability on NUMA architecture than the above classic methods. The flat-combining technique reduce contention by aggregating lock acquisitions in batch and processing them with a combiner thread. A key difference between this technique and our multi-resource lock is that our method aggregates lock acquisition requests for multiple resources from one thread, while the flat-combining lock gathers requests from multiple threads for one resource. Although the above queue-based locks could not solve the resource allocation problem on their own, they share the same inspiration with our method: using queue to reduce contention and provide FIFO fairness.

5   Performance Evaluation

In this section, we assess the overhead, scalability and performance consistency of our multi-resource lock (MRLock) and compare it with the std::lock function from GCC 4.7 (STDLock), the boost::lock function from Boost library 1.49 (BSTLock), the resource hierarchy scheme combined with std::mutex (RHSTD), and the extended TATAS lock (ETATAS) described in Algorithm 1. We use std::mutex as the underlying lockable object for std::lock, and boost::mutex for boost::lock. These alternative approaches have been widely used in practice and highly optimized for our testing system.

We employ a micro-benchmark to evaluate the performance of these approaches for multiple resource allocation. It consists of a tight loop that acquires and releases a predetermined number of locks. The loop increments a set of integer counters, where each counter represents a resource. The counters are not atomic, so without the use locks their value will be incorrect due to data races. When the micro-benchmark's execution is complete, we check each counter's value to verify the absence of data races and validate the correctness of our lock implementations. All tests are conducted on a 64-core ThinkMate RAX QS5-4410 server running Ubuntu 12.04 LTS. It is a NUMA system with four AMD Opteron 6272 CPUs (16 cores per chip @2.1 GHz) and 64 GB of shared memory (16 × 4GB PC3-12800 DIMM). Both the micro-benchmark and the lock implementations are compiled with GCC 4.7 (with the options -o1 -std=c++0x to enable level 1 optimization and C++ 11 support).

When evaluating classic mutual exclusion locks, one may increase the number of concurrent threads to investigate their scalability. Since all threads contend for a single critical section, the contention level scales linearly with the number of threads. However, the amount of contention in the resource allocation problem can be raised by either increasing the number of threads or the size of the resource request per thread. Given k total resources with each thread requesting h of them, we denote the *resource contention* by the fraction h/k or its quotient in percentage. This notation reveals that *resource contention* may be comparable even though the total number of resources is different. For example, 8/64 or 12.5% means each request needs 8 resources out of 64, which produces about the same amount of contention as 4/32. We show benchmark timing results in Section 5.2 that verifies this hypothesis. The product of the thread number p and *resource contention* level roughly represents the overall contention level.

To fully understand the efficiency and scalability in these two dimensions, we test the locks in a wide range of parameter combinations: for thread number $2 \leq p \leq 64$ and for resource number $4 \leq k \leq 64$ each thread requests the same number of resources $2 \leq h \leq k$. We set the loop iteration in the micro-benchmark to 10,000 and get the average time out of 10 runs for each configuration.

5.1   Single-Thread Overhead

To measure the lock overhead in the absence of contention, we run the micro-benchmark with a single thread requesting two resources and subtract the loop overhead from the results. Table 1 shows the total timing for the completion of a pair of lock and unlock operations. In this scenario MRLock is slightly slower than ETATAS because of the extra queue traversing operation. The other four methods take about twice the time of MRLock because each of them takes at a minimum two lock operations to solve a non-trivial resource allocation problem. Although std::mutex and boost::mutex does not solve the resource allocation problem, we compare against them as a baseline performance metric.

Table 1. Lock overhead obtained without contention

MRLock	STDLock	BSTLock	RHLock	ETATAS	std::mutex	boost::mutex
$42ns$	$95ns$	$105ns$	$88ns$	$34ns$	$35ns$	$35ns$

5.2   Resource Scalability

Our performance evaluation exploring the scalability of the tested approaches when increasing the level of *resource contention* is shown in Figures 2a, 2b and 2c. The y-axis represents the total time needed to complete the micro-benchmark in a logarithmic scale, where a smaller value indicates better performance. The x-axis represents levels of resource contention, and it is divided by five tick marks into six sections. Each tick mark on the x-axis represents the beginning of the section to its right, and the tick mark label denotes the total number of resources in that section. For example, the section between Tick 32 and 64 has a total of 32 resources, while the section to the right of Tick 64 has 64 resources. Within each section, the level of contention increases from 1% to 100%. We observe a saw pattern because the resource contention level alternates as we move along the x-axis. In addition, we observe that the timing pattern is similar among different sections, supporting our argument that the contention is proportional to the quotient of the request size divided by total number of resources.

When increasing the number of requested resources per thread, the probability of threads requesting the same resources increases. This poses scalability challenges for both two-phase locks and the resource hierarchy implementations because they rely on a certain protocol to acquire the requested locks one by one. As the request size increases, the acquiring protocol is prolonged thus prone to failure and retry. At high levels of contention, such as the case with 64 threads (Figure 2c) when the level contention exceeds 75%, STDLock is more than 50 times slower when compared to MRLock. BSTLock exhibits the same problem, and its observed performance closely resembles STDLock's performance. Unlike the above two methods, RHLock acquires locks in a fixed order, and it does not release current locks if a required resource is not available. This hold-and-wait paradigm helps stabilize the timings and reduce the overall contention. RHLock resembles the performance of STDLock in the two thread scenario (Figure 2a), but it outperforms both BSTLock and STDLock by about three times under 50% resource contention on 16 threads (Figure 2b).

While the time of all alternative methods show linear growth with respect to resource contention, MRLock remains constant throughout all scenarios. In the case of 64 threads and request size of 32, MRLock achieves a 20 times speedup over STDLock, 10 times performance gain over BSTLock and 2.5 times performance increase over RHLock. The fact that MRLock provides a centralized manager to respond the lock requests from threads in one batch contributes to this high degree of scalability. ETATAS also adopts the same all-or-nothing scheme, thus it could be seen as an MRLock algorithm with a queue size of

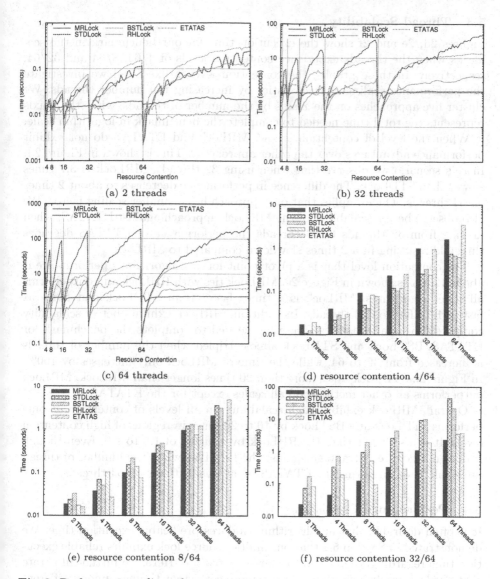

Fig. 2. Performance scaling when increasing resource contention (2a, 2a, and 2a) and the number of threads (2d, 2e, and 2f)

one. It outperforms MRLock on two threads by about 40% (Figure 2a), and almost ties with MRLock on 32 threads. However, MRLock is 1.7 time faster on 64 threads, because the queuing mechanism relieves the contention of the CAS loop.

5.3   Thread Scalability

Figures 2d, 2e and 2f show the execution time for our benchmark in the scenarios when the threads experience contention levels of 4/64, 8/64 and 32/64, respectively. In these graphs, the contention level is fixed and we investigate the performance scaling characteristics by increasing the number threads. We cluster five approaches on the x-axis by the number of threads, while the y-axis represents the total time needed to complete the benchmark in logarithm scale.

When the level of contention is low, MRLock and ETATAS do not exhibit performance advantages over the other approaches. This is shown in Figure 2d. In this scenario we observe that when using 32 threads, MRLock is 3.7 times slower than STDLock. The difference in performance decreases to about 2 times on 64 threads, which implies that our approach has a smaller scaling factor. We also observe better scalability of the MRLock approach against ETATAS; when moving from 32 threads to 64 threads the performance of ETATAS degrades threefold resulting in a 2 times slowdown compared to MRLock.

The contention level that is a pivot point for our algorithm's performance is about 12.5% as shown in Figure 2e. MRLock ties with RHLock and outperforms all other algorithms. MRLock is 4 times faster than STDLock and twice as fast as ETATAS on 64 threads. In addition, MRLock exhibits better scalability compared to its alternatives. The time needed to complete the benchmark for ETATAS, BSTLock and STDLock almost tripled when the number of threads is increased from 32 to 64, while the time of MRLock only increases by 100%. In Figure 2f, STDLock takes more than 20 times longer than MRLock. MRLock outperforms all other methods on all scales except for the ETATAS.

Overall, MRLock exhibits high scalability on all levels of contention, it outperforms STDLock and BSTLock by 10 to 20 times in regions of high contention levels. It is also faster than the RHLock by a factor of 1.5 to 2.5. Even though it does not hold an advantage against the ETATAS when the number of thread are small, it outperforms the ETATAS by at least 2 times on 64 threads.

5.4   Performance Consistency

It is often desirable that an algorithm produces predicable execution time. We demonstrated in Section 5.2 that our multi-resource lock exhibits reliable execution time regardless the level of resource contention. Here, we further illustrate that our lock implementation achieves more consistent timings among different runs when compared to the competing implementations. Figures 3b and 3a display the percentage deviation of execution times from 10 different runs. Since we generate randomized resource requests at the beginning of each test run, the resource conflicts is different for each run. We show the deviation normalized by the average execution time of each approach on the y-axis, and the number of threads on the x-axis. Overall, MRLock produces the smallest deviation that is often within 2% or its executing time. Notably ETATAS, which adopts the same batch request handling approach as MRLock, reached a maximum deviation of 18%. This indicates that the incorporation of a FIFO queue alleviated the contention and stabilized our lock algorithm.

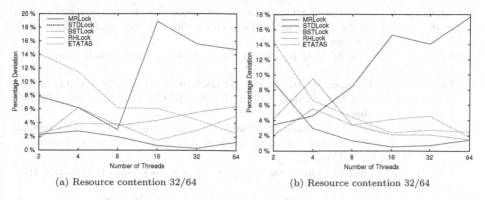

(a) Resource contention 32/64   (b) Resource contention 32/64

Fig. 3. Normalized deviation out of 10 runs

6 Conclusion and Future Work

Our multi-resource lock algorithm (MRLock) provides a robust solution to the resource allocation problem on shared-memory multiprocessors. The MRLock algorithm provides FIFO fairness for contending threads, and is scalable with minimal overhead increase over the best available solutions. As demonstrated by our performance evaluation, the MRLock algorithm exhibits reliability and scalability that can be beneficial to applications with high contention or when system scalability is desired.

Possible extension for this algorithm includes creating a NUMA aware algorithm by adopting the hierarchical queue structure and an adaptive method to choose from several locking algorithms based on the level of contention in a system.

Acknowledgment. This material is based upon work supported by the National Science Foundation under CCF Award No.1218100. The authors would like to thank Dimitry Vyukov for providing insightful implementation tips on the non-blocking queue.

References

1. Anderson, J., Kim, Y., Herman, T.: Shared-memory mutual exclusion: Major research trends since 1986. Distributed Computing 16(2), 75–110 (2003)
2. Anderson, T.E.: The performance of spin lock alternatives for shared-money multiprocessors. IEEE Transactions on Parallel and Distributed Systems 1(1), 6–16 (1990)
3. Awerbuch, B., Saks, M.: A dining philosophers algorithm with polynomial response time. In: Proceedings of the 31st Annual Symposium on Foundations of Computer Science, pp. 65–74. IEEE (1990)
4. Bar-Ilan, J., Peleg, D.: Distributed resource allocation algorithms. In: Segall, A., Zaks, S. (eds.) WDAG 1992. LNCS, vol. 647, pp. 277–291. Springer, Heidelberg (1992)

5. Bernstein, P., Goodman, N.: Timestamp based algorithms for concurrency control in distributed database systems. In: Proceedings 6th International Conference on Very Large Data Bases (1980)
6. Craig, T.: Building fifo and priorityqueuing spin locks from atomic swap. Tech. rep., Citeseer (1994)
7. Dechev, D., Pirkelbauer, P., Stroustrup, B.: Lock-free dynamically resizable arrays. In: Shvartsman, M.M.A.A. (ed.) OPODIS 2006. LNCS, vol. 4305, pp. 142–156. Springer, Heidelberg (2006)
8. Dice, D., Marathe, V.J., Shavit, N.: Flat-combining numa locks. In: Proceedings of the 23rd ACM Symposium on Parallelism in Algorithms and Architectures, pp. 65–74. ACM (2011)
9. Dijkstra, E.: Hierarchical ordering of sequential processes. Acta Informatica 1(2), 115–138 (1971)
10. Eswaran, K., Gray, J., Lorie, R., Traiger, I.: The notions of consistency and predicate locks in a database system. Communications of the ACM 19(11), 624–633 (1976)
11. Fischer, M.J., Lynch, N.A., Burns, J.E., Borodin, A.: Distributed fifo allocation of identical resources using small shared space. ACM Transactions on Programming Languages and Systems (TOPLAS) 11(1), 90–114 (1989)
12. Fischer, M., Lynch, N., Burns, J., Borodin, A.: Resource allocation with immunity to limited process failure. In: 20th Annual Symposium on Foundations of Computer Science, pp. 234–254. IEEE (1979)
13. Harris, T.L., Fraser, K., Pratt, I.A.: A practical multi-word compare-and-swap operation. In: Malkhi, D. (ed.) DISC 2002. LNCS, vol. 2508, pp. 265–279. Springer, Heidelberg (2002)
14. Herlihy, M., Moss, J.E.B.: Transactional memory: architectural support for lock-free data structures. SIGARCH Comput. Archit. News 21(2), 289–300 (1993)
15. Herlihy, M., Shavit, N.: The Art of Multiprocessor Programming, Revised Reprint. Morgan Kaufmann (2012)
16. Johnson, R., Pandis, I., Hardavellas, N., Ailamaki, A., Falsafi, B.: Shore-mt: a scalable storage manager for the multicore era. In: Proceedings of the 12th International Conference on Extending Database Technology: Advances in Database Technology, pp. 24–35. ACM (2009)
17. Lynch, N.: Fast allocation of nearby resources in a distributed system. In: Proceedings of the Twelfth Annual ACM Symposium on Theory of Computing, pp. 70–81. ACM (1980)
18. Mellor-Crummey, J., Scott, M.: Algorithms for scalable synchronization on shared-memory multiprocessors. ACM Transactions on Computer Systems (TOCS) 9(1), 21–65 (1991)
19. Michael, M., Scott, M.: Simple, fast, and practical non-blocking and blocking concurrent queue algorithms. In: Proceedings of the Fifteenth Annual ACM Symposium on Principles of Distributed Computing, pp. 267–275. ACM (1996)
20. Raynal, M.: A distributed solution to the k-out of-m resources allocation problem. In: Dehne, F., Fiala, F., Koczkodaj, W.W. (eds.) ICCI 1991. LNCS, vol. 497, pp. 599–609. Springer, Heidelberg (1991)
21. Raynal, M., Beeson, D.: Algorithms for mutual exclusion. MIT Press (1986)
22. Scott, M.L., Scherer, W.N.: Scalable queue-based spin locks with timeout. In: Proceedings of the Eighth ACM SIGPLAN Symposium on Principles and Practices of Parallel Programming, PPoPP 2001, pp. 44–52. ACM (2001)

Author Index